AF616221

Integrated approaches to health

Integrated approaches to health

A handbook for the evaluation of One Health

edited by:
Simon R. Rüegg
Barbara Häsler
Jakob Zinsstag

ISBN: 978-90-8686-324-2
e-ISBN: 978-90-8686-875-9
EAN: 9789086863242
e-EAN: 9789086868759
DOI: 10.3920/978-90-8686-875-9

First published, 2018

Wageningen Academic Publishers,
P.O. Box 220,
NL-6700 AE Wageningen,
The Netherlands.
www.WageningenAcademic.com
copyright@WageningenAcademic.com

**To the inspired and inspiring
who work together with passion,
respect and vision to achieve
sustainable health for all.**

This publication is based upon work from COST Action 'Network for Evaluation of One Health' (TD1404), supported by COST (European Cooperation in Science and Technology).

COST (European Cooperation in Science and Technology) is a funding agency for research and innovation networks. Our Actions help connect research initiatives across Europe and enable scientists to grow their ideas by sharing them with their peers. This boosts their research, career and innovation.

www.cost.eu

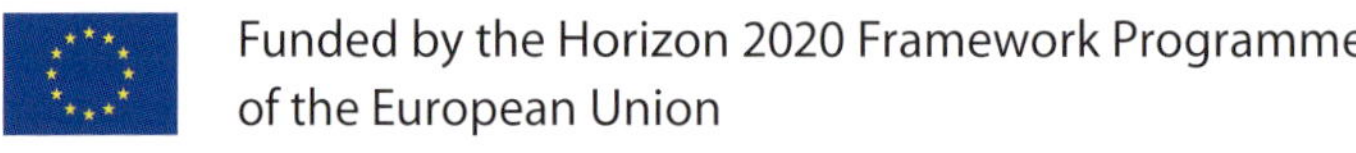

Acknowledgements

This book was produced by the COST funded Action TD1404, Network for Evaluation of One Health (NEOH). The Network brings together researchers, practitioners, decision-makers and other stakeholders with an interest in One Health and evaluation to develop and use methods and frameworks for improved decision making.

We are greatly indebted to the COST programme, for its financial support and its key role in facilitating networking and exchange, thereby enabling innovation and generation of new knowledge. Without the support of COST, we would not have been able to engage as many people from different countries and create this dynamic learning network. Special thanks go to the COST Scientific Officer Mafalda Quintas and the Administrative Officers Ange Marie Ina Uwase, Aranzazu Sanchez and Marie-Astrid Mukankusi who accompanied our Network. Their professional, patient, competent and friendly support helped to make our Action a success.

We also express our heartfelt gratitude to all participants in NEOH for their enthusiasm, expertise, motivation, manifold contributions, willingness to share and discuss ideas, exchange of materials, collaboration, reflection, and engagement. Many working hours in-kind and spare time have gone into developing, testing, refining and designing of this book and we thank them for joining forces and continuing to be inspired and inspiring!

Finally, we thank the grant holder institution, the Royal Veterinary College in London, United Kingdom, for their unsparing support in the administrative and financial management of NEOH. We are particularly grateful to the RVC's Vice-Principal for Research and Innovation and all members of the Research Office; the Director, Assistant Directors and all members of the Finance Department; and the management of the Department for Pathobiology and Population Sciences for their willingness to take on this responsibility and the diligent and skilled execution of tasks.

Table of contents

Preface

Recent financial, economic, social, environmental and health crises have led to the renewed recognition that collaborative approaches between disciplines and sectors are needed to address such wicked problems. Antibiotic resistance or outbreaks of highly infectious diseases, e.g. highly pathogenic avian influenza (HPAI), Ebola, severe acute respiratory syndrome (SARS), Zika virus disease, but also obesity, food security and green urbanisation are examples where integrated approaches to health such as One Health can be employed. One Health emphasises the commonalities of human, animal, plant and ecosystem health. In this perspective, the term can be used in lieu of many other integrated approaches to health across these highly interlinked components[1]. While there is considerable literature describing the characteristics of integrated approaches to health, there are no recognised guidelines so far on how to evaluate to what extent the underlying integration contributes to constructive management of complex health problems. There is thus a need to provide evidence on the added value of One Health to governments, researchers, funding bodies and other stakeholders, and to explore how to evaluate integrated approaches to health. The Network for Evaluation of One Health (NEOH) (http://neoh.onehealthglobal.net) is an initiative funded by the European Cooperation in Science and Technology (COST) that has tackled this challenge by bringing together over 250 scientists and One Health practitioners from more than 30 countries globally. One important result of our collaborative efforts is this handbook for the evaluation of One Health.

Integrated approaches to health are challenging because they represent complex systems of communication and collaboration that are difficult to delimit. In addition they are embedded in an ecological and cultural context where no element can be considered independently. Evaluation under such conditions requires a good understanding of the dynamics within the system and its tangible as well as intangible elements. For example, cultural practices may have a major impact on alimentary habits, which may in turn affect the prevalence of diabetes. This framework has therefore taken a systems approach to evaluation and employs qualitative and quantitative techniques and models developed in systems science. Moreover, we have considered outcomes of integrated health initiatives in the three pillars of sustainable

[1] Related examples are Ecohealth, Global Health, Planetary Health, Ecological Public Health, or Health in scaled Social-Ecological Systems.

development – ecology, society, and economy. The relevant chapters on these topics give the reader an introduction to the relevant theory, metrics, and methods used in these fields and help to understand possibilities emerging from collaboration and integration of disciplines. Further, we provide some novel insights into the governance of One Health and make a call for suitable governance mechanisms in this evolving field.

This handbook evolved over four years in an iterative process between conceptualisation, application and feedback. The concepts are thus framed to fit the many ways in which One Health can be put into practice. The result is a comprehensive overview on what integration can mean, how we can measure it, what outcomes such integrated approaches can have and how we can assess those. It is therefore suitable not only for One Health practitioners and evaluators who want to enhance their knowledge on One Health evaluation, methods, and metrics, but also for policy makers or funding bodies who are considering to support integrated health initiatives. The content and tools are broad and can be adapted to individual needs. We hope that you will find them useful to assess what we implicitly expect when employing an integrated approach to health. First applications of the framework have also been published in a special issue of Frontiers (https://www.frontiersin.org/research-topics/5479).

The development and the application of the handbook has provoked many discussions in the Network on aspects of health which we did not expect to encounter. We kept our minds open and tried to be open to suggestions from all. In this spirit, we hope to inspire you to an open-minded reflection on health and evaluation of integrated approaches to health. The handbook should provide a frame for this reflection and allow a systematic approach. We hope you will enjoy this book as much as we all enjoyed exploring these topics and writing the chapters. We also hope that you will feel encouraged to apply our concepts and tools in your work as you continue to contribute to the field.

The editors
Simon Rüegg, Barbara Häsler and Jakob Zinsstag

Chapter 1

Evaluation of integrated approaches to health with a focus on One Health

Photo: Kevin Queenan

Jonathan Rushton[1*], Liza Rosenbaum Nielsen[2], Laura Cornelsen[3], Kevin Queenan[4], Simon R. Rüegg[5] and Barbara Häsler[4]

[1]Institute of Infection and Global Health, University of Liverpool, IC2 Building, 146 Brownlow Hill, Liverpool L3 5RF, United Kingdom; [2]Faculty of Health and Medical Sciences, University of Copenhagen, Grønnegårdsvej 8, 1870 Frederiksberg C, Denmark; [3]Department of Global Health and Development, London School of Hygiene and Tropical Medicine, 15-17 Tavistock Place, London, WC1H9SH, United Kingdom; [4]Department of Pathobiology and Population Sciences, Veterinary Epidemiology Economics and Public Health Group, Royal Veterinary College, Hawkshead Lane, North Mymms, Hatfield, Hertfordshire, AL9 7TA, United Kingdom; [5]Section of Epidemiology, Vetsuisse Faculty, University of Zürich, Winterthurerstrasse 270, 8057 Zürich, Switzerland; j.rushton@liverpool.ac.uk

Abstract

One Health is a relatively novel term rooted in long held understandings of the link between diseases shared between humans and animals (zoonoses), and that underlying biological and physiological processes are found across species. Despite these understandings, health provision and research have increasingly become separated into areas of human, animal and environmental health. However, recent emergence of diseases such as BSE, SARs and highly pathogenic avian influenza has raised the need to look at health in a more holistic manner and apply principles of transdisciplinarity to difficult health problems. In some circumstances One Health has come to the fore with the understanding that addressing the health of species and the environment with an intersectoral and transdisciplinary approach will provide additional benefits. The frameworks to assess One Health programmes and projects are not well developed, and the guide this chapter introduces outlines an evaluation framework for One Health activities such as the provision of services, research and education.

Keywords: One Health, intersectoral, transdisciplinarity, evaluation

1.1 Background

The term 'One Health' is relatively novel. However, the concept has long been recognised. For instance, in the 1800s Rudolf Virchow, a German physician, coined the term 'zoonosis' indicating that there were links between humans and animals regarding infectious diseases. In reality, before the 1800's, the medical profession dealt with animal diseases that became epidemic problems, and the veterinary profession emerged out of a societal need towards the end of the 1800s (Wilkinson, 1992). Prior to this period there was the emergence of specialist veterinary schools and much experimental work conducted in physiology and microbiology that cut across species groups. However by the 20th century an increasing separation of medicines was seen (Zinsstag *et al.*, 2005), albeit with the same core underlying biology principles and with some interest in comparative medicine. Possibly the reasons for the separation of medicines has been the ability for disease control to be conducted species by species, and the increasing specialisation of human as well as veterinary medicine in all aspects of specific disease management and individual treatment, population and public health medicine. Chapter 2 provides a good overview of these changes in how One Health is perceived and what it includes.

In the last two decades, there has been a re-emergence of the recognition that a combined approach to health issues is needed, together with an increasing awareness that environmental health affects the health and livelihood of humans, domestic animals and wildlife, and is an important component for sustainability and resilience of the planet. This recognition has led to the emergence of the One Health, EcoHealth, Planetary Health and other integrated health movements, which are discussed in Chapter 2.[2] The drivers of emergence and spread of diseases are rooted in the way we organise our production and use of food,

[2] While we use the term One Health in this book, the principles discussed are relevant for any integrated, systems-based approaches to health.

feed, water and energy for a growing world population (FAO, 2013). One Health initially gained momentum triggered by the threat of major food borne disease problems such as salmonella and bovine spongiform encephalopathy (BSE) and more recently with the zoonotic pandemics threats such as severe acute respiratory syndrome (SARS), cross species influenza and Ebola. Antimicrobial resistance across species and within the general environment is also a growing concern (O'Neill, 2016). This has brought to the fore the need for medical doctors, veterinarians and human and animal health professionals to collaborate. Indeed, the need for wider interdisciplinary collaboration is increasingly recognised in order to address the complex interplay between humans, animals and the environment in the context of disease prevention and control and incorporating both health and welfare issues. In addition to human and animal health practitioners, the role of wildlife biologists, environmentalists, ecologists, anthropologists, economists and social scientists amongst others were included when designing One Health approaches for holistic disease prevention and mitigation strategies (Zinsstag *et al.*, 2015). The use of One Health approaches with a stronger emphasis on the environmental component is growing due to the rapid development of environmental change including exponentially growing world population, urbanisation, deforestation, wildlife and plant species extinctions and global warming.

The strategic direction of One Health is to assess actions and interventions that aim to promote health through common aims and collaboration between disciplines across different species and their environment. The transdisciplinarity of One Health brings with it the challenge of harmonising the definition of health across disciplines and sectors with an underlying core that concerns the importance of people affected by health outcomes be they animal, environmental or human. In this context the definitions of human health outcomes becomes critical and the definition of what is the state of human health that is desirable changes as society evolves and our understanding of our needs is better understood. Commonly used metrics to assess the disease burden for humans both qualitatively and quantitatively, have been the use of either quality adjusted life years (QALY) or disability adjusted life years (DALY) (Murray, 1994), with judgements on interventions on the cost per DALY avoided as a basis for policy change (Drummond *et al.*, 2005). On the other hand, health in domestic or production animals tend to have a strong focus on absence of disease compared to human health due to the links between health and productivity, which have important societal and economic value. The animal health issues can be reduced to monetary values whereas change or intervention can be modelled with a cost-benefit analysis framework similar to other types of investment in society (Gittinger, 1982). Some may contest that not all aspects of livestock and domesticated animals can be monetarised, yet economics has developed methods to place prices on outputs that have no markets, and the zDALY attempts to place the issue of zoonoses into a new framework to capture impact across species (Torgerson *et al.*, 2017). In some countries, welfare measures for animals are employed in legislation and daily management of domesticated animals. These welfare measures have been defined as the five freedoms: (1) freedom from hunger and thirst; (2) freedom from discomfort; (3) freedom from pain, injury and disease; (4) freedom to express (most) normal behaviour; and (5) freedom from fear and distress (FAWC, 2012). Similarly, plant and aquatic animal health can be reduced to monetary terms, yet it has rarely been included or even thought about, even within a One Health context. Finally, environmental or ecosystem health measurement and assessment have been the most creative in the development of methods to define the value or prices of goods and services that have no market (TEEB, 2015; Winpenny, 1991). Ecosystems services refer the direct and indirect contributions of healthy ecosystems to human well-being. There are three categories: provisioning ecosystem services, regulating and maintenance ecosystem services, and cultural ecosystem services (TEEB, 2015). However this area of environmental health has a relatively poorly developed set of metrics within the human-animal-environment disease triad. Costanza *et al.* (1992) defined ecosystem health as the occurrence of normal ecosystem processes and functions, with a system being free from distress and degradation, which maintains its organisation and autonomy over time and is resilient to stress. The concept of ecosystem health depends on human-social values and desires, and therefore, integrates numerous ecological, social, economic and political factors (Tzoulas *et al.*, 2007). Charron (2012) has explored the how the ecosystem interacts with and established methods for such assessment, and Zinsstag *et al.* (2017) propose extending the Health Impact Assessment framework to a One Health format. These developments are useful processes of innovation, the big step is to incorporate them into legislation systems for the public sector as is the case for cost-benefit analysis or to encourage their use in social charters for private companies.

Across the human-animal-environment system there is a lack of universally accepted methods and metrics to evaluate problems and interventions. In turn this generates a problem of how the added value of One Health actions can be measured (Babo Martins *et al.*, 2015; Cleaveland *et al.*, 2006; Coker *et al.*, 2011) and also how costs to achieve a better societal outcome are borne across society. Capturing changes in human and animal welfare, environment services and economic returns provide a major challenge (Häsler *et al.*, 2014; Manlove *et al.*, 2016). The transdisciplinary nature of One Health makes it difficult to fund as most research funding is focused on specific diseases or disease mechanisms, and redirecting funds rarely takes place unless faced with a crisis emergency situation such as the one lived in the recent Ebola epidemic. Manlove *et al.* (2016) in their analysis of what has so far been published detect separation across different areas of One Health with clustering of activity in ecology and veterinary science and some more diverse work. However, research is expanding in this area at a faster rate than other life sciences when measured by publication output. Yet the question remains whether this is an old approach with a new badge and therefore the need for evaluation methods and metrics to test how holistic and interdisciplinary the research has been.

The increase in research output labelled as One Health reflects a change in funding focus of the major organisations. For example the EU have funded ICONZ[3]; USAID support the four-part emerging pandemic threat programmes[4] PREDICT, PREVENT, IDENTIFY and RESPOND; and the British research councils have funded a zoonotic disease programme (ZELS[5]) with a One Health focus. One Health funding from the private sector is also increasing. For instance, the UK based Wellcome Trust includes a strategic funding section under the name 'Our Planet, Our Health' that supports transdisciplinary research that connects the environment and health. Furthermore, the Bill and Melinda Gates Foundation also funded One Health

[3] https://www.ed.ac.uk/global-health/research/project-profiles/one-health/zoonotic-diseases/iconz.
[4] https://www.usaid.gov/news-information/fact-sheets/emerging-pandemic-threats-program.
[5] https://bbsrc.ukri.org/research/international/engagement/global-challenges/zels/.

projects under the call 'The One Health Concept: Bringing Together Human and Animal Health for New Solutions'. These research funding projects are in stark contrast to the lack of government institutional change and delivery of One Health services and mechanisms. The policy change at present is largely on funding research, rather than changing practices in the delivery of health services in a multi-species and trans-sectoral manner, yet the joint publication by the World Bank and Ecohealth Alliance on operationalization of health systems would indicate that this may be changing (World Bank, 2018).

These initiatives are a start to bring about investments across all species that reflect their relative importance in terms of health outcomes, including humans and the environment. Such an approach needs to recognise that market prices are not a good measure of determining how to achieve a balance across species. Yet measuring the added value from a One Health approach requires both clarity of what should be measured as well as the reason why a change in resource use should be valued. This guide sets out to develop a protocol for the evaluation of One Health and a series of methods to solve this problem and thereby adding greater certainty on the when, where and how One Health activities are needed to promote health in the global society.

1.2 Structure of the guide

The guide contains seven core chapters, which can be read either in isolation or in combination. In order to help the reader the following gives a brief overview of each chapter:

- Chapter 2 describes the existing separate health disciplines and approaches and their dependency on high tech, linear solutions, which are becoming less effective and less sustainable in solving increasingly complex problems. The sustainable development goals are presented as a unique opportunity for a paradigm shift to a fully integrated approach to health. A convergence of the various movements that support this, including One Health, EcoHealth, Planetary Health and Ecological Public health, is called for.
- Chapter 3 gives a step-by-step protocol to be used when designing evaluations for One Health initiatives with key steps of:
  - defining the system/context;
  - describing and characterising the One Health Initiative;
  - describing the theory of change and expected/unexpected outcomes;
  - selecting the outcomes and metrics;
  - assessment of One Health-ness including One Health planning, working, systemic organisation, learning infrastructure, sharing infrastructure and the One Health index;
  - reviewing, planning and conducting the evaluation.
- Chapters 4-6 investigate methods and metrics utilized for the relevant outcomes of interest for three dimensions of ecology, society and the economy.
- Chapter 7 examines the integration of outcomes in the various dimensions and discusses the governance of One Health focusing in particular on knowledge integration within One Health policy cycles.

Our intention of this guide is to provide a methodology that will be regularly used with results reported with a common format. Once results and publications begin to flow it will be possible to establish a longitudinal database of that can be further analysed for trends in the value of One Health over time and in different regions.

Acknowledgements

This chapter is based upon work from COST Action 'Network for Evaluation of One Health' (TD1404), supported by COST (European Cooperation in Science and Technology).

References

Babo Martins, S., Rushton, J. and Staerk, K.D.C., 2015. Economic assessment of zoonoses surveillance in a 'One Health' context: a conceptual framework. Zoonoses Public Health 63(5): 386-395.

Charron, D.F.E., 2012. Ecohealth research in practice. Springer, New York, NY, USA.

Cleaveland, S., Kaare, M., Knobel, D. and Laurenson, M.K., 2006. Canine vaccination – providing broader benefits for disease control. Vet Microbiol 117(1): 43-50.

Coker, R., Rushton, J., Mounier-Jack, S., Karimuribo, E., Lutumba, P., Kambarage, D., Pfeiffer, D., Stärk, K. and Rweyemamu, M., 2011. Towards a conceptual framework to support 'One Health' research and policy: health systems and the human – animal interface. Lancet Infect Dis 11(4): 326-331.

Costanza, R., Norton, B. and Haskell, B.J., 1992. Ecosystem health: new goals for environmental management. Island Press, Washington, DC, USA.

Drummond, M.F., Sculpher, M.J., Torrance, G.W., O'Brien, B. and Stoddart, G.L., 2005. Methods for the economic evaluation of health care, 3rd edition. Oxford University Press, Oxford, UK.

FAO, 2013. World livestock 2013: changing disease landscapes. FAO, Rome, Italy. Available at: www.fao.org/docrep/019/i3440e/i3440e.pdf.

Farm Animal Welfare Council (FAWC), 2012. Five freedoms. Available at: https://tinyurl.com/y9r8wzjz.

Gittinger, J.P., 1982. Economic analysis of agricultural projects, 2nd edition. World Bank, Washington, DC, USA, 505 pp.

Häsler, B., Hiby, E., Gilbert, W., Obeyesekere, N., Bennani, H. and Rushton, J., 2014. A one health framework for the evaluation of rabies control programmes: a case study from Colombo City, Sri Lanka. PLoS Negl Trop Dis 8(10): e3270.

Manlove, K.R., Walker, J.G., Craft, M.E., Huyvaert, K.P., Joseph, M.B., Miller, R.S., Pauline Nol, Patyk, K.A., O'Brien, D., Walsh, D.P. and Cross, P.C., 2016. 'One Health' or three? Publication silos among the one health disciplines. PLoS Biol 14(4): e1002448.

Murray, C., 1994. Quantifying the burden of disease: the technical basis for disability-adjusted life years. Bull World Health Organ 72(3): 429-445.

O'Neill, J., 2016. Tackling drug-resistance infection globally: final report and recommendations. The review of antimicrobial resistance. Available at: http://tinyurl.com/zxxjmww.

TEEB, 2015. The economics of ecosystems and biodiversity for agriculture and food: towards a global study on the economics of eco-agri-food systems. Available at: http://tinyurl.com/ydewwye9.

Torgerson, P.R., Rüegg, S., Devleesschauwer, B., Abela-Ridder, B., Havelaar, A.H., Shaw, A.P.M., Rushton, J. and Speybroeck, N., 2017. zDALY: an adjusted indicator to estimate the burden of zoonotic diseases. One Health 5: 40-45.

Tzoulas, K., Korpela, K., Venn, S., Yli-Pelkonen, V., Ka´zmierczak, A., Niemela, J. and James, P., 2007. Promoting ecosystem and human health in urban areas using Green Infrastructure: a literature review. Landsc Urban Plan 81: 167-178.

Wilkinson, L., 1992. Animals & disease. An introduction to the history of comparative medicine. Cambridge University Press, Cambridge, UK, 272 pp.

Winpenny, J.T., 1991. Values for the environment. A guide to economic appraisal. Overseas Development Institute, HMSO, London, UK, 277 pp.

World Bank, 2018. One Health: operational framework for strengthening human, animal, and environmental public health systems at their interface. World Bank, Washington, DC, USA, 138 pp.

Zinsstag, J., Schelling, E., Waltner-Toews, D., Whittaker, M. and Tanner, M., 2015. One Health: the theory and practice of integrated health approaches. CABI, Wallingford, Oxfordshire, UK.

Zinsstag, J., Schelling, E., Wyss, K. and Bechir, M., 2005. Potential of cooperation between human and animal health to strengthen health systems. Lancet 366(9503): 2142-2145.

Zinsstag, J., Crump, L. and Winkler, L.S., 2017. Biological threats from a 'One Health' perspective. Rev Sci Tech Off Int Epiz 36(2): 671-680.

Chapter 2

Health solutions: theoretical foundations of the shift from sectoral to integrated systems

Photo: Kevin Queenan

Richard Kock[1], Kevin Queenan[1*], Julie Garnier[2], Liza Rosenbaum Nielsen[3], Sandra Buttigieg[4], Daniele de Meneghi[5], Martin Holmberg[6], Jakob Zinsstag[7,8], Simon Rüegg[9] and Barbara Häsler[1]

[1]Department of Pathobiology and Population Sciences, Royal Veterinary College, Hawkshead Lane, North Mymms, Hatfield, Hertfordshire AL9 7TA, United Kingdom; [2]Odyssey Conservation Trust, Bakewell, Derbyshire DE45 1LA, United Kingdom; [3]Department of Veterinary and Animal Science, Faculty of Health and Medical Sciences, University of Copenhagen, 1870 Frederiksberg, Denmark; [4]Department of Health Services Management, Faculty of Health Sciences, University of Malta, Msida, Malta; [5]Department Veterinary Science, University of Turin, L.go P. Braccini 2, 10095 Grugliasco (Torino), Italy; [6]Department of Culture and Media Studies, Umeå University, 901 87 Umeå, Sweden; [7]Swiss Tropical and Public Health Institute, Socinstrasse 57, 4002 Basel, Switzerland; [8]University of Basel, Petersplatz 1, 4003 Basel, Switzerland; [9]Section of Epidemiology, Vetsuisse Faculty, University of Zurich, Zurich, Switzerland; kqueenan3@rvc.ac.uk

Abstract

The current fragmented framework of health governance for humans, animals and environment, together with the conventional linear approach to solving current health problems, is failing to meet today's complex health challenges and is proving unsustainable. Advances in healthcare depend increasingly on intensive interventions, technological developments and expensive pharmaceuticals. The disconnect grows between human health, animal health and environmental and ecosystems health. Human development gains have come with often unrecognised negative externalities affecting ecosystems, notably loss of resilience, mostly through biodiversity loss and land degradation. Reduced capacity of the ecosystem to serve humanity threatens to reverse the health gains of the last century. A paradigm shift is urgently required to de-sectoralise human, animal, plant and ecosystem health and to take a more integrated approach to health, One Health (OH). The sustainable development goals (SDGs) offer a framework and unique opportunity for this and we argue the need of an OH approach towards achieving them. Feasibility assessments and outcome evaluations are often constrained by sectoral politics within a national framework, historic possession of expertise, as well as tried and tested metrics. OH calls for a better understanding, acceptance and use of a broader and transdisciplinary set of evaluation approaches and associated metrics, which is a key objective of NEOH. We need to shift our current sectoralised, linear focus to a more visible balanced health investment with more global benefits to all species. This is encapsulated in the movements for OH, EcoHealth, Planetary Health and Ecological Public Health, which are essentially converging towards a paradigm shift for a more integrated approach to health.

Keywords: One Health, health governance, health policy, sustainable development goals, ecosystems health, global health, planetary health

2.1 Introduction

One health is a paradigm shift from mechanistic determinism in health sciences to post-normal science. Can we therefore, through a One Health approach, deal with a seemingly insoluble set of wicked problems through systems science and inter-/transdisciplinarity? Could the future direction in health be restoration of healthy lives in healthy ecosystems? The choice is ours but governance of this process is key.

The development of human and animal health, as well as environmental and ecosystems health continues within a governance and policy framework which remains highly sectoralised and structural despite calls for an integrated and transdisciplinary approach (Karesh *et al.*, 2014; Lee and Brumme, 2013; Valeix *et al.*, 2016; Wallace *et al.*, 2014). The medicalisation of health within conventional health systems and their increasing intensification and dependence on advances ensues in often highly profitable technological innovations and expensive pharmaceuticals. This occurs whilst neglecting drivers and preventive interventions, and has contributed to unhealthy practices (e.g. antibiotic misuse) that are now proving too expensive to maintain (Wallace *et al.*, 2015).

Earlier gains in human health are now looking vulnerable, with widening global health inequalities and increasing number of emerging and re-emerging diseases (Rabinowitz *et al.*, 2013). Many 20th century advances in human health and development came with a delayed unexpected/unforeseen cost to ecosystems, the consequences of which are now increasingly a cause for deterioration in human health (Prüss-Ustün *et al.*, 2016; UNEP/UNECE, 2016). Neglect of environmental or ecosystems health and associated loss of biodiversity is now at a critical point, threatening 'Planetary Health' and the fundamental processes on which life depends (Whitmee *et al.*, 2015). Human health is also vulnerable to consequences of concurrent underinvestment in the health and productivity of livestock (NAS, 2015) and plants (Chakraborty and Newton, 2011). This is particularly true within the context of climate change (Porter *et al.*, 2014). Poorer disease control and reduced productivity of livestock and crops will affect food security and livelihoods and indirectly human health.

Maintaining individual and public health in the ever changing, complex adaptive socio-ecological system that we form part of, requires us to think foresightedly and creatively, while remaining flexible and contributive. The same goes for maintaining the health or survival of an individual animal or single species population. Houle (2015) (p. 401), questioned whether the concepts upon which we base our understanding of health (within the disciplines of epidemiology, pathology, etc.) are themselves 'unhealthy and maladaptive' and that we should acknowledge our dependency, passivity, weakness and vulnerability as features of our human existence. Rook (2013) argues that microbial symbionts and commensals should be seen as a neglected ecosystem service, essential for the development of our immune systems and our well-being.

A paradigm shift is needed towards a fully integrated approach to health; a system(s) approach with a focus on restoring resilience of biological systems at all scales, including humans, animals and plants (Kock, 2015; Rabinowitz *et al.*, 2013; Zinsstag *et al.*, 2015), an approach known as One Health (OH), derived from the One World One Health concept which emerged in the first decade of the 21st Century (Anon, 2009; WCS, 2004).

When defining OH, Zinsstag *et al.* (2015) focused on the added value that could be achieved (improved health, financial savings and environmental services) through cooperation of human and veterinary medicine rather than having these disciplines functioning separately. Whilst integration of human and animal health, without specific consideration of socio-ecological factors, takes some steps towards inter-sectoral collaboration, it fails to address the many structural and environmental issues critical to health. By contrast, Wallace *et al.* (2014) (p. 1) state 'It (OH) redresses an epistemological alienation at the heart of much modern population health, which has long segregated studies by species. To this point OH research, however, has also omitted addressing fundamental structural causes underlying collapsing health ecologies.' Furthermore, 'ecosystem approaches to health' or 'EcoHealth' considers inextricable linkages between sustainable ecosystems, society and health of animals and humans (Rapport *et al.*, 1998, 1999). One Health and EcoHealth thinking converge strongly, especially through OH's recognition of health as an outcome of social-ecological systems and its implication for sustainability (Zinsstag, 2012; Zinsstag *et al.*, 2011, 2012). The term OH is used in this discussion because of its high acceptance, whilst we clearly recognise that 'ecosystems approaches to health' (Charron, 2012) (p. 257) and 'health in social-ecological systems' (Zinsstag *et al.*, 2011) (p. 2) are imbedded in the One Health approach to complex systems.

2.2 Names, definitions and hierarchy

Several uncomfortable truths confront human development and all its potential that was achieved in the 20th Century. The environment and biodiversity are rapidly declining, whilst ecosystem services, namely those benefits that humans derive from the dynamic system of plants, animals and microorganisms, such as clean air and water, fertile soils and timber as well as recreational and spiritual benefits (Millennium Ecosystem Assessment, 2005), are in themselves, metaphorically speaking unhealthy (Lu *et al.*, 2015). Earlier gains in human health are threatened by several emerging and multidirectional health and disease threats, including effects of climate change, novel pathogens, growing mental health issues, obesity and hunger, micronutrient deficiencies and ecotoxicologies. To compound this, global health (which focused only on humanity) seems more disconnected than ever, despite estimates that 23% of global human premature deaths representing 12.6 million deaths every year are attributed to modifiable environmental factors (Prüss-Ustün *et al.*, 2016).

How did these problems and these disparities emerge despite the growing understanding of and investment in health across all sectors? To answer this, we review the definitions of health around which the sectoralised health systems have developed.

2.2.1 Human health

When we think of health, we think firstly of the health of individual humans and communities. The WHO defined (human) health as 'a complete state of physical, mental and social well-being, and not merely the absence of disease or infirmity' (WHO, 1946). It has subsequently evolved to account for the rights (Saracci, 1997) and changing needs of the individual in relation to age, culture and personal responsibility (Bircher and Kuruvilla, 2014). The significance of health in underpinning development and the socio-ecological determinants of health are increasingly recognised (Dora *et al.*, 2015). Population health was presented by Frankish *et al.* (1996) (p. 6) as 'the capacity of people to adapt to, respond to, or control life's challenges and changes.' The term Global Health (frequently confused with OH) remains human centric, defined as 'an area for study, research and practice that places a priority on improving health and achieving equity in health for all people worldwide' (Koplan *et al.*, 2009).

2.2.2 Animal health

Animal health is much more disintegrated, in that notifiable diseases, welfare, terrestrial and aquatic wild animal health are often addressed in separate laws. For example, recent legislation (British Columbia Government, 2014) in British Columbia, Canada (British Columbia Government, 2014) defines it as 'the health of a population or subpopulation of animals and includes the preservation of a population or subpopulation of animals that is at risk of being exposed to or affected by a notifiable or reportable disease'. Animal health typically focusses on the control of domestic animal infectious diseases that impact on humans, either directly as zoonotic diseases or indirectly through economic losses. Only recently, the World Animal Health Organisation (OIE) added wildlife diseases to its listed diseases[1] and now also includes health as an aspect of its definition of animal welfare[2].

2.2.3 Wildlife health

Wildlife health is a fairly recent concept without any formal sectoral responsibility. It is often covered under environmental and biodiversity legislation and under legislation for zoonoses in public health or diseases of concern for domestic animal health. A working definition of wildlife health is needed; one that recognizes that the major threats to wildlife are not diseases but rather anthropogenic impacts through so-called development. Stephen (2014) states that a modern definition of wildlife health should emphasize that: (1) health is the result of interacting biologic, social, and environmental determinants that interact to affect capacity to cope with change; (2) health cannot be measured solely by what is absent but rather by characteristics of the animals and their ecosystem that affect their vulnerability and resilience; and (3) wildlife health is not a biologic state but rather a dynamic social construct based on human expectations and knowledge. Conservationists have recognised and promoted what are known as the 'Manhattan Principles'[3], that the health and sustainable maintenance of wildlife in natural reserves are mutually interdependent with the health of communities and the livestock surrounding them (Osofsky *et al.*, 2005).

2.2.4 Plant health

Plant health, much like animal health, is primarily understood in the context of plants' contribution to the food sector for humans and to livestock feeds, rather than in the context of their contribution to biodiversity and overall health of the ecosystem. More recently however, climate change has drawn attention to global plant population health as part of the solution to global warming (CBD, 2015). The links between plant health and their contribution to food security of animals and humans and determination of human health are recognised in the Three Health model (Boa *et al.*, 2015).

2.2.5 Ecosystems health/health in social-ecological systems

Ecosystem approaches to health concerns is embedded in the United Nations Environment Programme (UNEP). Recently UNEP has diverged from static reports on chemicals, waste, air, water, biodiversity and soils to a more holistic view of the health paradigm as evidenced in their 'Healthy Planet Healthy People' report (UNEP/UNECE, 2016). The theory and practice of understanding and managing human activities in the context of social-ecological systems has been well-developed by members of The Resilience Alliance[4] and was also used extensively in the Millennium Ecosystem Assessment[5] in its work on human wellbeing

---

[1] http://www.oie.int/animal-health-in-the-world/oie-listed-diseases-2016/.

[2] http://www.oie.int/doc/ged/d5517.pdf.

[3] http://www.oneworldonehealth.org/.

[4] https://www.resalliance.org/.

[5] http://www.millenniumassessment.org/en/index.html.

outcomes. It is therefore not difficult to relate human health (and similarly animal health) to social-ecological systems (SES) as health in social-ecological systems (HSES) (Zinsstag *et al.*, 2011). These systems relate outcomes to systemic interactions within related ecosystems, which are primarily influenced by resources, governance and users in a given social, economic and political setting.

2.3 Towards integration of health systems

The recent global changes in the social-ecological systems (urbanisation, globalisation, human population growth, increasing consumption, climate change and loss of habitat and biodiversity, etc.) favour the rapid and often global transmission of emerging and re-emerging pathogens (Jones *et al.*, 2017). The complexity of some of these recent global infectious disease threats (SARS, H5N1, ZIKA and Ebola) encouraged a lowering of sectoral walls and a more integrated approach to finding health solutions at an international level in principle (e.g. tripartite agreement between WHO, OIE and Food and Agriculture Organisation of the United Nations-FAO). However, at a national level in almost all countries, ministries remain separate and sectoralised, with their own budgets and agendas without integration of health programmes (Häsler *et al.*, 2013). Efforts and progress towards OH are still restricted by the inertia of long established divisions, institutional and logistical barriers to sharing data and information across institutions (World Bank, 2010), and power and leadership struggles with failure to agree on task and resource allocation issues (Rushton *et al.*, 2012). Besides a few studies on joint health service delivery (Schelling *et al.*, 2005), brucellosis (Roth *et al.*, 2003), rabies control (Zinsstag *et al.*, 2009) and laboratory infrastructure (World Bank, 2012), there is a lack of economic evidence and metrics to measure OH gains (Häsler *et al.*, 2014).

Beyond the paradigm shift called for by OH, Wallace *et al.* (2015) reinvigorate the notion of specifically focusing on the wider context which lies behind emerging health problems, including the geopolitical, economic and societal global crises and the unsustainability of natural resource use and current global economic systems. Structural OH is said to 'empirically formalise the connections among capital-led changes in the landscape and shifts in wildlife, agricultural and human health' (Kock, 2015). It requires a shift from linear thinking and simplistic medicalisation of health, to systemic transdisciplinary approaches with contributions from a wide range of professionals such as ecologists, agriculturalists, engineers, architects and also social scientists, including political scientists, economists, anthropologists and behavioural scientists, as well as from the stakeholder community and its representatives (Zinsstag *et al.*, 2015).

The United Nations (UN) community continues to develop policy and political instruments to drive change. In 2015, the 2030 Agenda for Sustainable Development set new goals (SDGs) to guide global development over the 15 years to 2030. The SDGs have a strong focus on equity and are described as being 'integrated and indivisible, global in nature and universally applicable' (United Nations, 2015). The new 2030 Agenda calls for a new cooperative paradigm based on the concept of 'full global partnership'. The need to 'think differently' to address the deep systemic changes required by this new Agenda has also been recognized at intergovernmental level (Giovannini *et al.*, 2015). We see the SDGs as a unique opportunity for change with a OH Agenda for 2030 (Queenan *et al.*, 2017).

2.4 The SDGs: opportunities for change

The latest WHO assessment of health in the SDGs acknowledges 'that the SDGs, by contrast to the Millennium Development Goals (MDGs), reflect a far wider range of environmental, economic and societal concerns. All the SDGs are designed to be cross-cutting and the inter-linkages and networks within the SDGs are as important as the individual goals themselves' (WHO, 2015a). Health, instead of being based as in the MDGs on three narrow targets in isolation from the other goals, is now recognised as a precondition, an outcome and an indicator of sustainable development (UNEP/UNECE, 2016), and is now one target embedded in the others. There is at least a current acceptance that health depends on many factors outside of human control and that only by attending to the health of other biological and physical elements of the planet, will this be sustained (Demaio and Rockstrom, 2015; Whitmee *et al.*, 2015).

2.5 The Interactive Web of SDGs

Waage *et al.* (2015) noted that total sustainable development is more than the sum of its parts and 'is an outcome of positive synergies between multiple elements and may be undermined by negative trade-offs between them' and criticise the SDGs for being developed within different sectors without recognising the interactions, both positive and negative, between them. To demonstrate, they positioned the SDGs in a framework of three concentric levels depending on their intended outcomes and argued that 'governance within silos is no longer tenable'. The inner level of 'well-being', which includes 'people-centred' goals such as health, education and nutrition (SDGs 1, 3, 4, 5, 10 and 16), were noted as providing opportunities for synergies. The middle level, infrastructure relate to those goals perceived as essential for a modern society to function (SDGs 2, 6, 7, 8, 9, 11 and 12) and are closely linked with those in the inner level. The outer level, 'environment' contains goals which relate to the management of natural resources and the provision of ecosystem services and life-supporting systems (SDGs 13, 14, 15). These were noted as having been largely ignored and seriously compromised. Achieving the goals in the infrastructure level must be done so without compromising those in the outer and inner levels.

We have adapted this framework further and added an all-inclusive level of OH which extends to include the SDG 17 for global partnerships, a cornerstone of the SDG's and of OH (Figure 2.1). We have also highlighted three of the infrastructure goals relating to economic growth, industrialisation and production and consumption (SDGs 8, 9 and 12). These goals have an antagonistic relationship with other goals, especially under current political economies (see Structural OH above). A comprehensive effort to apply the principles of New Institutional Economics, (Ostrom, 2007) could provide a global shift to decouple the dependency of economic growth on resource use (UNEP/UNECE, 2016) and move towards linking economic performance with sustainable practices; the only resource available in the future will be a renewable resource used in a greener, circular economy.

The SDGs provide a key entry point for a One Health approach to drive a paradigm shift in policy and practice towards a fully integrated approach to health in social-ecological systems (Zinsstag *et al.*, 2011). Due to the political consensus and momentum behind the SDGs as well as the recent frequent global reports on health concerns, this is a historic opportunity.

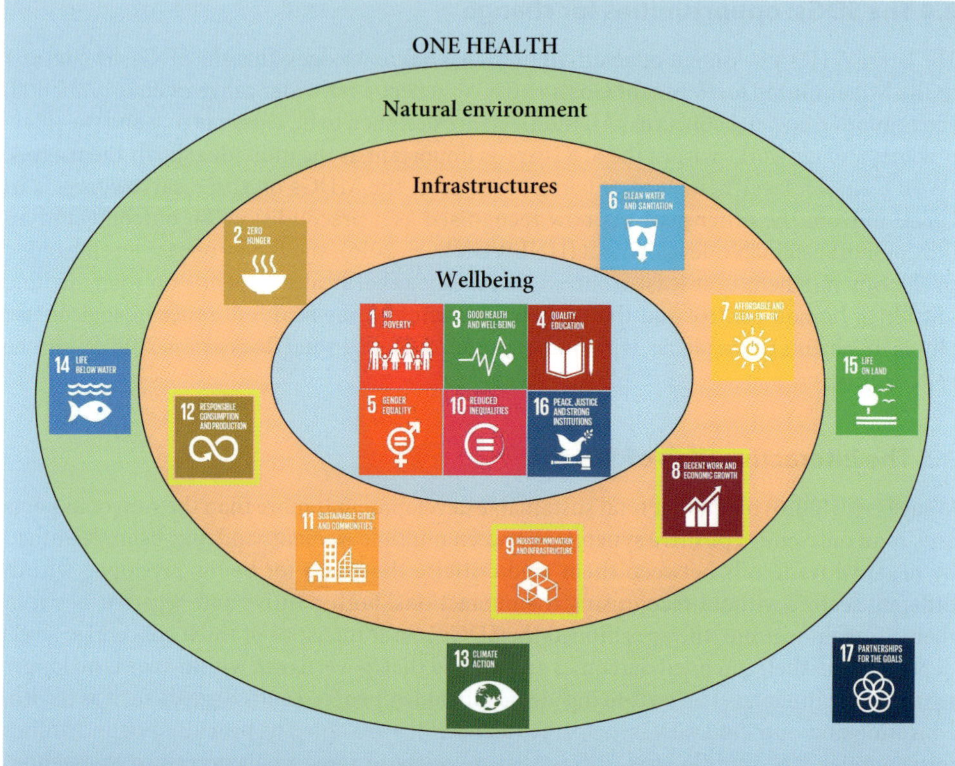

Figure 2.1. A framework grouping the sustainable development goals (SDGs) based on their intended outcomes, highlighting goals (in yellow) with antagonistic relationships with other goals (adapted from Waage *et al.*, 2015).

2.6 What have health assessments taught us?

Current health governance remains segregated in local, national and international institutions, which lack the authority and tools to prevent emerging health threats at various scales. Recent global threats like Ebola and Zika viruses provided valuable lessons, whilst the implementation of International Health Regulations have improved coordination and internationalisation of interventions (Gostin *et al.*, 2015; Heymann *et al.*, 2015; Moon *et al.*, 2015). In addition, governance is no longer dominated by health organisations but influenced by many actors, including UN agencies (WHO, UNICEF) and multinational agencies (World Bank), national governments, civil society organisations, multinational corporations and academic institutions, etc. (Frenk and Moon, 2013). Animal health and environmental health governance are in a similar state. With a better acceptance of the interconnectedness and the multiple determinants of health and the different sectors and actors involved, Frenk and Moon (2013) suggest using the more inclusive term 'global governance for health' to open health governance to others beyond health professionals.

As part of governance, priority setting and budget allocation is based on priority disease lists regularly provided by the WHO and OIE. These priority lists lack a OH assessment, despite the obvious linkages with the zoonotic diseases and less obvious environmental, socio-economic or structural drivers. For example, the WHO's 'top emerging diseases likely to cause major epidemics' includes diseases described as serious and requiring immediate action (WHO, 2015b). Despite all six diseases being zoonoses, with arguably strong environmental and socio-economic drivers, the list of experts responsible for prioritising these does not include veterinarians, ecologists, social scientists or other stakeholders. Although WHO and OIE are advocating a transdisciplinary approach there is little evidence yet of this in practice.

Our analysis of the drivers and risk factors for prioritised diseases listed by the WHO i.e. neglected tropical diseases, neglected zoonotic diseases, pandemic and epidemic diseases and the top ten causes of death globally, showed 98% of them could be classified as benefitting from a OH, systems thinking approach. A similar analysis of the OIE's 118 listed diseases was performed. This list has a focus on economically significant livestock diseases, however more recently, they have included wildlife diseases, including those of insects and amphibians. We analysed each disease to assess whether it had either a significant impact on producers' livelihoods (mass losses, culls or trade restrictions), on farmed and wild species populations, had a vector distribution affected by climate change, was zoonotic or caused biodiversity loss within natural ecosystems. On this basis, we advocate a One Health approach would be called for in all 118 OIE listed diseases.

Feasibility studies for policy making in society are frequently based on five elements; technical, economic, legal, operational and scheduling, with the economic element (cost benefit analysis) often having the most leverage. This is not always the case in human or animal health where political and technical considerations are primary. However, complex problems such as new emerging diseases, climate change and antimicrobial resistance create new challenges when assessing their feasibility for control. Current commonly used economic models, metrics and analyses often fail to capture the full extent of costs and benefits produced by health interventions. A sound assessment must be based on scientific evaluation and must combine economic, social, and ecological aspects (Häsler *et al.*, 2011, 2014). Predictions in complex problems are heavily dependent on modelling, whilst benefits may take many years to accrue, which increases confounding and makes a traditional cost benefit analysis difficult. Predicting human behaviour and how it may change over time, is an additional challenge. A OH approach, based on complex or wicked problem solving methods (Brown *et al.*, 2010) with transdisciplinary collaboration, warrants a better understanding, acceptance, integration and use of a broader set of evaluation metrics, as promoted by NEOH (Haxton *et al.*, 2015).

2.7 But is there proof of concept for a One Health Approach and its added value?

Policy decisions under challenging economic conditions rely not only on sound scientific evidence but on economic evidence too. Several authors have presented evidence of the feasibility and argued for the added value of a OH approach compared to isolated and linear approaches to disease prediction and control (Guimaraes and Mergler, 2012; Harris *et al.*, 2012; Monroy *et al.*, 2009; Queenan *et al.*, 2016; Rabinowitz *et al.*, 2013; Rushton *et al.*, 2012;

Schelling *et al.*, 2005; Valeix *et al.*, 2016; Zinsstag *et al.*, 2005, 2009, 2015; World Bank, 2012). The World Bank estimated the annual funding required to build capacity of human and animal health systems in developing countries (with high risk of zoonotic disease prevalence) to WHO and OIE standards was approximately US$3.4 billion (World Bank, 2012). They estimate that such annual investment would expect global benefits of US$30 billion each year. However, many examples lack the consideration of environment, ecosystems and structural elements of health in the interventions and benefit assessments.

Parallels between OH and sustainability (built on the pillars of society, environment and economy) have been identified and can be used to broaden the assessment of the added value of OH (Rüegg *et al.*, 2017). In particular, the economic dimensions require a wide assessment beyond the obvious cost benefit analysis to include the less tangible benefits to human and animal health and welfare (Babo Martins *et al.*, 2015; Queenan *et al.*, 2016; Rüegg *et al.*, 2017).

The objective of NEOH is to provide guidance on metrics and evaluation of OH for use into the future. Once established they will help to build confidence in the approach with scientific method in assessing the benefits to individuals up to planetary systems.

2.8 Conclusion

This chapter has described the current definitions of health, the segregation of health systems and the opportunities for change. We propose that considering animal, human and environmental or ecosystems health separately within narrow perspectives is no longer valid. This is based on the increasing evidence of deterioration in biodiversity, ecosystem services and function, and trends towards a reversal in human and animal health gains of the past century. Whilst business as usual may continue to achieve some apparent gains in human and domestic animal health (through technological advancement at high cost), failing to adopt integrated approaches to address structural issues will make the current health model increasingly unsustainable. The challenges faced by the continuously rising healthcare costs are already high on the political agenda in many developed countries. For example, the United Kingdom is financially burdened with a National Health Service (NHS), which is the 5th largest global employer accounting for ~9.75% of GDP (OECD, 2015). Although there is much to commend advanced social health systems such as the UK's NHS, being relatively more efficient than nearly half OECD countries, is it not also an indication of the parlous state of human health and the reactive rather than preventive focus of healthcare systems? Perhaps even more significant in this debate, is that the much admired NHS is in danger of collapse (Iacobucci, 2016) whilst in the USA, expenditure continues to increase in the expansion of the 'Obamacare' social health system, causing significant political and financial angst (Congressional Budget Office, 2016). Significant per capita expenditure on health in the high income economies has had historic benefits, with improving longevity a key metric. However, this trend is tailing off in some countries e.g. England (Office for National Statistics, 2016), and many other health gains threatened by resurgence of bacterial infections associated with antimicrobial resistance and emergent novel pathogens and non-communicable diseases such as obesity. Kock (2013) in a prescient piece stated that 'Awareness of the decline in ecosystem, human and animal health, reversing the hitherto positive trends in human longevity, well-being and economy might be a more effective means of achieving a new political economy.';

this being necessary to shift the current development pathways, which seem increasingly associated with these trend shifts. Ironically, low-income countries in some ways are more resilient to these changes, for example Kenya is maintaining economic growth (6% GDP) and improving health and longevity (WHO, 2017) despite investment being as low as $169 per capita (World Bank, 2017), yet some countries spending as little as $32 per capita remain starkly disproportionate in terms of burden of disease.

Whatever the theoretical foundations are, so as to effectively implement this change towards a fully integrated approach to health, the added value will need to be demonstrated. However, we clearly need to shift our current sectoralised, linear focus to a more visible balanced health investment with more global benefits to all species. This is encapsulated in the movements for OH, EcoHealth, Planetary Health and Ecological public health which are essentially converging towards a paradigm shift for a more integrated approach to health.

Acknowledgements

This chapter is based upon work from COST Action 'Network for Evaluation of One Health' (TD1404), supported by COST (European Cooperation in Science and Technology).

References

Anonymous, 2009. One world one health™: from ideas to action. Report of the expert consultation. March 16-19, 2009. Winnipeg, Manitoba, Canada.

Babo Martins, S., Rushton, J. and Stark, K.D., 2015. Economic assessment of zoonoses surveillance in a 'One Health' context: a conceptual framework. Zoonoses Public Health 63(5): 386-395.

Bircher, J. and Kuruvilla, S., 2014. Defining health by addressing individual, social, and environmental determinants: new opportunities for health care and public health. J Publ Health Pol 35: 363-386.

Boa, E., Danielsen, S. and Haesen, S., 2015. Better together: identifying the benefits of a closer integration between plant health, agriculture and one health. In: Zinsstag, J., Schelling, E., Waltner-Toews, D., Whittaker, M. and Tanner, M. (eds.) One Health: the theory and practice of integrated health approaches. CABI, Wallingford, Oxfordshire, UK.

British Columbia Government, 2014. Animal Health Act, Chapter 16. The Queen's Printer, Victoria, British Columbia, Canada. Available at: https://tinyurl.com/yd8ahh2w.

Brown, V.A., Harris, J.A. and Russell, J.Y.E., 2010. Tackling wicked problems through the transdisciplinary imagination. Earthscan, Washington, DC, USA.

Convention on Biological Diversity (CBD), 2015. Connecting global priorities: biodiversity and human health: a state of knowledge review. World Health Organisation, Geneva, Switzerland.

Chakraborty, S. and Newton, A.C., 2011. Climate change, plant diseases and food security: an overview. Plant Pathol 60: 2-14.

Charron, D.F., 2012. Ecosystem approaches to health for a global sustainability agenda. Ecohealth 9: 256-266.

Congressional Budget Office (CBO), 2016. Federal subsidies for health insurance coverage for people under age 65: 2016 to 2026. Available at: https://tinyurl.com/yan74ysq.

Demaio, A.R. and Rockstrom, J., 2015. Human and planetary health: towards a common language. Lancet 386: e36-7.

Dora, C., Haines, A., Balbus, J., Fletcher, E., Adair-Rohani, H., Alabaster, G., Hossain, R., De Onis, M., Branca, F. and Neira, M., 2015. Indicators linking health and sustainability in the post-2015 development agenda. Lancet 385: 380-391.

Frankish, C.J., Green, L.W., Ratner, P.A., Chomik, T. and Larsen, C., 1996. Health impact assessment as a tool for population health promotion and public policy. Report submitted to the health promotion development division of Health Canada. Institute of Health Promotion Research, University of British Columbia, Vancouver, Canada.

Frenk, J. and Moon, S., 2013. Governance challenges in global health. N Engl J Med 368: 936-942.

Giovannini, E., Niestroy, I., Nilsson, M., Roure, F. and Spanos, M., 2015. The role of science, technology and innovation policies to foster the implementation of the sustainable development goals report of the expert group 'follow-up to rio+20, notably the SDGs'. European Commission, Brussels, Belgium.

Gostin, L.O., Debartolo, M.C. and Friedman, E.A., 2015. The international health regulations 10 years on: the governing framework for global health security. Lancet 386: 2222-2226.

Guimaraes, J.R.D. and Mergler, D., 2012. A virtuous cycle in the amazon: reducing mercury exposure from fish consumption requires sustainable agriculture. In: Charron, D.F.I.D.R.C. (ed.) Ecohealth research in practice: innovative applications of an ecosystem approach to health. Springer, New York, NY, USA.

Harris, A., Mohan, V., Flanagan, M. and Hill, R., 2012. Integrating family planning service provision into community-based marine conservation. Oryx 46: 179-186.

Häsler, B., Cornelsen, L., Bennani, H. and Rushton, J., 2014. A review of the metrics for one health benefits. Rev Sci Tech Off Int Epiz 33: 453-464.

Häsler, B., Gilbert, W., Jones, B.A., Pfeiffer, D.U., Rushton, J. and Otte, M.J., 2013. The economic value of one health in relation to the mitigation of zoonotic disease risks. Curr Top Microbiol Immunol 365: 127-151.

Häsler, B., Howe, K.S. and Stark, K.D., 2011. Conceptualising the technical relationship of animal disease surveillance to intervention and mitigation as a basis for economic analysis. BMC Health Serv Res 11: 225.

Haxton, E., Šinigoj, Š. and Rivière-Cinnamond, A., 2015. The network for evaluation of one health: evidence-based added value of one health. Infect Ecol Epidemiol 5: 1-2.

Heymann, D.L., Chen, L., Takemi, K., Fidler, D.P., Tappero, J.W., Thomas, M.J., Kenyon, T.A., Frieden, T.R., Yach, D., Nishtar, S., Kalache, A., Olliaro, P.L., Horby, P., Torreele, E., Gostin, L.O., Ndomondo-Sigonda, M., Carpenter, D., Rushton, S., Lillywhite, L., Devkota, B., Koser, K., Yates, R., Dhillon, R.S. and Rannan-Eliya, R.P., 2015. Global health security: the wider lessons from the West African Ebola virus disease epidemic. Lancet 385: 1884-1901.

Houle, K.L.F., 2015. Towards a healthy concept of health. In: Zinsstag, J., Schelling, E., Waltner-Toews, D., Whittaker, M. and Tanner, M. (eds.) One Health: the theory and practice of integrated health approaches. CABI, Wallingford, Oxfordshire, UK.

Iacobucci, G., 2016. Former ministers call for commission to review nhs and social care funding. BMJ 352: i95.

Jones, B.A., Betson, M. and Pfeiffer, D.U., 2017. Eco-social processes influencing infectious disease emergence and spread. Parasitology 144: 26-36.

Karesh, W.B., Mazet, J.A.K. and Clements, A., 2014. Upstream surveillance – scanning for zoonoses in wildlife. Intern J Infect Dis 21: 28.

Kock, R., 2013. Will the damage be done before we feel the heat? Infectious disease emergence and human response. Anim Health Res Rev 14: 127-132.

Kock, R., 2015. Structural One Health – are we there yet? Vet Rec 176: 140-142.

Koplan, J.P., Bond, T.C., Merson, M.H., Reddy, K.S., Rodriguez, M.H., Sewankambo, N.K. and Wasserheit, J.N., 2009. Towards a common definition of global health. Lancet 373: 1993-1995.

Lee, K. and Brumme, Z.L., 2013. Operationalizing the one health approach: the global governance challenges. Health Pol Plan 28: 778-785.

Lu, Y., Wang, R., Zhang, Y., Su, H., Wang, P., Jenkins, A., Ferrier, R.C., Bailey, M. and Squire, G., 2015. Ecosystem health towards sustainability. Ecosyst Health Sustain 1: 1-15.

Millennium Ecosystem Assessment, 2005. Ecosystems and human well-being: synthesis. Millennium Ecosystem Assessment, Washington, DC, USA.

Monroy, C., Bustamante, D.M., Pineda, S., Rodas, A., Castro, X., Ayala, V., Quinones, J. and Moguel, B., 2009. House improvements and community participation in the control of triatoma dimidiata re-infestation in Jutiapa, Guatemala. Cad Saude Publica 25, Suppl. 1: s168-178.

Moon, S., Sridhar, D., Pate, M.A., Jha, A.K., Clinton, C., Delaunay, S., Edwin, V., Fallah, M., Fidler, D.P., Garrett, L., Goosby, E., Gostin, L.O., Heymann, D.L., Lee, K., Leung, G.M., Morrison, J.S., Saavedra, J., Tanner, M., Leigh, J.A., Hawkins, B., Woskie, L.R. and Piot, P., 2015. Will Ebola change the game? Ten essential reforms before the next pandemic. The report of the Harvard-lshtm independent panel on the global response to Ebola. Lancet 386: 2204-2221.

National Academy of Sciences (NAS), 2015. Global considerations for animal agriculture research. Critical role of animal science research in food security and sustainability. National Academies Press, Washinton, DC, USA.

Organisation for Economic Cooperation and Development (OECD), 2015. Health at a glance 2015: OECD indicators. OECD Publishing, Paris, France.

Office for National Statistics, 2016. National life tables, UK: 2013-2015. Available at: https://tinyurl.com/m5u3hw2.

Osofsky, S.A., Cleaveland, S., Karesh, W.B., Kock, M.D., Nyhus, P.J., Starr, L. and Yang, A., 2005. Conservation and development interventions at the wildlife/livestock interface: implications for wildlife, livestock and human health. IUCN, Gland, Switzerland.

Ostrom, E., 2007. A diagnostic approach going beyond panaceas. Proc Nat Acad Sci 104: 15181-15187.

Porter, J.R., Xie, L., Challinor, A.J., Cochrane, K., Howden, S.M., Iqbal, M.M., Lobell, D.B. and Travasso, M.I., 2014. Food security and food production systems. In: Field, C.B., Barros, V.R., Dokken, D.J., Mach, K.J., Mastrandrea, M.D., Bilir, T.E., Chatterjee, M., Ebi, K.L., Estrada, Y.O., Genova, R.C., Girma, B., Kissel, E.S., Levy, A.N., Maccracken, S., Mastrandrea, P.R. and White, L.L. (eds.) Climate change 2014: impacts, adaptation, and vulnerability. Part a: global and sectoral aspects. Contribution of working group ii to the fifth assessment report of the Intergovernmental Panel on Climate Change. Cambridge University Press, Cambridge, UK.

Prüss-Ustün, A., Wolf, J., Corvalán, C., Bos, R. and Neira, M., 2016. Preventing disease through healthy environments: a global assessment of the burden of disease from environmental risks. WHO, Geneva, Switzerland.

Queenan, K., Garnier, J., Nielsen, L.R., Buttigieg, S., Meneghi, D.D., Holmberg, M., Zinsstag, J., Rüegg, S., Häsler, B. and Kock, R., 2017. Roadmap to a One Health agenda 2030. CAB Rev 2017 12: 14.

Queenan, K., Häsler, B. and Rushton, J., 2016. A One Health approach to antimicrobial resistance surveillance: is there a business case for it? Intern J Antimicrob Agent 48: 422-427.

Rabinowitz, P.M., Kock, R., Kachani, M., Kunkel, R., Thomas, J. and Gilbert, J.E.A., 2013. Toward proof of concept of a one health approach to disease prediction and control. Emerg Infect Dis 19: 12.

Rapport, D.J., Böhm, G., Buckingham, D., Cairns, J., Costanza, R., Karr, J.R., De Kruijf, H.A.M., Levins, R., Mcmichael, A.J., Nielsen, N.O. and Whitford, W.G., 1999. Ecosystem health: the concept, the ISEH, and the important tasks ahead. Ecosyst Health 5: 82-90.

Rapport, D.J., Costanza, R. and Mcmichael, A.J., 1998. Assessing ecosystem health. Tree 13: 397-402.

Rook, G.A., 2013. Regulation of the immune system by biodiversity from the natural environment: an ecosystem service essential to health. Proc Natl Acad Sci USA 110: 18360-18367.

Roth, F., Zinsstag, J., Orkhon, D., Chimed-Ochir, G., Hutton, G., Cosivi, O., Carrin, G. and Otte, J., 2003. Human health benefits from livestock vaccination for brucellosis: case study. Bull WHO 81: 867-876.

Rüegg, S.R., Mcmahon, B.J., Häsler, B., Esposito, R., Nielsen, L.R., Ifejika Speranza, C., Ehlinger, T., Peyre, M., Aragrande, M., Zinsstag, J., Davies, P., Mihalca, A.D., Buttigieg, S.C., Rushton, J., Carmo, L.P., De Meneghi, D., Canali, M., Filippitzi, M.E., Goutard, F.L., Ilieski, V., Milićević, D., O'shea, H., Radeski, M., Kock, R., Staines, A. and Lindberg, A., 2017. A blueprint to evaluate one health. Front Publ Health 5: 1-5.

Rushton, J., Hasler, B., De Haan, N. and Rushton, R., 2012. Economic benefits or drivers of a 'One Health' approach: why should anyone invest? Onderstepoort J Vet Res 79: 461.

Saracci, R., 1997. The World Health Organisation needs to reconsider its definition of health. BMJ 314: 1409-1410.

Schelling, E., Wyss, K.B., Chir, M., Daugla, D. and Zinsstag, J., 2005. Synergy between public health and veterinary services to deliver human and animal health interventions in rural low income settings. BMJ 331: 1264.

Stephen, C., 2014. Toward a modernized definition of wildlife health. J Wildlife Dis 50: 427-430.

UNEP/UNECE, 2016. Geo-6 assessment for the Pan-European region. United Nations Environment Programme, Nairobi, Kenya.

United Nations, 2015. Transforming our world: the 2030 agenda for sustainable development. General assembly 17th session, agenda items 15 and 116. UN, New York, NY, USA.

Valeix, S., Stein, C. and Bardosh, K., 2016. Knowledge flows in one health. The evolution of scientific collaboration networks. In: Bardosh, K. (ed.) One Health. Science, politics and zoonotic disease in Africa. Routledge, Abingdon, UK.

Waage, J., Yap, C., Bell, S., Levy, C., Mace, G., Pegram, T., Unterhalter, E., Dasandi, N., Hudson, D., Kock, R., Mayhew, S.H., Marx, C. and Poole, N., 2015. Governing sustainable development goals; interactions, infrastructures, and institutions. In: Waage, J. and Yap, C. (eds.) Thinking beyond sectors for sustainable development. Ubiquity Press, London, UK.

Wallace, R.G., Bergmann, L., Kock, R., Gilbert, M., Hogerwerf, L., Wallace, R. and Holmberg, M., 2015. The dawn of structural one health: a new science tracking disease emergence along circuits of capital. Soc Sci Med 129: 68-77.

Wallace, R.G., Gilbert, M., Wallace, R., Pittiglio, C., Mattioli, R. and Kock, R., 2014. Did Ebola emerge in West Africa by a policy-driven phase change in agroecology? Environ Planning 46: 2533-2542.

Wildlife Conservation Society (WCS), 2004. One world, one health: building interdisciplinary bridges to health in a globalized world. Available at: https://tinyurl.com/yckbxd5m.

Whitmee, S., Haines, A., Beyrer, C., Boltz, F., Capon, A.G., De Souza Dias, B.F., Ezeh, A., Frumkin, H., Gong, P., Head, P., Horton, R., Mace, G.M., Marten, R., Myers, S.S., Nishtar, S., Osofsky, S.A., Pattanayak, S.K., Pongsiri, M.J., Romanelli, C., Soucat, A., Vega, J. and Yach, D., 2015. Safeguarding human health in the anthropocene epoch: report of the rockefeller foundation-lancet commission on planetary health. Lancet 386: 1973-2028.

World Health Organisation (WHO), 1946. Preamble to the constitution of the World Health Organisation as adopted by the International Health Conference. June 19-22, 1946. New York, NY, USA.

World Health Organisation (WHO), 2015a. Health in 2015: from MDGS, millennium development goals to SDGS, sustainable development goals. WHO, Geneva, Switzerland.

World Health Organisation (WHO), 2015b. WHO publishes list of top emerging diseases likely to cause major epidemics. WHO, Geneva, Switzerland. Available at: https://tinyurl.com/obb3ndg.

World Health Organisation (WHO), 2017. Member counties: Kenya. Available at: https://tinyurl.com/y7qclkja.

World Bank, 2010. People, pathogens and our planet. Towards a One Health approach for controlling zoonotic diseases: the World Bank agriculture and rural development health, nutrition and population. World Bank, Washington, DC, USA.

World Bank, 2012. People, pathogens and our planet. Vol. 2. The economics of One Health. World Bank, Washington, DC, USA.

World Bank, 2017. Health expenditure per capita. World Bank, Washington, DC, USA.

Zinsstag, J., 2012. Convergence of ecohealth and One Health. Ecohealth 9: 371-373.

Zinsstag, J., Durr, S., Penny, M.A., Mindekem, R., Roth, F., Gonzalez, S.M., Naissengar, S. and Hattendorf, J., 2009. Transmission dynamics and economics of rabies control in dogs and humans in an African city. Proc Nat Acad Sci USA 106: 14996-15001.

Zinsstag, J., Mackenzie, J.S., Jeggo, M., Heymann, D.L., Patz, J.A. and Daszak, P., 2012. Mainstreaming One Health. Ecohealth 9: 107-110.

Zinsstag, J., Roth, F., Orkhon, D., Chimed-Ochir, G., Nansalmaa, M., Kolar, J. and Vounatsou, P., 2005. A model of animal-human Brucellosis transmission in Mongolia. Prev Vet Med 69: 77-95.

Zinsstag, J., Schelling, E., Waltner-Toews, D. and Tanner, M., 2011. From 'One Medicine' to 'One Health' and systemic approaches to health and well-being. Prev Vet Med 101: 148-156.

Zinsstag, J., Schelling, E., Waltner-Toews, D., Whittaker, M. and Tanner, M., 2015. One Health: the theory and practice of integrated health approaches. CABI, Wallingford, Oxfordshire, UK.

Chapter 3

A One Health evaluation framework

Photo: Rawpixel.com/Shutterstock.com

Simon R. Rüegg[1*], Barbara Häsler[2*], Liza Rosenbaum Nielsen[3*], Sandra C. Buttigieg[4], Mijalche Santa[5], Maurizio Aragrande[6], Massimo Canali[6], Timothy Ehlinger[7], Kevin Queenan[2], Ilias Chantziaras[8], Elena Boriani[9], Miroslav Radeski[10], Mieghan Bruce[11], Hans Keune[12,13], Houda Bennani[2], Chinwe Ifejika Speranza[14,15], Luís P. Carmo[16], Roberto Esposito[17], Maria-Eleni Filippitzi[8,18], K. Marie McIntyre[19], Barry J. McMahon[20], Marisa Peyre[21], Laura C. Falzon[22], Kevin L. Bardosh[23], Chiara Frazzoli[24], Tine Hald[25], Grace Marcus[2] and Jakob Zinsstag[26]

[1]Section of Epidemiology, Vetsuisse Faculty, University of Zürich, Winterthurerstrasse 270, 8057 Zürich, Switzerland; [2]Department of Pathobiology and Population Sciences, Veterinary Epidemiology Economics and Public Health Group, Royal Veterinary College, Hawkshead Lane, North Mymms, Hatfield, Hertfordshire, AL9 7TA, United Kingdom; [3]Department of Veterinary and Animal Sciences, Faculty of Health and Medical Sciences, University of Copenhagen, Grønnegårdsvej 8, 1870 Frederiksberg C, Denmark; [4]Department of Health Services Management, Faculty of Health Sciences, University of Malta, MSD2080, Msida, Malta; [5]Faculty of Economics – Skopje, Saints Cyril and Methodius University, Blvd Goce Delcev 9V, 1000 Skopje, FYR Macedonia; [6]Department of Agricultural and Food Sciences, University of Bologna, Viale Giuseppe Fanin 44, 40127 Bologna, Italy; [7]Center for Global Health Equity, University of Wisconsin Milwaukee, P.O. Box 413, Milwaukee, WI 53201, USA; [8]Faculty of Veterinary Medicine, Ghent University, Salisburylaan 133, 9820, Merelbeke, Belgium; [9]Global Decision Support Initiative (GDSI) and National Food Institute, Technical University of Denmark, Bygningstorvet, Building 115,2800 Kongens Lyngby, Denmark; [10]Faculty of Veterinary Medicine, Saints Cyril and Methodius University, Lazar Pop Trajkov 5-7, 1000 Skopje, FYR Macedonia; [11]School of Veterinary and Life Sciences, Murdoch University, 90 South Street, Perth, 6150, Australia; [12]Belgian Biodiversity Platform, Research Institute Nature & Forest (INBO), Herman Teirlinckgebouw, Havenlaan 88 bus 73, 1000 Brussels, Belgium; [13]University of Antwerp, Campus Drie Eiken, gebouw R R.3.07. Universiteitsplein 1, 2610 Wilrijk, Belgium; [14]Institute of Geography, University of Bern, Hallerstrasse 12, 3012 Bern, Switzerland; [15]Centre for Development and Environment, University of Bern, Mittelstrasse 43, 3012 Bern, Switzerland; [16]Veterinary Public Health Institute, Vetsuisse Faculty, University of Bern, Schwarzenburgstrasse 155, 3097 Liebefeld, Bern, Switzerland; [17]External Relations Office, Istituto Superiore di Sanità, Via Giano della Bella 34, 00199 Rome, Italy; [18]Faculty of Veterinary Medicine, Federal Research Institute Sciensano, Ernest Blerotstraat 1, Anderlecht, 1070, Brussels, Belgium; [19]Department of Epidemiology and Population Health, Institute of Infection and Global Health, University of Liverpool, Leahurst Campus, Neston, Cheshire CH64 7TE, United Kingdom; [20]UCD School of Agriculture and Food Science, University College Dublin, Belfield, Dublin 4, Ireland; [21]CIRAD, Avenue Agropolis, TA 178/04, 34398 Montpellier Cedex 5, France; [22]Institute of Infection and Global Health, University of Liverpool, 8 West Derby Street, Liverpool, L69 7BE, United Kingdom; [23]Department of Anthropology and Emerging Pathogens Institute, University of Florida, Turlington Hall, Room 1112, Gainesville, FL 32611, USA; [24]Department of Cardiovascular, Dysmetabolic and Aging-Associated Diseases, Istituto Superiore di Sanità, Via Giano della Bella 34, 00199 Rome, Italy; [25]National Food Institute, Technical University of Denmark, B204, Kemitorvet, 2800 Kgs. Lyngby, Denmark; [26]Department of Epidemiology and Public Health, Swiss Tropical and Public Health Institute, University of Basel, P.O. Box, 4002 Basel, Switzerland; srueegg@vetclinics.uzh.ch; bhaesler@rvc.ac.uk; liza@sund.ku.dk

Abstract

Challenges calling for integrated approaches to health, such as the One Health (OH) approach, typically arise from the intertwined spheres of humans and animals, and the ecosystems constituting their environment. Initiatives addressing such wicked problems commonly consist of complex structures and dynamics. The Network for Evaluation of One Health (NEOH) proposes an evaluation framework anchored in systems theory to address the intrinsic complexity of OH initiatives and regards them as subsystems of the context within which they operate. Typically, they intend to influence a system with a view to improve human, animal, and environmental health. The NEOH evaluation framework consists of four overarching elements, namely: (1) the definition of the OH initiative and its context; (2) the description of its theory of change with an assessment of expected and unexpected outcomes; (3) the process evaluation of operational and supporting infrastructures (the 'OH-ness'); and (4) an assessment of the association(s) between the process evaluation and the outcomes produced. It relies on a mixed-methods approach by combining a descriptive and qualitative assessment with a semi-quantitative scoring for the evaluation of the degree and structural balance of 'OH-ness' (summarised in an OH-index and OH-ratio, respectively) and conventional metrics for different outcomes in a multi-criteria-decision analysis. We provide the methodology for all elements, including ready-to-use Microsoft Excel spread-sheets for the assessment of the 'OH-ness' (Element 3) and further helpful worksheets as electronic supplements. Element 4 connects the results from the assessment of the 'OH-ness' to the methods and metrics described in Chapters 4 to 6 in this handbook. Finally, we offer some guidance on how to produce recommendations based on the results. The presented approach helps researchers, practitioners, policy makers and evaluators to conceptualise and conduct evaluations of integrated approaches to health and enables comparison and learning across different OH activities, thereby facilitating decisions on strategy and resource allocation. Examples of the application of this framework have been described in eight case studies, published in a dedicated Frontiers Research Topic (https://www.frontiersin.org/research-topics/5479).

Keywords: One Health, transdisciplinary, integrated approaches to health, evaluation framework, theory of change

3.1 Introduction

Many current health challenges, such as spread of zoonotic infectious diseases, environmental pollutants, antimicrobial resistance, climate or market-driven food system changes with consequences on food and feed supplies, malnutrition including obesity and many more arise from the intertwined spheres of humans, animals and the ecosystems constituting their environment (FAO, 2013; Jones *et al.*, 2008). They are recognised to be wicked problems and need to be tackled using integrated approaches to health (Pfeiffer, 2014; Romanelli *et al.*, 2015; Whitmee *et al.*, 2015). Here, we conceptualise integration as inter- or transdisciplinary approaches. Such approaches consider the needs, values and opinions of multiple disciplines, sectors and stakeholders. They also bring together the scientific and non-scientific communities, influencing, or influenced by, the challenge and their combined know-how and resources (Rüegg *et al.*, 2017; Stokols *et al.*, 2013; Zinsstag *et al.*, 2011). Due to the existing, historically contingent, separation of sectors and disciplines, developing integrated approaches is difficult and the realisation of benefits can be delayed. There is a need to provide evidence on the added value of these integrated and transdisciplinary approaches to governments, researchers, funding bodies, and stakeholders (Ledford, 2015; Rabinowitz *et al.*, 2013; Stokols *et al.*, 2003).

The NEOH evaluation framework uses a systems approach and regards the context of a OH initiative as the system within which it operates, and the initiative itself as a subsystem, which has a potential to affect the system to a smaller or larger degree. Drivers, operations, supporting infrastructure and outcomes were identified as fundamental characteristics of OH initiatives (Rüegg *et al.*, 2017). The NEOH evaluation framework relates the aspects of operations (i.e. OH thinking, OH planning and OH working) and supporting infrastructure (i.e. systemic organisation, learning and sharing) summarised as OH process characteristics ('One Health-ness'), to changes and outcomes evoked by a specific initiative. This is an important step towards identifying added value arising from integration across disciplines and sectors (i.e. transdisciplinarity). Figure 3.1 illustrates the relations between drivers, operations, supporting infrastructure and outcomes of OH and how the system evolves when a OH approach is engaged (Rüegg *et al.*, 2017).

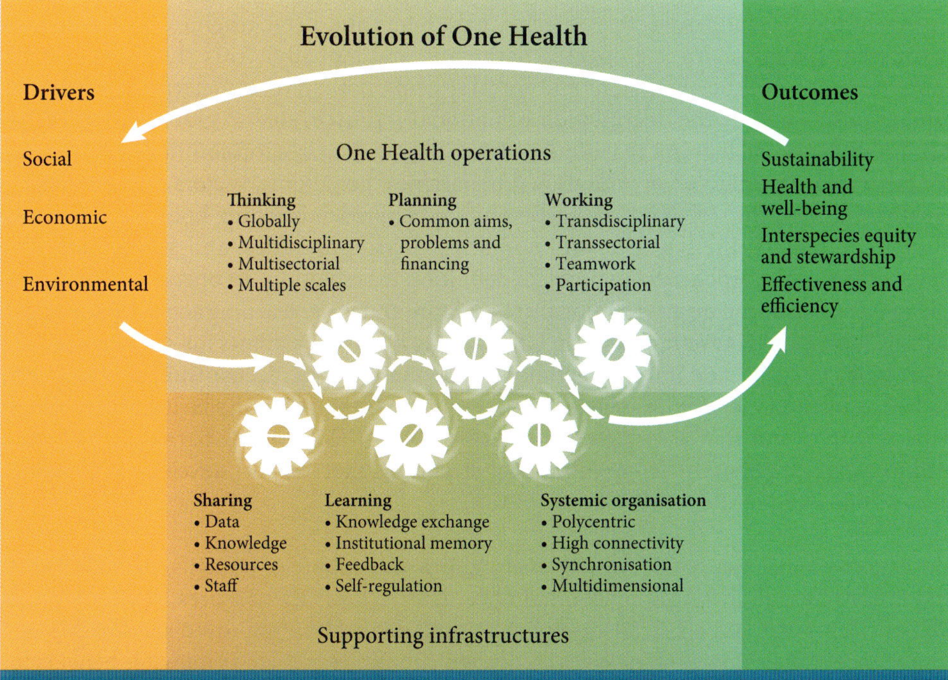

Figure 3.1. One Health characteristics identified during a workshop held in Cluj, Romania, June 2015, by members of the COST Action TD1404: Network for Evaluation of One Health. Published in Rüegg *et al.* (2017).

In brief, drivers refer to a collective perception of multiple and complex origins behind health problems, such as social (Commission on Social Determinants of Health, 2008) and environmental determinants of health (Lang and Rayner, 2012), as well as economic drivers (Woodward *et al.*, 2001). Social drivers include lack of participation or well-being, as well as the presence of ignorance, poverty, poor governance, mental and physical illness, or high risks for these. Environmental drivers include climate change, land degradation, and other ecosystem changes rooted in both natural phenomena as well as human actions. Economic drivers are mostly related to the globalisation process, dominated by market deregulation and financial capital, and largely irrespective of social needs at the local level (Rayner and Lang, 2012; Woodward *et al.*, 2001). These examples are by no means exhaustive and there is clearly an interplay between different drivers. For example, increased poverty in conjunction with close contact to previously unexploited environments puts human and animal health at risk (Pfeiffer, 2014). Similarly, economic crises and financial deregulation reduce public resources for interventions, reinforcing negative environmental, economic and social drivers, and exacerbating negative health outcomes (Khanal and Bhattarai, 2016).

As a response to these drivers, OH initiatives can range from development projects to educational programmes, research projects and inter-governmental strategies. Although disparate, these initiatives often have specific operating principles, characterised by a way of thinking, planning and working. 'OH thinking' is holistic, inclusive, respectful and tolerant, as opposed to approaches that are specific, reductionist, with a tendency to focus on single or limited outcomes that impact positively on few people only. It considers multiple scales (levels) of life, disciplines, sectors, species, paradigms and demographics, and integrates at different spatial scales (e.g. locally, nationally and globally). This should reflect the connected nature of social relations and social systems, both in their material and symbolic dimensions as well as the degradation of national resources due to globalisation (Wolf, 2015). 'OH planning' requires that aims, problem formulation, responsibilities and financing are organised regardless of organisational hierarchies, paradigms, sectors and disciplines. Finally, 'OH working' relies on a transdisciplinary approach bridging knowledge between disciplines, sectors, the scientific and non-scientific communities, and actively includes stakeholders in the process, from problem definition to resolution. To operate as conceived, OH must rely on adequate information infrastructure and foster learning across all scales and fields (Ciborra and Hanseth, 1998). An OH learning framework allows for stakeholders and institutions to evolve and improve autonomously, and requires mechanisms for knowledge exchange, institutional memory, feedback and regulation. This relies on sharing of knowledge, data, resources and staff across sectors and disciplines. This working paradigm will often lead to complex, poly-centric organisational structures that support development towards sustainability and resilience (Retief *et al.*, 2016).

The expected outcomes of OH initiatives are health and welfare of humans, animals, plants and ecosystems, all managed by common health strategies. This ensures healthy food, as well as clean water and air. Transdisciplinarity should result in improved stewardship and compliance, and promote interspecies equity, which would facilitate sustainable benefits for humans from other species (domestic and wild) and their habitats. Furthermore, OH should improve effectiveness across different sectors and at multiple scales. It relies on and results in more efficient communication, thereby generating a higher degree of awareness that can enable rapid detection of illness and consequent action. By having a more inclusive voice for neglected human populations, animals and ecosystems, OH is intended to widen our usual anthropocentric perspectives, and to simultaneously enhance human health. The expected outcomes of OH approaches contribute to the three pillars of sustainability, namely society, environment and economy.

3.2 Evaluation framework and steps

Figure 3.2 provides an overview of the NEOH evaluation framework. There are four overarching Elements in the evaluation process:

- Element 1: defining and describing the OH initiative and its context (i.e. the system, its boundaries, and the OH initiative as a subsystem), providing information for the further Elements.
- Element 2: assessing expected outcomes based on the theory of change (TOC) of the initiative, and collecting unexpected outcomes emerging in the context of the initiative.
- Element 3: assessing the 'One Health-ness', i.e. the implementation of operations and infrastructure contributing to the OH initiative.
- Element 4: comparing the degree of 'One Health-ness' and the outcomes produced.

The framework relies on a mixed methods approach that combines a descriptive and qualitative assessment with a semi-quantitative evaluation (scoring) for the evaluation of the 'One Health-ness' with a OH-index, while including conventional metrics for outcomes in a multi-criteria-decision-analysis.

The following chapters translate the schematic into distinct steps to be considered from defining the system to characterising the OH initiative to elaborating a TOC to identifying and selecting the evaluation type and metrics for outcomes.

The framework can be used for external or self-evaluation. It is recommended that the evaluator is comfortable with systems thinking (Trochim *et al.*, 2006; Whitehead *et al.*, 2015) to approach the complex structures and dynamics of OH initiatives and their context. Data and information can be gathered from actors and stakeholders using methods such as open or semi-structured interviews, focus group discussions or other qualitative data collection approaches, from resources used or produced by the initiative (Garcia and Zazueta, 2015), and related (external) primary or secondary datasets.

For examples that apply the method presented here, the readers can refer to the case studies included in the Frontiers research topic 'Concepts and experiences in framing, integration and evaluation of One Health and EcoHealth' (https://www.frontiersin.org/research-topics/5479). Paternoster and co-workers evaluated integrated surveillance of West-Nile virus (Paternoster *et al.*, 2017), Radeski and co-authors applied the framework to an animal welfare centre (Radeski *et al.*, 2018), Léger and co-workers evaluated a research project on antimicrobial resistance involving four faculties, the industry and health authorities (Léger *et al.*, in press), Buttigieg and collaborators compared control strategies for brucellosis in Serbia and Malta

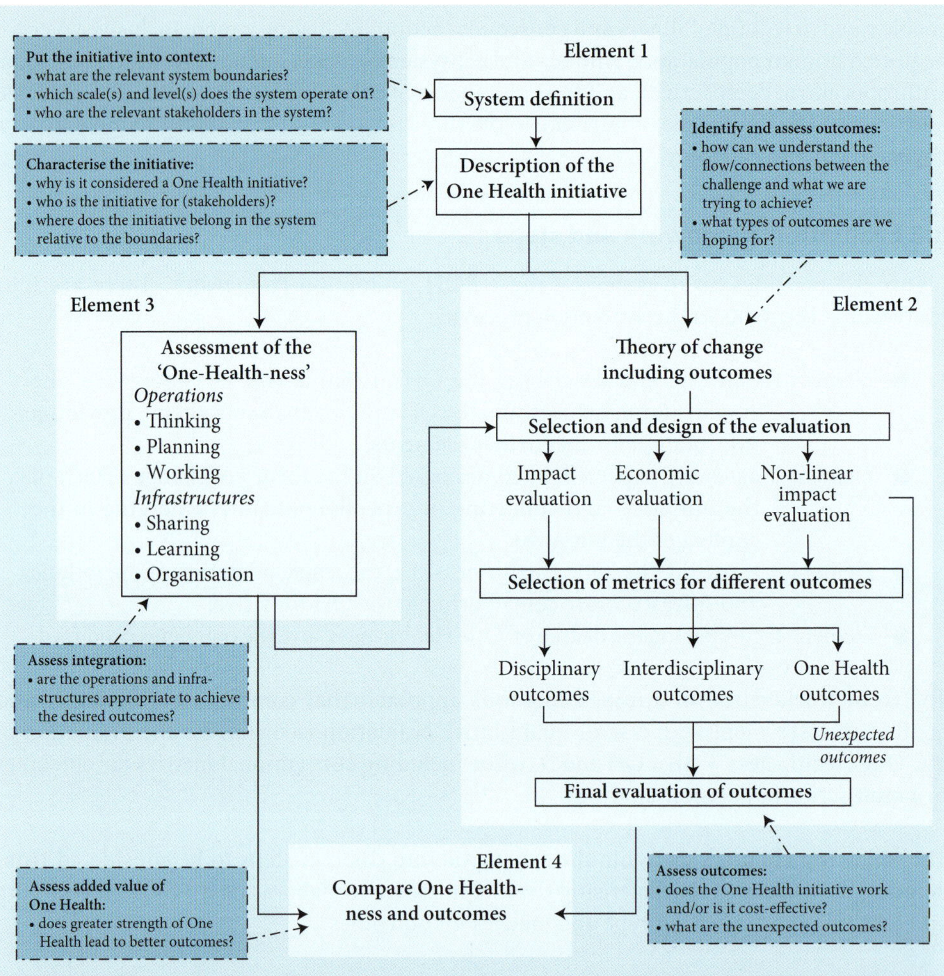

Figure 3.2. Flow chart of elements to be considered during a One Health evaluation (in grey) with their purpose and the associated questions to be answered (blue boxes). In Element 1, the initiative and its context are described to inform Element 2 and 3. Element 2 relies on a Theory of Change to identify expected outcomes and collects unexpected outcomes through non-linear impact assessment. In Element 3 the implementation of operations and infrastructure contributing to the One Health initiative is assessed. The two assessments are compared in Element 4. Published in Rüegg *et al.* (2018).

(Buttigieg *et al.*, 2018), Muñoz-Prieto and co-workers assessed a study on factors affecting obesity in dogs and dog-owners (Muñoz-Prieto *et al.*, 2018), Laing and collaborators evaluated a project mitigating the effects of the unexpected domestic re-use of containers employed for organophosphates in a tick control programme (Laing *et al.*, 2018), Fonseca and co-authors applied the framework to evaluate a cross-sectoral observatory of taeniasis and cysticercosis

(Fonseca *et al.*, 2018), and finally Hanin and collaborators evaluated an international and inter-sectoral centre for infectious disease surveillance (Hanin *et al.*, 2018).

3.3 Element 1: definition of the OH initiative and its context

Element 1 of the evaluation framework (Figure 3.2) consists of a general overview (Section 3.3.1), a visual representation and a textual description of the system in which the initiative operates (Section 3.3.2), and an analogous illustration and description of the initiative within this context (Section 3.3.3). They do not need to be developed in sequence, but may evolve iteratively, and may be developed by a group of evaluators, by the stakeholders of the initiative, or by these two groups in collaboration.

Before designing an evaluation, the evaluation question(s) must be clearly stated. To answer these questions and to select an adequate evaluation design, it is important to gain a principle understanding and overview of the activities to be evaluated (Williams, 2016). The framework presented here uses a systems approach and regards the context of an OH initiative as the system within which it operates, and the initiative itself as a subsystem conceived to induce change in this context. Systems have been defined in many different disciplines and frameworks e.g. (Anderson and Johnson, 1997; Ifejika Speranza *et al.*, 2014; Meadows, 2008; Whitehead *et al.*, 2015; WHO, 2009). A fundamental feature is that systems are composed of a set of interacting or interdependent components that form a complex whole (Anderson and Johnson, 1997). This implies a hierarchical organisation and a concept of levels or scales within different dimensions (Pumain *et al.*, 2006). Although the term 'level' is used ambiguously in science, the concept used here is that of 'grades of being ordered', which captures what biologists and social scientists refer to as 'levels of organisation' (Bunge, 1960). Three such grades or levels can be identified at which OH outcomes are usually measured: individual level of health, population level of health and ecosystem level of health (Lerner and Berg, 2015). Systems can be considered as a network of components, which can be tangible (e.g. humans, animals, forests, lakes) or intangible (e.g. cultural behaviours, values, norms, language expressions) and which are linked by interactions (Anderson and Johnson, 1997; WHO, 2009). The system's components depend on the perspective and determine its boundaries, which are important for evaluation (Garcia and Zazueta, 2015). While the perspectives of stakeholders (and thus system boundaries) may differ, the stakeholders may become agents of change or part of a pathway towards successful solutions (Ostrom, 2009; WHO, 2009; Williams, 2016). OH initiatives might create additional opportunities to produce relevant – expected as well as unexpected – outcomes by including stakeholders and system boundaries explicitly (Figure 3.2).

3.3.1 The general overview

For the general overview, the evaluator should put together a concise description of the background, objectives, key features and rationale of the OH initiative under evaluation so that the user is aware of the important characteristics that can affect the evaluation.

3.3.2 Visual representation and textual description of the context

Here the focus is specifically on the system targeted by the OH initiative; in other words the wider context within which the initiative operates. We will describe the initiative itself later. For the visual representation of the system (Figure 3.3), we propose a combination of the socio-ecological system framework by Ostrom and a causal loop diagram (Anderson and Johnson, 1997; Ostrom, 2009).

To capture the socio-ecological system, three core subsystems are plotted first (Figure 3.3): the resource systems (blue ovals), the resource units they provide (blue boxes), and the governing systems (grey boxes). In the next step, further tangible and intangible components relevant to the system (white ovals, e.g. use of antibiotics, effectiveness of antimicrobials) are added. For legibility of the graph it is recommended to use nouns that fit into phrases such as 'the level of…', to avoid verbs and to use neutral terms, e.g. 'use of antimicrobials' rather than 'increase of antimicrobial use'. Finally, relationships are added as arrows: governance relations (grey), membership relations (black) and causal relations (blue). For causal relations, it is useful to note the relation using S for same direction change and O for opposite direction change, in order to identify reinforcing and balancing loops at a later stage. Subscripts and explanatory text as well as annotations of time delays can be convenient for later reference.

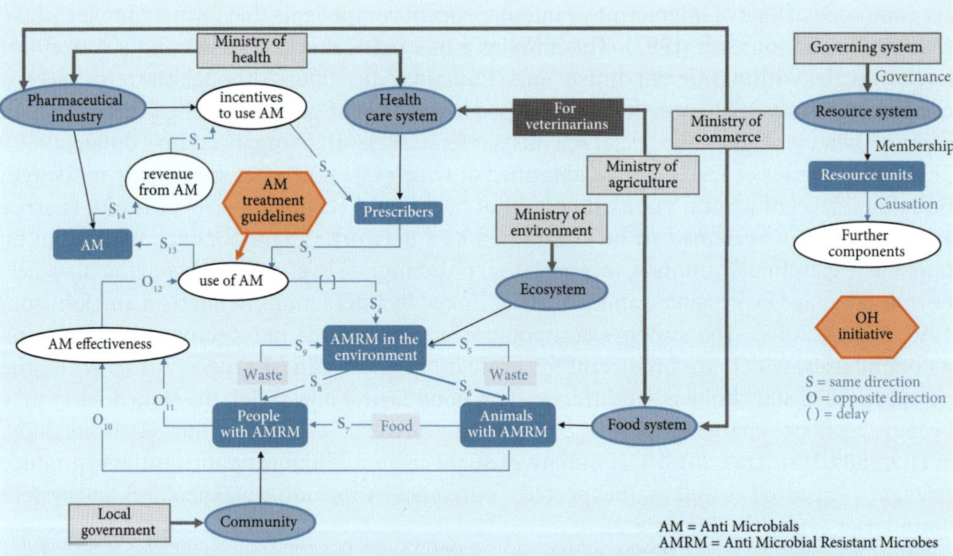

Figure 3.3. Example for visual representation of an initiative in its context exemplified by occurrence of antimicrobial resistance within a given system: resource systems (blue ovals), resource units (blue boxes), and governance systems (grey boxes) within which an initiative operates. Furthermore, tangible and intangible components (white ovals) are included. Relationships (arrows) are classified as governance (grey), membership (black) and causal interactions (blue) with explanatory text. Letters designate changes of two components in the same (S) or opposite (O) direction, respectively. The red hexagon represents the initiative with arrows where it impacts the system. Published in (Rüegg *et al.*, 2018).

Visual representation is powerful, but lacks any dimension beyond the plane and therefore hinders the depiction of overlapping sub-systems or nested hierarchies. Hence, to explore further the system in which the OH initiative operates, the textual description is guided by three questions formulated by Williams (2016): (1) to understand interrelationships: What is the reality we are dealing with?; (2) to engage with perspectives: How do we understand/how do we see that reality?; and (3) to reflect on boundaries: How do we decide to do what needs to be done? (Williams, 2016). In Table 3.1 we adapted the tabular system description by Boriani *et al.* (2017) for a broader application. It allows capturing aspects complementary to the graph and sometimes overlapping, namely the aim of the system, the stakeholders and actors and their interactions, the system dimensions with corresponding boundaries, and the system evolution.

The aim and/or indicators of the system are not to be confused with the aim of the initiative and should answer the question 'why does the system exist?' or 'what does it produce?', e.g. the result of a food chain may be to 'produce Salami'. A social-ecological system may not have an explicit aim, but it can be characterised by indicators that allow the description of selected attributes, such as resilience, productivity or health. In this evaluation framework, we differentiate between the declared aim by the system and the observed, enacted and the perceived aims. The declared aim of a veterinary practice may be to provide animal health services. However, this will be enacted within a socio-economic context, which may result in therapeutic choices that prioritize practice income over animal welfare. These actions may be observed by a subset of clients, while others do not notice them. Each stakeholder may have a different perception of the declared aim and again, each of them can have a different way of interpreting how the system is performing in relation to its aim (Anderson and Johnson, 1997). In socio-ecological systems the perceptions differ mainly in regard to the way one verifies if the system is intact/healthy. This is important as it explains the motivational background of the concerned stakeholders. If the system has an explicit aim, specific indicators should be identified and compared to indicators used by stakeholders to assess their perceived aim(s), thereby shedding light on discrepancies and identifying ways of resolving them.

Following the interactive terminology for Europe (Anonymous, 1999), we define stakeholders as 'any individual, group or organisation who may affect, be affected by, or perceive themselves to be affected by a decision or activity', while actors are a subgroup of stakeholders such as 'any individual, group or organisation who acts, or takes part' in system activities. To gain clarity about roles of stakeholders, we recommend referring to the visual representation of the system exemplified in Figure 3.3 and probe for 'who is involved in the system as an actor and who is merely affected?'. For example, the pharmaceutical industry produces a certain compound, people can decide whether to take that compound or not, while animals are affected by a certain preparation distributed to them by an actor in the system (e.g. veterinarian or owner). An overview of relevant actors and stakeholders allows further delimiting the system under evaluation. Stakeholders could be actors at the same time, and in these situations, the capacity that a group is stakeholder or actor, respectively, should be differentiated.

In order to understand the context of the OH initiative, it is important to understand how the components of the system are arranged or interact (Williams, 2016). There are four aspects of relationships that should be considered and described: (1) the structure or arrangement

Table 3.1. An overview of how to describe the system at which the One Health initiative is targeted, i.e. the context of the initiative (adapted from Boriani *et al.*, 2017).

Aspect	Description	Secondary questions	Evolution
Aims	What is the context of the OH initiative - why does this system exist? What does it produce? For social-ecological systems that have no explicit aim, what are indicators that the system is intact/ healthy?	Perspectives • What does the system aim to do? Are there different declarations? • What do the actors and stakeholders perceive the system does and how do those perceptions differ? (For social-ecological systems: how do the actors and stakeholders perceive/evaluate that the system is intact/operational?) • Are there measurable outcomes/ indicators of the system? • How do the declared, perceived and measured aims/outcomes relate?	Do the various aims/ indicators change as the system evolves with time?
Actors	Who are the actors? Who acts within the system?	Relationships • How do they affect the other actors/ stakeholders and the aim of the system? • How are they affected by the other actors/stakeholders and the aim/ indicators of the system? • How are the relationships distributed/ arranged? • Which are the most important links? • What are the processes between the related components? • How can the links be characterised (slow/fast, strong/weak)?	Do the actors change their activity and behaviours as the system evolves (new trade-offs)? Does the system have secondary effects on the actors?
Stakeholders	Who are the stakeholders? Who is affected by the system?	Relationships • How are they affected by the actors and the dynamics of the system? • How are the relationships distributed/ arranged? • Which are the most important links? • What is the nature of the processes between the related components? • How can the links be characterised (slow/fast, strong/weak)?	Does the system have secondary effects on the stakeholders?

Table 3.1. Continued.

Aspect	Description	Secondary questions	Evolution
Geographical dimension	Which geographical space does the system occupy and where is it situated (surface concerned, climate, location)?	Boundaries • How is the system delimited in geographical area? • How do these boundaries affect the system aims/indicators and dynamics?	Does the system have secondary effects in geographical space within the boundaries? Does the system produce 'externalities' in geographical space?
Temporal dimension	Which is the most important time scale in which events are happening in the system (e.g. minutes, months, years)? Are there other important time scales?	Boundaries • How is the system delimited in time? Is it infinite, terminated, transient? • How does this time limit affect the system aims/indicators?	Does the system affect the frequency of events or its own time limit? Does the system produce 'externalities' in time (accelerating or slowing down external systems)?
Governance/ institutional dimension	Which governance entities/levels are involved (shire, agglomeration, state, nation, or international space)? What institutional structures (companies, corporations, organisations) play a role?	Boundaries • How is the system delimited in the governance/institutional dimension? • How do these boundaries affect the system aims/indicators?	Does the system have secondary effects in the governance/ institutional dimension within the boundaries? Does the system produce 'externalities' in the governance/ institutional dimension?
Further dimensions	How does the system extend within this dimension and how many levels of this dimension are part of the system?	Boundaries • How are these dimensions delimited? • How do these boundaries affect the system aims/indicators?	Does the system have secondary effects in these dimensions within the boundaries? Does the system produce 'externalities' in these dimensions?

of the links between the components (topology); (2) the nature of the processes between the components (e.g. information flow, transfer of goods, etc.); (3) the characteristics of the links (slow/fast, strong/weak, antagonistic/synergistic, etc.); and (4) identifying the links that are most important in the system.

Dimensions are defined as spaces in which levels of organisation according to Bunge occur (Bunge, 1960). In other words, entities within a dimension feature the same quality (e.g. metric) but to a different degree. Examples include geographical space, time, governance/institutional, economic, linguistic, faith and value dimensions. Within these dimensions we consider scales or levels of analysis, e.g. cell – organism – population in the dimension of life (Pumain *et al.*, 2006, pp. 39-70). These levels are important, because they will determine the relationship between the resolution of the analysis and the resolution of observations and what can be measured or evaluated in the system in a particular dimension. Due to their importance, geographical, temporal and governance/institutional dimensions are included.

Time, in particular, is related to the scale in other dimensions, i.e. the larger the system the larger its characteristic time, which is the time at which average change occurs (e.g. cells react within milliseconds, individuals between minutes and hours, ecosystems between years and decades; the same applies to the adaptability of laws at different scales or the frequency that vocabulary is used in a language) (Pumain *et al.*, 2006). Together with geographical space, time is a particularly important dimension, because it will characterize if the system is evolving over seconds, hours, days, years, decades or even longer. It can be considered in the past, present or future, and opportunities to affect the system are highly dependent on time due to the system disposition (the same intervention may have different effects when applied at different times). Furthermore, causes and effects may occur in different time scales, where short actions may result in effects with a time lag of years. The governance/institutional dimension will determine which organisational levels (ranging from international governance mechanisms to household structures) are represented and addressed in an initiative. Considering scales is important, because initiatives may aim to change systems at several different levels according to the most promising leverage points. Consequently, well intended initiatives may remain ineffective if they do not address all appropriate levels.

Further dimensions are the 'dimension of life' (or 'biology') comprising nested living entities from cells to biosphere with levels such as 'cell', 'organ' and 'individual', the 'economic dimension' defined by rules and institutions involved in production, trade and exchange of goods and services, the 'linguistic dimension' delimited by languages and dialects used, and the 'faith/value dimension', which represents the values and beliefs underlying the system. Other dimensions may also be relevant to the system, such as communication, transportation, legal frame, socio-cultural dimensions and many others.

The primary importance of a systems approach to evaluation implies less the idea of being comprehensive, but rather being 'thoughtful, smart and aware about what you are leaving out' (Williams, 2016). The evaluator(s) will need to be transparent about the consequences of choices and declare their relation to the initiative, the system and the evaluation per se. Although the dynamics, boundaries and stakeholders of a system are clear, they will be constrained by physical limits (e.g. a mountain range, river), social limits (e.g. country, community), regulations (e.g. quotas, prohibitions) and/or other norms (e.g. social norms, religious norms) that are either imposed by the systems' nature or selected by the evaluators (Garcia and Zazueta, 2015). Many restricting factors will be found in the system dimensions identified earlier. For example, a food system can be limited due to production regulations (e.g. the previous milk quotas system in Europe), food hygiene standards (e.g. restrictions on raw milk consumption), or cultural practices (e.g. no pork consumption in certain faith groups). The system boundaries characterise the interaction between the context of the initiative with the broader world in which it is imbedded, and determine how this affects the aim of the system (Garcia and Zazueta, 2015). Finally, dimensions can also interact and may even be closely correlated, to the extent that it may not be useful to differentiate them (e.g. when religious beliefs are prescribed by the law).

The evolution of a system can be regarded as interaction of time with other dimensions in terms of iterations and pathways along those dimensions and time. Apart from the aim of the system, the interactions in the system may produce secondary effects within the system and 'externalities' beyond the boundaries as it evolves. Highly self-organising systems may even change their (aim) dynamics and boundaries as time goes by.

3.3.3 Illustration and description of the OH initiative within the context

In a next step, the OH initiative can be added to the visual representation of the context to illustrate its effects on various components and their interactions (Figure 3.3). If an affected component is missing, it is added and the system graph is corrected accordingly. In the example in Figure 3.3, we have included a hypothetical OH initiative that involves new antimicrobial treatment guidelines for veterinarians and general practitioners (prescribers) that are assumed to impact directly on the amount and distributions of types of antimicrobials used in the system.

The user should now have a clear understanding of the system in which the OH initiative is situated. Next, the initiative itself is described using the template in Table 3.1 in analogy, namely as a nested subsystem of the context which it aims to change. Many elements may be congruent, but the boundaries of the initiative will inevitably be smaller and there will be fewer actors, stakeholders and more limitations than in the description of the system. Care should be taken, as actors and stakeholders and their particular roles may not be identical in the initiative and in the wider system. The initiative may be likely to consider fewer dimensions compared to the system, but it is important to identify how it will influence the context and what the limitation of the actions are. A key question in this description is: How is OH conceptualised by the various participants and is there a common understanding?

3.4 Element 2: the theory of change, outcomes, evaluation design and implementation

Element 2 involves an elaboration of the TOC, which helps to explain how an initiative is intended to produce the desired (or expected) outcomes. This is an important step to define the evaluation question and to choose the evaluation methods and metrics. It entails generating hypotheses about the causal mechanisms by which the components and activities of the initiative produce outcomes by asking pertinent questions about: (1) why people expect the initiative to bring about the change(s) and the outcome(s) they seek, (2) to question their assumptions about how the change process will unfold, and (3) to be clear about how they are selecting outcomes to focus on, in the evaluation. Identifying and developing a theoretical understanding of the likely process of change is a key task to evaluate successfully complex initiatives (Craig *et al.*, 2013). It also provides an opportunity for stakeholders to assess what they can influence, what impact they can have, and whether it is realistic to expect their goals to be reached with the time and resources they have available.

Measuring (or assessing) change in multiple outcomes, facilitates the evaluation of whether the OH initiative works as intended and whether it is cost-effective. In addition, unexpected outcomes may arise from an OH initiative. A good description and understanding of the system and OH initiative in Element 1 facilitates the identification of interactions and dynamics that may lead to unexpected and indirect outcomes not specified by the TOC. This framework standardises the evaluation through a systematic approach based on the TOC, while explicitly remaining open for potentially emerging systemic effects through non-linear impact evaluation (Figure 3.2). During the implementation of an initiative, the TOC can be reviewed based on progress. Retrospectively, it helps to inform a reflective process of learning about what has worked and why, as part of an evaluation process (Taplin *et al.*, 2013).

3.4.1 Description of the theory of change

Essentially, the TOC presents a roadmap with all the building blocks required to bring about a desired (long-term) goal; it hence, spells out the logic behind the initiative. The presentation of the TOC can be assisted by a graphical presentation (e.g. Figure 3.4), or the TOC description can refer back to the illustration of the system used in Element 1.

The impact is defined as the long-term effects (or goals) to be induced by an OH initiative. It is a change that continues to exist after the end of the initiative, and can be a direct (first order) or indirect (second order) impact. Outcomes are changes (e.g. improvement, learning) resulting from the initiative that can be considered to be stepping stones for progress towards the longer-term goals. In a transdisciplinary process, the outcomes are situated in societal and scientific practice and can be of multiple natures (e.g. technical, economic, social, sanitary, political) (Lang *et al.*, 2012). Outputs are products, goods and services, which result from the transdisciplinary process of an OH initiative and are necessary for the achievement of outcomes. For illustration, we use an example from a fictive research project aiming to produce new knowledge and methods to combat the development of antimicrobial resistance (Figure 3.4): OH research outputs (new data and knowledge) result in new treatment guidelines (outcome) leading to new regulations restricting (and hence lowering) the use of specific antimicrobials in farmed animals (first order impact of political nature), which then may reduce the development of antimicrobial resistance in farmed animals and the associated transmission to people (second order societal impact). The impacts can be realised at different political levels (e.g. individual, institutional, regional, national, international) and can consist of different types of effects (positive or negative; direct or indirect). Outcomes for societal and scientific practice (e.g. an improved integrated surveillance programme for antimicrobial resistance or a new simulation model, respectively) are disseminated, adapted and applied by other actors, resulting in societal impact or scientific progress. Between the initial problem formulation and the expected impact(s), new inputs might be required as a result of intermediary outcomes and will feed a further iteration of knowledge co-production. An example could be new research collaborations such as the outcome of an OH initiative, which may lead to new knowledge or tools for improved control of infectious diseases in

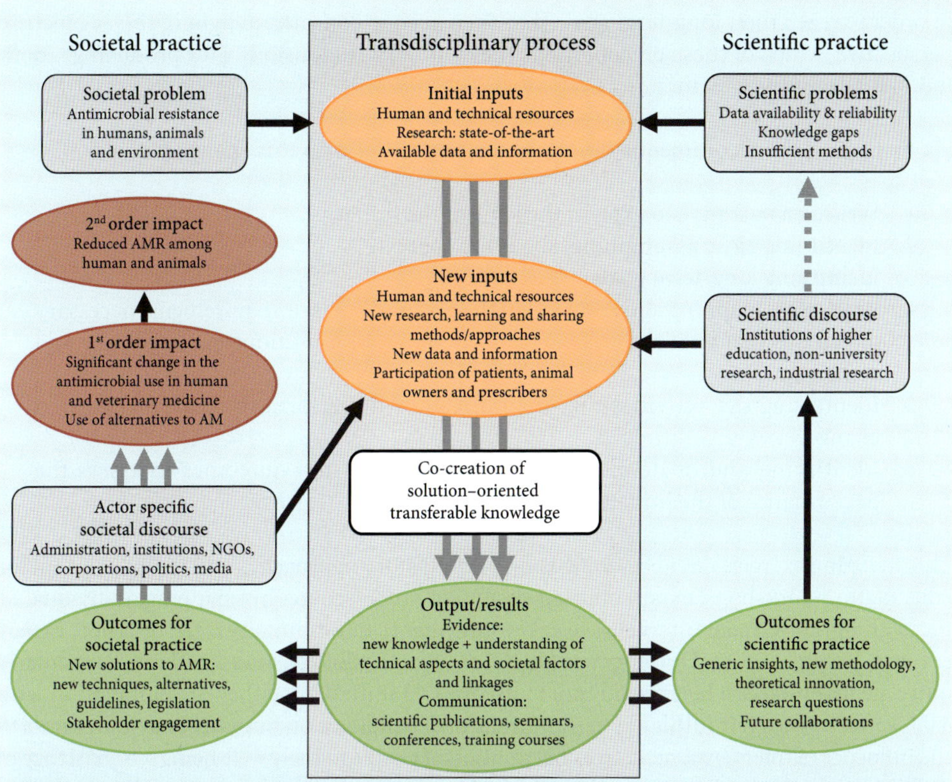

Figure 3.4. The change pathway for a fictive One Health research initiative aiming to mitigate the development of antimicrobial resistance in a transdisciplinary process. It illustrates the inputs from science and society to co-produce outputs that are taken up by society and the scientific community and disseminated through a specific discourse before resulting in first and second order impacts and scientific progress. On the way to impact(s), several iterations with new inputs and outputs of the transdisciplinary process may be needed. Published in (Rüegg *et al.*, 2018).

a second initiative. The sequence of inputs (i.e. resources needed to perform the actions), outputs, outcomes and impact can be graphically represented by a change pathway also known as an impact pathway (Taplin *et al.*, 2013) or a logical framework or logic model, which presents the flows in a 'logical', sequential way (Brown, 2016). Importantly, the classification into outputs, outcomes and impacts depend on the perspective that is taken for the evaluation and may differ among stakeholders (INTRAC, 2015). It is therefore important to elaborate the TOC in collaboration with the entity contracting the evaluation.

To generate a TOC, stakeholders must be clear about what they want to achieve with their initiative. In a OH team it is likely that the group members often have very different ideas about what they are working towards and are lacking a joint understanding. Therefore, everybody involved should agree on the preconditions – the building blocks – that must exist in order to reach their long-term goal. They then need to consider, in light of this big picture perspective, which of these preconditions they will take responsibility for producing – both individually and as a team.

Six main steps are recommended in the evaluation to build up this change pathway:

1. Writing a narrative to explain the logic of the initiative.
2. Identifying basic assumptions about the context.
3. Identifying long-term goals.
4. Backwards mapping and connecting the preconditions or requirements necessary to achieve that goal and explaining why these preconditions are necessary and sufficient.
5. Identifying the activities that the initiative will perform to create the desired change.
6. Identifying and/or developing indicators to measure outcomes to assess the performance of the initiative.

This mapping exercise could be done using participatory approaches and tools such as actor consultation workshops; expert opinion elicitation process; outcome mapping; individual or focus group; convergent interviews (e.g. key informant), questionnaires (e.g. internet), expert reviews, Delphi studies, Dotmocracy, ORID, or Q methodology, among others. Particularly outcome mapping can be a useful tool to use for OH initiatives, either in combination with TOC or on its own if it fulfils key assumptions of dependence on human behaviour, limits to the influence of interventions, active contribution of people to their well-being, co-existence of differing yet valid perspectives, and resilience dependent on interrelationships (Deprez, 2014).

Usually there is just a subset of outcomes that OH collaborators can influence. Some preconditions are beyond the sphere of influence of any single initiative, such as needing a stable economy to produce enough jobs to reach an employment goal. Others may be beyond a programme's influence, but stakeholders could suggest ways that a particular programme may be able to influence other programmes, or they could identify areas for strategic collaboration or partnerships. Combining different options during the process can provide more insightful understandings by: (1) identifying issues or obtaining information on variables not obtained by quantitative surveys; (2) generating hypotheses to be tested through the quantitative approach; (3) understanding unanticipated results from quantitative data; or (4) verifying or rejecting results (triangulation).

3.4.2 Expected outcomes and impacts

The description and definition of outcomes and impacts are dependent on the problem the OH initiative is addressing and the associated boundaries of the system, objective, rationale and consequently the resulting TOC. Given the diversity of OH initiatives, there is no single outcome that summarises OH endeavours, but rather a wide range of different outcomes (Baum *et al.*, 2016; Falzon *et al.*, in press; Häsler *et al.*, 2014a). However, at the longer-term impact level, there are commonalities that OH endeavours to strive for (Rüegg *et al.*, 2017). The outcomes and impacts to be measured need to be selected as a best fit for the specific OH initiative and its TOC. Because of their nature, OH initiatives will commonly span different sectors and disciplines and therefore are likely to produce disciplinary, interdisciplinary and OH outcomes and impacts. Evaluators consequently need to be aware of disciplinary paradigms, data and approaches as well as methods of combining outcomes from different disciplines. A range of outcomes used in the fields of social, ecological and economic assessments are presented in the following chapters. Here we limit ourselves to the distinction between disciplinary, interdisciplinary and systemic OH outcomes.

Disciplinary outcomes relate to outcomes that are measurable within a distinct discipline or sub-speciality within the natural or social sciences. Examples of disciplinary outcomes include health outcomes such as decreased levels of non-communicable or infectious diseases; nutrition outcomes such as reduced levels of undernutrition or obesity; economic outcomes such as increased productivity or savings in the health care system; social outcomes such as improved societal stability; and ecological outcomes such as slower rates of biodiversity reduction or improved water or air quality. Importantly, these outcomes can be achieved in disciplinary or sectoral approaches (e.g. promotion of a new anti-diabetes treatment or childhood vaccination in a national health service), but more often, they are the results of collaborations across disciplines and sectors. Interdisciplinary activities, by definition, have an impact on multiple fields or disciplines and produce results that feed back into and enhance disciplinary or sectoral work. In these instances, the pathway to the outcome may be characterised by collaboration and contributions from different disciplines and sectors, but the outcomes may still be conceptualised (and consequently measured) at the level of a field or discipline. Combining these disciplinary outcomes in methods such as multi-criteria decision analysis gives a solid basis for an assessment of the achievements of the OH initiative. In interdisciplinary outcomes, the efforts realised by individuals from different disciplines getting together to create new knowledge and understanding through sharing of ideas and bringing together different perspectives result in a product or measure, which explicitly reflects the shared responsibility among disciplines for outcomes (Strang and McLeish, 2015; Trochim *et al.*, 2006; WHO, 2009). Consequently, interdisciplinary outcomes occur in the realm of at least two disciplines simultaneously, e.g. food security as an interdisciplinary outcome of successful alignment of multiple sectors (i.e. food availability, food access and food utilisation), which contribute different skills and expertise (Ingram and White, 2015). Other examples are the Human Development Index, the Environmental Performance Index, and the Planetary Boundaries, which combine a diversity of indicators into a single or a few measure(s). An improvement in the index cannot be achieved with a disciplinary approach, but needs activities in health (e.g. investment in health service capacity, public awareness campaigns), education (e.g. build infrastructure, attract talented teachers, provide incentives for school attendance), social protection (e.g. policies to reduce poverty and vulnerability of disadvantaged population groups), and economics (e.g. promotion of efficient labour markets, robust governance). Interdisciplinary outcomes are ideally measured in a common metric, i.e. they should rely on a consensus on how to assess and weigh the particular outcomes. Such metrics are even more policy relevant and effective if they are produced and measured in a transdisciplinary process, which transcends both horizontal boundaries between scientific disciplines, and vertical boundaries between science and other societal fields (private sector, public agencies and civil society) (Lélé and Norgaard, 2005). Through the process stakeholders share different perspectives and can therefore improve the contextualization of the problem and its potential solutions and targets (Hirsch Hadorn *et al.*, 2008).

OH outcomes or impacts occur as result from a broader integration of activities in the system at stake. The main domains of OH outcomes are the three pillars of sustainability, i.e. society, environment and economy. Typical examples are interspecies equity, health stewardship, human and animal welfare, efficiency and effectiveness (Rüegg *et al.*, 2017). Clear causal attribution to the OH initiative may be difficult, but a contribution of the OH initiative can be assessed. An overview of the links between the OH characteristics (Figure 3.1) and some OH outcomes is available as supplementary online material (ESM-1).

Given the perspective chosen and the resource availability for the evaluation, the description of the TOC and the selection of associated outcomes may be more or less comprehensive and complex. However, the evaluator should make sure to pay careful attention to the contributions from different disciplines and sectors, their integration and the resulting positive and negative effects.

While One Health appears to be an endeavour towards sustainability and resilience relying on the three pillars of society, economy and the environment (http://www.un.org/sustainabledevelopment/), deficiencies in any of these aspects is obviously a reason to engage in OH (Rüegg *et al.*, 2017). Similarly, any driver for OH can be understood as the negative expression of the desired outcome, e.g. disparity versus equity, illness versus health, etc. Consequently, any driver identified earlier can be measured as an outcome of the OH initiative, and progress over time may convert what was considered to be a driver (a problem) into some form of improvement (a positive outcome).

3.4.3 Unexpected outcomes and impacts

By definition unexpected outcomes and impacts cannot be planned or covered by a TOC, even though attempts are sometimes made to capture a wide range of eventualities. Throughout a OH initiative within its system, interactions among components and feedback loops frequently produce rapid, non-linear and unanticipated changes (Fath *et al.*, 2015; Garcia and Zazueta, 2015; Reynolds, 2015). Typically, integrated approaches in complex systems generate unexpected added value, e.g. a new stakeholder organisation, but may also result in unexpected negative impacts, e.g. discrimination among stakeholders (Garcia and Zazueta, 2015), which is why capturing unexpected outcomes constitutes an essential process of OH evaluation. Other examples would be emerging diseases due to new contact rates or closer contact between previously isolated populations, or due to new social behaviours in urbanised environments (Wallace and Wallace, 2016). If unexpected outcomes are not captured, evaluation fails in informing adaptive management that seeks to improve outcomes in complex dynamic environments (Mowles, 2014). Some exemplary methods to capture unexpected outcomes and impacts are presented in the section on non-linear impact assessment (Section 3.4.4.2).

3.4.4 Evaluation design and selection of outcomes

3.4.4.1 Consider/select evaluation question(s)

It is important to select the appropriate evaluation questions before conducting the evaluation to avoid wasting scarce resources by evaluating aspects that are not of interest to end-users. During the planning it is therefore recommended to look at the TOC and to reflect on what exactly stakeholders want to know about the initiative. This should clarify why the evaluation is conducted and why the community of interest, the team, the funding bodies or other stakeholders may be interested in the evaluation. Different types of evaluation questions may be important, which will also influence the selection of the evaluation type. Adding questions during the evaluation may be possible (e.g. non-linear impact assessment), but may be difficult for others with more rigid evaluation designs (e.g. impact evaluation). It may be useful to include a brainstorming sessions with all stakeholders to come up with a full list of questions and then refine it based on priorities and resources available.

If the purpose of the evaluation is about learning and finding out how to improve the programme, the following questions may be important:

- Are the activities being implemented as planned?
- What works and what does not work?
- What are the strengths and weaknesses?
- What are participants' reactions?
- What works for whom in what ways and under what conditions?
- How can outcomes and impacts be increased?

If the purpose is about the performance, the following questions may apply:

- Does the programme meet participants' needs?
- Is there a gap between the intended and actual population served?
- How can quality be enhanced?
- Does the programme work as intended?
- To what extent can outcomes be attributed to the intervention?
- Is the programme theory clear and supported by findings?

If the purpose of the evaluation is about economic efficiency, the following may be relevant:

- How can costs be reduced?
- Does the programme deliver value for money?
- Could a higher outcome be achieved at the same cost?
- Is one strategy more beneficial than the other one?
- How do outcomes and costs compare with other options?

3.4.4.2 Select evaluation type

Taking into account the information gathered so far, the user needs to make a decision on the evaluation type to be used taking into account the complexity of the OH initiative, its rationale, and the scope and purpose of the evaluation. There are three main evaluation types that need to be considered in this process, namely impact evaluation, non-linear impact assessment and economic evaluation, which are briefly explained in the following sections.

Impact evaluation

Impact evaluation (IE) seeks to show that intended results are achieved as a result of a programme's activities, directly or indirectly. In other words, IE tries to identify whether a programme or policy as a cause can be linked to identifiable and intended results. This is often described as making a 'causal claim'.

Impact evaluation belongs to the broader agenda of evidence-based policy making. By making programme processes and resulting effects more transparent, IE proves or disproves accountability to funders and policy makers. It is concerned with both demonstrating and measuring effects as well as explaining these effects, to be able to answer 'how' and 'why' questions. It can also help us understand how to do things better and more accurately in the future. The need to explain the effects highlights the importance of theory and of context, in order to address questions of generalisability beyond a particular programme evaluation.

To decide whether to perform an impact evaluation, it is generally worth performing first a preliminary assessment to collect information on the topic of interest, the relevance of the intervention programme (e.g. what is the innovative and influential potential; what is the number of people who are or will be affected by it) and the feasibility of the impact evaluation (e.g. financial resources and logistics; ethical, political or other constraints prohibiting randomisation in a controlled trial; incomplete baseline data to allow for comparison with and without the intervention). Based on this information, a decision can be made on whether a full-scale impact evaluation needs to or can be conducted.

Once it is decided to conduct an impact evaluation, the further design implies important decisions which are determined by the hallmark of IE, i.e. the focus on causality and attribution. Three basic factors need to be taken into account when deciding on a suitable IE design: (1) the evaluation questions to be answered; (2) the 'attributes' of the programmes to be evaluated; and (3) the realistic capabilities of available designs. Many decisions related to those factors are interconnected.

Evaluation questions

The selected evaluation question may need to be refined further to capture the essence of an impact evaluation. Four typical questions in impact evaluation are the following:

- To what extent can a specific impact be attributed to the intervention?
- Did the intervention make a difference?
- How has the intervention made a difference?
- Will the intervention work elsewhere?

Because pre-existing theory rarely exists for OH initiatives, it is important to take into account the elaborated TOC (Section 3.4.1) to capture the expected dynamics. Additional questions that are likely relevant for the impact evaluation of OH initiatives include the following:

- Is the work consonant with/grounded in its source disciplines/methodologies or is it likely to develop novel methodological approaches?
- Has the work added or will it add to knowledge, even in a non-conventional way?

Programme attributes

The attributes of programmes, including their purpose, form, location, inter-relationship results and duration, can highly vary. These attributes affect the impact evaluation design and the questions. Many OH initiatives are likely to be in areas of limited understanding or they overlap with other interventions with similar aims and their results are difficult to measure. Consequently, precise attribution questions will increase the complexity of the evaluation design required and resources needed (including capacity).

Impact evaluation designs

In IE, a link between cause and effect needs to be established. This link can be established through comparison of: either two populations at the same time, with and without intervention, ensuring there is no mixing; or of the same population in time, before and after the intervention. The basic questions concerning an evaluator regarding the choice of the design are:

- What do we want to measure (e.g. a disease incidence rate)?
- How could we measure it (e.g. is an experimental approach feasible?)?
- What are assumptions on the measurement (e.g. is the way we detect cases stable over time)?

The key to useful IE is a sound methodological approach including high quality data, addressing issues of most interest for policy and programme makers (it may be advisable to focus on fewer or one particular question to be addressed) and to acknowledge the limitations of the factual analysis of the causal chain and its assumptions. For many OH initiatives, it may be more appropriate to combine the effort with a robust non-linear impact assessment (see next section). Given that no single approach seems to provide a complete picture, mixed designs (i.e. using a variety of methods, quantitative and qualitative) are most useful in strengthening confidence in conclusions. For instance, an IE could combine an experiment to assess the impacts of a programme, with a participatory design to ensure validity and relevance, and case-based, comparative studies to identify the implications of different contexts. In principle, IE for OH follow the generic guidelines, for instance explained in detail by Gertler *et al.* (2011) and Stern (2015). The main designs useful for IE, their variants and causal inference (i.e. way to show the link between cause and effect) are given in Table 3.2.

There is not always a need for a full-scale extensive impact evaluation. If a full impact evaluation is not deemed feasible, encouragement designs (e.g. a real-time, formative evaluation) can be used to test different approaches and to extract estimates of the programme's impact. Having to refer to approximations is quite likely, because OH outcomes and impacts are expected in society, ecosystems and economy, and hence the IE must be informed by the vast field of methods from social assessment, environmental and/or economic evaluation outlined in Chapters 4-6. The main issue here is that most of these investigations do not provide causal

Table 3.2. Main designs used in impact evaluation, their variants and causal inference (Stern, 2015).[1]

Design approaches	Variants/methods	Basis for causal inference
Experimental	Randomised controlled trials Quasi experiments Natural experiments	Counterfactuals: the difference between two otherwise identical cases – the manipulated and the controlled; the co-presence of cause and effects.
Statistical	Statistical modelling Longitudinal studies Econometrics	Regularity: Correlation between cause and effect or between variables, influence of (usually) isolatable multiple causes on a single effect. Control for 'confounders'.
Theory-based	Causal process designs: Theory of change, process tracing, contribution analysis, impact pathways. Causal mechanism designs: Realist evaluation, congruence analysis.	Generative causation: Identification and confirmation of causal processes or 'chains'. Supporting factors and mechanisms at work in context.
Case-based	Interpretative: Naturalistic, grounded theory, ethnography. Structured: Configurations, QCA, within-case-analysis, simulations and network analysis.	Multiple causation: Comparison across and within cases of combinations of causal factors. Analytic generalisation based on theory.
Participatory	Normative designs: Participatory or democratic evaluation, empowerment evaluation. Agency designs: Learning by doing, policy dialogue, collaborative action research.	Actor agency: Validation by participants that their actions and experienced effects are 'caused' by the programme. Adoption, customisation and commitment to a goal.
Synthesis studies	Meta-analysis, narrative synthesis, realist-based synthesis.	Accumulation and aggregation within a number of perspectives (statistical, theory based, ethnographic).

[1]Contains public sector information licensed under the Open Government Licence v3.0.

relationships, but they can be more informative when exposed to counterfactual thinking and quasi-experimental designs that collect data as to reveal hidden biases.

A list with references to detailed guidelines on impact evaluation for evaluators and risk managers, and databases with past and current development programmes, including health, is available as supplementary online material (ESM-2).

Non-linear impact assessment

An expanding array of methods for complexity-enabled monitoring, evaluation and learning (CeMEL) is available for use in the fields of development and peacebuilding (Befani *et al.*, 2015; Britt, 2016; Chigas *et al.*, 2014), many of which can be contextually adapted for One Health projects and programmes. A recent edition of the IDS Bulletin (Befani *et al.*, 2015) is entirely dedicated to methods, questions and approaches necessary to embrace complex systems. In the following sections, we will briefly describe some qualitative and quantitative methods and refer to more detailed sources.

Due to the complexity of OH initiatives, their diversity of stakeholders, actors and objectives in human, animal and ecosystem health, the use of CeMEL is almost imperative. We therefore recommend to implement at least one of the mentioned methods to remain aware and attentive to possible emerging features that result from such a holistic approach. This not only helps avoiding unintended negative consequences, but also contributes to demonstrating the added value of a holistic approach in contrast to a focussed initiative.

Qualitative methods embracing complexity

The advantage of using qualitative methods in CeMEL is that they are less constrained in measuring progress towards a predefined goal and can be used to engage stakeholders in participatory processes. A discussion note produced for the US Agency for International Development recommends five approaches for complexity-aware monitoring without claim for completeness (Britt, 2016):

1. Sentinel indicators are the most basic way to complement a TOC-based evaluation system with a complexity-aware approach (Britt, 2016). The concept is borrowed from ecology where it refers to an indicator which captures the essence of the process of change affecting a broad area of interest and which is also easily communicated. As such, a sentinel indicator facilitates monitoring and communicating about complex processes that are difficult to study within a OH initiative. As a proxy, however, this type of indicator provides incomplete information, and judgments about complex processes or entire social systems based on a single indicator can be dangerous. Therefore, a sentinel indicator should be used to trigger further observation or probes.
   The identification of sentinel indicators begins with a description of the system at stake or a system map. Sentinel indicators are critical points in the map to help monitor and inform the mutually influencing relationship between the initiative and its context. These critical points are similar to leverage points mentioned in Table 3.3. Effective sentinel indicators signal changes in the relationships among actors, represent key perspectives separate from those of the initiative, or are placed outside the boundaries of an initiative.
2. The most significant change (MSC) technique focuses first on collecting and selecting stakeholder accounts of significant changes that have occurred during a specified time period, then following a structured process in discerning which changes are the most significant and why (Davies and Dart, 2005). The MSC approach validates the stories provided by stakeholder process of cross-validation with other sources. But in its essence, it is an inductive, goal-free method

with no pre-determined notion of what impacts ought to have been achieved – both positive and negative. The foundation of the approach is the systematic collection and selection of a sample of significant change stories from people most directly involved with an intervention (such as participants, field staff, affected community members). They are asked a simple, open-ended question: 'what was the most significant change that took place for your community' (in a particular domain, such as relationships among people, over a particular period of time)? The most significant of the stories are selected through a multi-layered group process of review, selection – often (although not always) involving participants and community stakeholders. The deliberative engagement helps programme implementers learn how local stakeholders view their environment and the various aspects of change that occur over time and space. MSC is not intended as a stand-alone methodology for evaluating impact. It is an inductive method that is best utilized as an exploratory tool that can be combined with other methods to further evaluate reported changes.

3. Also 'outcome harvesting' uses a participatory process to identify, formulate, verify, and make sense of outcomes, relationships and causal pathways (Wilson-Grau and Britt, 2012). It is focused on establishing the story of how an intervention has contributed to changes in behaviour of and relationships among actors intended to be engaged and influenced by the project. The method is particularly useful when there is difficulty in attributing impact to a particular programme because of diverse interacting actors and factors. Outcome harvesting looks for results that occur 'upstream' from an anticipated impact by focusing on the changes that occur within a programme's sphere of influence. It draws attention to incremental, often subtle changes that are necessary to support the large-scale, more prominent impacts in the system. In short, rather than looking for measureable 'quick wins', this method looks for the smaller sustained systemic changes in key actors and system factors that are necessary to sustain longer term system improvement. Outcome harvesting works in reverse from most standard evaluation methods by collecting evidence of what has been achieved, and working backward to determine whether and how the project or intervention contributed to the change. It relies on six iterative steps, which are outlined in detail by Wilson-Grau and Britt (2012): Design the harvest, Review documentation and draft outcomes, Engage with informants, Substantiate, Analyse and interpret the findings, Support use of findings. The World Bank (Gold *et al.*, 2014) has recently published the results of a pilot project examining 10 cases that explore outcome harvesting in development processes.
4. Monitoring approaches that privilege feedback from stakeholders or make use of participatory methods are particularly valuable in complexity (Britt, 2016). Diverse perspectives are important for at least two reasons. First, in complexity, knowledge of the system is partial and predictability is low. Second, how actors perceive a situation motivates their behaviour. Understanding the system from different perspectives will help any single actor create a more holistic and useful picture.

   Examples of stakeholder feedback include citizen report cards, community scorecards, client surveys or other forms of collecting opinions. Alternatively, feedback may target those excluded from or marginalized by the initiative as a means of questioning whether the boundaries have been drawn in the most useful way. Sampling errors may include failure to properly identify the relationship between a respondent and an intervention, or capturing the responses of dominant individuals or groups only. Obtaining feedback may be costly and logistically or technically difficult to achieve. Measurements can be misunderstood and misreported.

5. Process monitoring of impacts (PMI) focuses on monitoring impact-producing processes (Britt, 2016). These describe how a result at one level is used by specific individuals or organizations to achieve results at the next level. In a sense, impact-producing processes take place between results. Like sentinel indicators and stakeholder feedback, PMI may be used to complement, rather than replace, performance monitoring systems.

Further approaches are described by the BetterEvaluation network (http://betterevaluation.org/en/approaches/), and organisations such as Cognitive Edge (www.cognitive-edge.com) provide tailor-made software and decision making tools based on micro-narratives in complex socio-ecological systems.

Quantitative methods embracing complexity

In circumstances where sufficient data exist, advances in computational analytics using non-linear modelling procedures and artificial neural networks have made it much easier to explore multivariate associations among indicators in complex systems. A straight forward alternative to linear models are Acyclic Bayesian Networks used to assess nested causality chains (Ward and Lewis, 2013). The method matches network models of causal factors to observational data in order to identify the most likely network which could have produced the observations. Although, allowing for much richer relationships among causal factors, the method does not consider feedback loops. An alternative approach is the risk propagation assessment as presented by Dellinger and Ehlinger, for example (Dellinger *et al.*, 2012). They used a 2-step methodology first employing self-organising feature maps (SOM, (Manolakos *et al.*, 2007; Novotny *et al.*, 2005)) to generate multidimensional clusters that visualize various outcome syndromes (e.g. ecological health, causes-of-death, and birth outcome metrics) and then applying supervised learning to identify key factors influencing dynamics of the system. The underlying propagation models typically consist of 4 layers (e.g. Figure 3.5) comprised of: (1) root stressors that act on a global, regional or local scale; (2) drivers of change that create exposure to risk factors; and (3) risk probabilities associated with exposures. These are typically expressed by numerical probabilities of undesirable hazards; and (4) impact endpoints reflect measures of system-related goods and services of value to the public.

Such risk profiles do not unequivocally demonstrate linear cause-effect relationships, but rather provide tools for identifying leverage points for targeted investigations and for risk-management prioritization. The profiles can be manipulated to simulate how the system changes in response to changing individual stressors alone or in combination. This process not only assists in producing a more robust understanding of the feedbacks, but also is an effective learning-enabling tool that helps facilitate situational-dependence, indirect effects and unanticipated consequences that single-issue interventions can have on the system. When

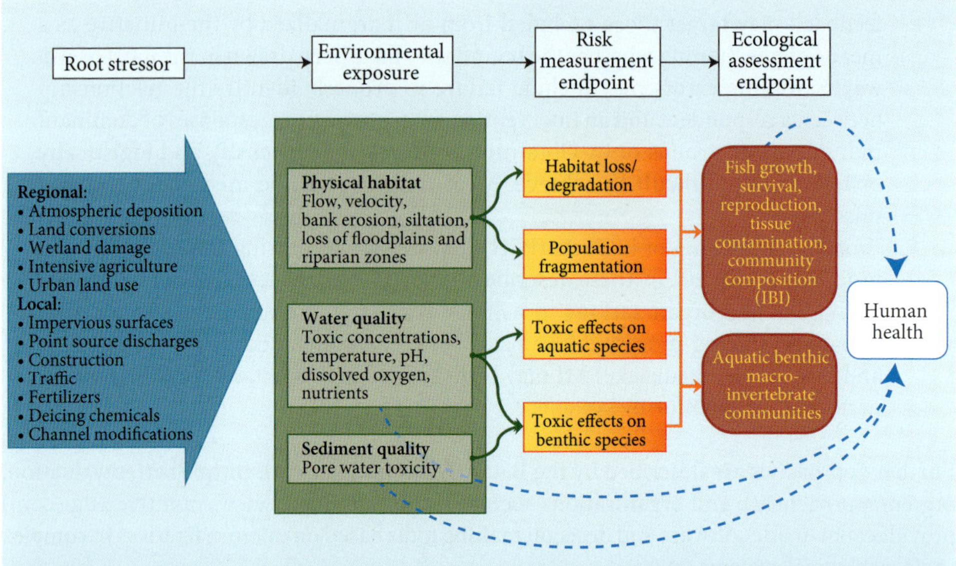

Figure 3.5. Example of a risk propagation model for an aquatic ecological risk assessment: Root Stressors act on a global, regional or local scale; Drivers of change create exposure to risk factors; Risk Probabilities are associated with exposures (typically expressed by numerical probabilities of undesirable hazards); and Impact Endpoints reflect measures of system-related goods and services of value to the public (courtesy of Timothy Ehlinger).

combined with systems mapping and critical examination of TOC process, risk profiling has the strong potential to identify key sentinel indicators (Britt, 2016).

Economic evaluation

An important claim of OH is that it generates an 'added value' through closer cooperation among professionals health, animal and environment sectors at all levels of organisation compared to uni-disciplinary or uni-sectoral approaches (Zinsstag *et al.*, 2012). Generally, there is an expectation that an integrated approach to the prevention and management of zoonotic disease risk leads to better disease control, prevention, and more efficient use of the scarce resources available (Rushton *et al.*, 2012). However, there is little evidence on the measured added value of OH in comparison to traditional approaches (Häsler *et al.*, 2014a), partly due to the complexity arising when value needs to be captured in humans, animals, society, and ecosystems. Expected benefits in OH include improved disease surveillance and control; better livelihood; more efficient production; greater health for humans, animals and ecosystems; food safety and food security; and avoidance of food scares (Häsler *et al.*, 2014a). However, greater integration, cooperation and collaboration in OH can also increase the resources needed for materials, operations, and labour. Time will be required to develop human, institutional and infrastructure capacities. These investments can be substantial, which brings up questions about who (private sector, public sector, NGOs), will be able to afford such an initial fixed capital investment (Häsler *et al.*, 2012). Consequently, the extra cost needs to be valued and compared to resulting benefits to determine whether an effective action is also economically viable and justifiable.

Economic evaluations assign a value to the resources used in a specific action, and the consequences of such action expressed in monetary or non-monetary values. If all economic resources needed and the resulting outcomes can be expressed in a common metric (e.g. money), an economic analysis can – in theory – span multiple sectors or take a whole society or multiple societies approach. However, reality is more often such, that multiple outcomes result thereby requiring different metrics and mixed methods, which makes interpretation difficult (Häsler *et al.*, 2014b). Moreover, resources are often not easily divisible and instantly available (Häsler *et al.*, 2012).

Chapter 6 sheds more light on the trade-off between OH complexity and the reductionist approaches of economists. It presents the main concepts and explains the cost-benefit analysis, and other methods that find wide application in health-related studies, i.e. the cost-effectiveness, the cost-utility, and the cost-consequence analysis. Then it summarizes the limitations and challenges of economic evaluation techniques in the context of OH and shows how the economic thought evolved to deal with complex phenomena. Finally, the chapter presents a variety of methods and models, mainly of systemic type, that can contribute to account the diversified and intangible values created by OH initiatives.

3.4.4.3 Select outcomes and metrics

With the OH initiative characterised, the TOC formulated, evaluation rationale and approach selected including relevant outcomes, it will be important to identify metrics suitable to measure the outcomes in question. For disciplinary outcomes, it is recommended to refer to relevant disciplinary literature. Interdisciplinary and OH outcome metrics deemed particularly relevant are described in detail in Chapters 4 to 6. Unexpected outcomes are primarily in the realm of emerging, qualitative information that should be captured through a non-linear impact assessment (Section 3.4.4.2), which we highly recommend as a complement to any other evaluation type. If this information is suited, it may result in new outcomes and metrics to monitor over the remaining time of the initiative.

3.4.5 Review and implementation

Once the evaluation plan is complete, it is recommended to review it carefully with the whole evaluation team and relevant stakeholders to determine whether the rationale, questions, evaluation type, metrics, intended activities, outputs and outcomes are relevant to the target population and the end-user of the evaluation. Moreover, it is advisable to consider if enough resources are available to conduct the evaluation as planned and if there is the relevant capacity. Additionally, it is worthwhile contemplating how the results will be communicated and set aside respective resources and capacity.

Once the evaluation plan has been reviewed and updated (if necessary), the data collection and analyses processes can be implemented. The evaluator must remain involved to monitor data collection as well as the implementation of activities and make sure that the integrity of data collection stays intact, that appropriate measurement instruments are used, reporting bias and similar is avoided and that the evaluation steps are well coordinated and documented.

3.5 Element 3: assessment of the One Health-ness

Aspects of implementation of initiatives (i.e. the structures, resources, and processes through which delivery is achieved, and the quantity and quality of what is delivered); mechanisms of impact (i.e. how activities, and participants' interactions with them, trigger change); and context (i.e. how external factors influence the delivery and functioning of activities) are examined through process evaluation (Moore *et al.*, 2014, 2015). Process evaluations allow seeing how an initiative develops, its structures, environment and associated activities like communications and marketing. Detailed generic guidelines for process evaluations are available (Anonymous, 2009; Moore *et al.*, 2014, 2015; Saunders, 2005), and provide methods to look at the processes of programme, management and infrastructure together, or, in other words, the capacity of a OH initiative to deliver on its promised outcomes. Critical aspects to be examined are (Moore *et al.*, 2014, 2015):

- Implementation: the structures, resources and processes through which delivery is achieved, and the quantity and quality of what is delivered.
- Mechanisms of impact: how intervention activities, and participants' interactions with them can trigger change.
- Context: how external factors influence the delivery and functioning of interventions. Process evaluations may be conducted within feasibility testing phases, alongside evaluations of effectiveness, or alongside post-evaluation scale-up.

The supplementary online material (ESM-3) contains a table summarising the key characteristics of these guidelines for process evaluation and links them to OH characteristics to facilitate the selection of available guidance. In the following, we describe a set of systematic assessment tools that contribute to a One Health Index (OH-index) as an indicator for the degree of integration of processes in an evaluated initiative.

An implicit characteristic of any OH initiative is its focus on sharing, exchanging, collaborating, learning (from each other), reflecting and generating change across disciplines and sectors in an enabling environment (Rüegg *et al.*, 2017). Consequently, this affects the delivery of an OH initiative (e.g. availability of training, learning about other fields, provision of resources), the mechanisms of impact (e.g. the responses of participants and their interactions with the initiative), and context factors (e.g. shaping of theories on how an initiative works). We refer to the sum of these characteristics as One Health-ness composed of six aspects outlined below and hypothesise that they need to be an integral element of any (process) evaluation in OH. We collate scores and indices that have been suggested in a variety of contexts, adapt them to OH and combine them in a One Health index (OHI) and ratio (OHR) for a holistic appreciation. The six assessment tools have been standardised for use and are made available together with the calculation of the indices and spider diagrams in an Excel workbook for download (ESM-4). Each assessment tool consists of a series of up to 17 questions to be answered and an associated scoring system with values between 0 and 1 as well as spider diagrams. The questions were developed by Working Group 1 of the NEOH and probe for the specificities of each aspect (outlined below) that can be captured in a semi-quantitative way. They are based on the concept of SMART goals (specific, measurable, achievable, relevant, timely) and wherever appropriate, were adapted from existing evaluation tools. They were then circulated in the NEOH community and revised in several workshops. The scoring recommendations were determined so that scores close to one reflect a high degree of realisation of the different OH characteristics. Here it must be emphasised that the authors do not presume that a high degree of implementation necessarily results in a high impact or effectiveness and underline that at this stage, the benchmark still needs to be established. Each question has the same weight, with exception of the learning assessment, where different levels of organisational learning are weighted according to their level of influence on institutional learning. Consequently, care was taken to balance the number of questions across all assessment tools to provide equal representation in the overall OH-index. The underlying assumption is that each question contains equivalent information to describe the OH initiative. However, because there is no measurable gold standard for each of the questions, the questionnaire and primarily the OH-index and OH-ratio are then assessed for their usefulness and representativeness using case studies as outlined in the overview and a meta-analysis of further published studies. Similar to Element 1, the assessment of the characteristics in this element should ideally be informed by a group of evaluators or (preferably) by relevant stakeholders identified in Element 1.

3.5.1 OH thinking: system thinking and match between context and initiative

OH as a systemic approach with corresponding methodology is of little worth if not based on a foundation of systems thinking (Whitehead *et al.*, 2015). This tool assesses how an OH initiative conceptualises the system in which it operates and in how far it considers features specific to complex adaptive systems. The fundamental idea is that a complex initiative addresses multiple dimensions of the system in which it operates (see Element 1 above). The first set of questions (ESM-4) measure the number of dimensions and the scales within each to gain a semi-quantitative appreciation of the context and the embedded OH initiative. Subsequent questions assess the match between the dimensions of the initiative and its context. Particular attention is given to the scales in different dimensions and whether the initiative reflects the reality of the context in which it operates. A third set of questions probes for concepts and thoughts typically contained in a systems approach (Anderson and Johnson, 1997; Meadows, 2008). To assess systems thinking in written documents, e.g. in a retrospective evaluation or in a proposal, we refer to a method based on statistical semantics proposed by Whitehead and Scherer (2015).

3.5.2 OH planning: cross-sectorial, integrated planning

OH planning is essentially the unfolding of the OH thinking into operational features of the initiative that should facilitate OH working towards achieving the aims and objectives during as well as after the OH initiative. The planning of OH initiatives goes beyond the type of planning that is required for disciplinary and inter-disciplinary projects in which it might be easier to maintain control of what tasks, engagement and resources are required. For instance, OH initiatives typically require human resources with competences in transdisciplinary working methods and excellent communication skills to bridge disciplines and sectors

(Stokols *et al.*, 2013). It is important that the planning includes appropriate methods to engage all of the essential actors and stakeholders, who should be aiming to reach a common goal. Part of the planning evaluation is to assess whether the planned structure, location and timing of the initiative support the OH outcomes aimed for. Due to the complex and trans-domain characteristics of OH challenges, another important aspect of OH initiatives is the ability to self-assess, learn, reflect and adapt to new knowledge and changing conditions, constraints and opportunities over time (Gunderson *et al.*, 2016). Therefore, adaptability features prominently in the evaluation of the planning of OH initiatives. Finally, the planning evaluation helps assessing the tasks and resources allocated to each task employed to achieve the specified objectives of the initiative. The questions in the supplementary online material (ESM-4) were developed to probe if the challenges of complex initiatives described here are addressed in the planning phase and if funding as well as organisational aspects are set up to accommodate adaptive behaviour by the participants. High scores are recommended for a strong support of adaptability and flexibility.

3.5.3 OH working: transdisciplinarity

Interdisciplinary collaboration brings together people with different skills and expertise to tackle complex problems, which often have a high societal stake and require an understanding of the human behaviour (Anonymous, 2005; Hirsch Hadorn *et al.*, 2008; Ledford, 2015). Appreciating potential contributions of multiple disciplines requires examining the limits imposed by a discipline, and rejecting or accepting different disciplinary theories based on their relevance and credibility in order to gain a new understanding about the defined challenge (Lattuca *et al.*, 2012; Nikitina, 2005). In the context of OH, interdisciplinarity has developed towards a participatory approach in the form of transdisciplinarity (Hirsch Hadorn *et al.*, 2008). Both inter- and transdisciplinarity rely on appropriate leadership and management to promote strategic dialogue and shared decision-making (Nancarrow *et al.*, 2013; Strang and McLeish, 2015), which in turn will foster a non-hierarchical relationship between the different disciplines and members within the team. It must also allow for self-reflection, flexibility and recursiveness (Aragrande and Canali, 2015; Hirsch Hadorn *et al.*, 2008; Lélé and Norgaard, 2005; Strang and McLeish, 2015), to be able to challenge and modify underlying assumptions and concepts and thereby enrich understanding. It must be emphasised that such transdisciplinary work demands a high level of commitment and collaboration of all participants to establish personal relationships founded within a climate of trust (Ledford, 2015; Lélé and Norgaard, 2005; Nancarrow *et al.*, 2013). The questions probing for transdisciplinarity (ESM-4) focus on disciplinary diversity, team building and adaptability and were adapted based on the work cited above.

Further aspects of trans- and interdisciplinarity may be assessed, namely for (A) evaluating (academic) participants; and (B) assessing scientific outputs of a OH initiative. However, because individuals may have different roles in an OH initiative, assessing their trans- and interdisciplinary capacity may not always be required or relevant. Also, printed scientific output may not be a primary objective of an OH initiative and occurs with some delay, thereby contributing more to the assessment of outputs than to the implementation per se:

A. The transdisciplinarity of (academic) participants may be assessed based on the interdisciplinarity of publications (see method B) below); interdisciplinarity of teaching, other academic activity (e.g. teaching experience in other disciplines than the own, co-teaching with experts from other disciplines/ sectors, etc.); previous experience with various non-academic communities (e.g. public debate, main stream media, sports and leisure organisations, politics, NGOs, volunteering, etc.); involvement in other disciplinary and interdisciplinary networks (e.g. social and natural science networks other than the own expertise, explicitly interdisciplinary initiatives, science policy, etc.).
B. A framework to evaluate the interdisciplinarity of knowledge production based on citation network analysis can be found here: https://www.mcgill.ca/msr/msr-volume-4/evaluating-knowledge-production-systems. It must be emphasised that this is only represents the written knowledge published in peer reviewed journals, which does not reflect the actual knowledge production occurring in the field.

3.5.4 Systemic organisation: adaptive and shared leadership

In many complex settings, change-oriented leadership has helped to overcome the fallacies of conventions, norms and traditions (Thygeson *et al.*, 2010; Yukl, 2012). Complex systems have leverage points where they can be influenced according to their potential to modify a systems behaviour (Meadows, 2008). The use of these points by an OH initiative determines the dimension(s) and scales at which the initiative is effective. However, in order to be effective, the implementation of the initiative needs to be facilitated by corresponding leadership behaviour. Yukl classifies leadership into four meta-categories with specific objectives (Yukl, 2012): (1) task-oriented behaviour, the primary objective is to accomplish work in an efficient and reliable way; (2) relations-oriented behaviour, the primary objective is to increase the quality of human resources and relations, which is sometimes called 'human capital'; (3) change-oriented behaviour, the primary objectives are to increase innovation, collective learning, and adaptation to the external environment; (4) external leadership behaviour, the primary objectives are to acquire necessary information and resources, and to promote and defend the interests of the team or organisation. These leadership behaviours can be related to the leverage points in a system according to their objectives (Table 3.3).

Yukl emphasises that all leadership behaviours and particularly their flexible applications are relevant for effective leadership. The table simply illustrates that the lack of a particular leadership behaviour may hamper the implementation of a well-conceived OH initiative. The effectiveness of leadership behaviours also depends on the extent to which the leader is trusted by people to be influenced. Most types of leadership behaviours can be used in ethical or un-ethical ways. Moreover, a leader, who is not trusted because of unethical behaviour will have less influence. Values, namely honesty, altruism, compassion, fairness, courage, and humility may further catalyse effects of good leadership behaviour. In contrast, excessive institutional structure and organisation can nullify these effects (Yukl, 2012). Rooke and Torbert identify further common personality traits of leaders that effectively manage wicked problems: They can challenge the prevailing view without provoking outrage or cynicism; they can act on the big and small picture at the same time, and change course if their chosen path turns

Table 3.3. Ranked list of leverage points at which to intervene in complex systems, from least to most effective, according to Meadows (2008), in relation to leadership behaviour according to Yukl (2012).

Leverage point	Leadership behaviour
• Constants, parameters, numbers (such as subsidies, taxes, standards) • The sizes of buffers and other stabilising stocks, relative to their flows. • The structure of material stocks and flows (such as transport networks, population age structures).	Task-oriented leadership: clarifying, planning, monitoring, problem solving
• The lengths of delays, relative to the rate of system change. • The strength of negative feedback loops, relative to the impacts they are trying to correct against. • The gain around driving positive feedback loops. • The structure of information flows (who does and does not have access to information). • The rules of the system (such as incentives, punishments, constraints).	Relation-oriented leadership: supporting, developing, recognising, empowering
• The power to add, change, evolve, or self-organise system structure. • The goals of the system	Change-oriented leadership: Advocating change, envisioning change, encouraging innovation, facilitating collective learning.
• The mindset or paradigm out of which the system (its goals, structure, rules, delays, parameters) arises. • The power to transcend paradigms.	Change-oriented, and external leadership: Networking, external monitoring, representing

out to be incorrect; and they lead with inquiry as well as advocacy, with engagement as well as command, operating all from a deeply held humility and respect for others (Rooke and Torbert, 2005).

A further challenge for leading OH projects is that there may be less interest, commitment, and collaboration if one discipline dominates. Consequently, other disciplines may retract their activity and reinforce the disciplinary silo mentality. To ensure that disciplines are effectively engaged and involved in decision-making from the planning to the implementation stages of projects, shared/distributed leadership and governance should be implemented involving all stakeholders (Houghton *et al.*, 2015; Scott and Caress, 2005).

Consequently, the selection of questions for the systemic organisation of OH initiatives focuses on the structure of teams, as well as management, social and leadership skills of key players and its implementation (ESM-4). The questions were taken from the leadership assessment tools and the published questionnaires on team work and transdisciplinarity described in Section 3.5.3. High scores were recommended for strong teams, change-oriented leadership skills, clear competences, goals and criteria of success.

3.5.5 Learning infrastructure

Learning is a change in cognition, potential behaviour or actual behaviour through better knowledge and understanding (Fiol and Lyles, 1985; Tsang, 1997). Organisations, such as OH initiatives, learn when they 'encode inferences from history into routines that guide behaviour' (Levitt and March, 1988). This is achieved when discoveries, evaluations and insights by individuals are successfully embedded in the organisation's mental models or cognitive systems and memories (Argyris, 1999). This requires that organisational learning takes into account the learning that takes place at the individual, group, and organisational levels (Giesecke and McNeil, 2004) and the interplay between them (Argyris, 1999). The three levels of learning work together and influence each other and are thus not clearly distinct and mutually exclusive (Redding and Catalanello, 1994). Nevertheless, each level of learning has its characteristics for evaluation.

Individuals can engage in single-loop or double-loop learning. Single-loop learning happens when the output is corrected or existing competences, procedures, technologies and paradigms are improved, without necessarily examining or challenging the underlying beliefs and assumptions. In contrast, double loop learning involves seeing beyond the situation and questioning operating norms. It results in modification of the organisation's underlying norms, policies and objectives.

Individual learning is not a sufficient condition for organisational learning (Gould, 2000). Teams enable the interplay between individual and organisational learning, because they can better share the knowledge (Gould, 2000; Guns, 1998; Watkins and Marsick, 1993) and include more people in the learning process. As a result, team members share awareness of each individual member's expertise, knowledge, and skills, and build a transactive memory system (Stokols *et al.*, 2013). Thus, the evaluation should examine the knowledge shared through teams, to what extent it is shared and how it is shared. The conclusion should show whether the teams provide the appropriate interplay between the individual and the OH initiative. Without supporting the development of a transactive memory system within and across teams, the initiative may have individuals who learn, but it cannot engage in organisational learning (Garvin, 2000). It is important to assess how knowledge is gathered, stored and distributed within a OH initiative (Huysman, 1999), and if and how it provides working environments, technology, rewards, systems, structures, and policies that will support learning (Watkins and Marsick, 1993).

Finally, the context in which the OH initiative is located has influence on the organisational learning (Santa, 2015). The context can be divided in the direct system in which it operates and general environment (Santa, 2014a). The direct system consists of other components with which the initiative interacts, e.g. actors and stakeholders with various relationships. The general environment consists of less specific elements that might affect learning like economic, technological, sociocultural and other factors. The questions probing for learning are taken from a tool to change organisations towards learning organisations (Santa, 2014b) and focus on the frequency single-loop and double loop learning occur at the level of individuals, teams and the OH initiative, as well as how the system and broader environment support learning (ESM-4).

3.5.6 Sharing infrastructure and processes

In a broad sense, data and information sharing is a catalyser of knowledge generation (Piwowar *et al.*, 2007). Data are often a pre-requisite for the operational gears to function. In OH initiatives, data and information are often the 'raw material' that ultimately will lead to better understanding and a more inclusive and sustainable way of tackling the challenge. If managed appropriately, data and unbiased information sharing can foster trust between participants, as well as minimise misconduct in data management and reporting (Schelling and Zinsstag, 2015; Walter *et al.*, 2007). Additionally, this process can avoid duplication of data collection, ensuring an optimisation of resources (Tenopir *et al.*, 2011).

A central benefit of data sharing is that the data can be analysed to a much greater extent than if only the data owner examines them. This brings benefits to the data owners themselves, as the analysis of others might lead them to further develop their knowledge on the systems the data originated from or the strengths and limitations of their datasets, as well as raising the awareness of the existence of the data in the wider community (Piwowar and Chapman, 2010; Piwowar *et al.*, 2008, 2007). Despite these benefits, data and information sharing often lead to barriers for establishing collaborations (Chokshi *et al.*, 2006) and are hampered by confidentiality issues, time delays and even mistrust in established collaborations. Consequently, data sharing is not as frequent as desirable, and needs to be incentivised to become a natural part of the science and governance cultures. For example, in some countries research relies on a tripartite agreement to share information and collaborate between academia, government institutions and industry, but public access to data may also be reinforced through legislation.

A frequent barrier to data procurement is the bureaucratic process to access data, particularly its complexity and duration. Moreover, fees and technical constraints may arise (Houe *et al.*, 2011), and often too little resources are set aside to for data extraction from databases. Data accessibility and ownership are further critical factors, with data owned by collaborating parties contributing more to knowledge generation than public data or data owned by third parties. Data confidentiality may affect its sharing, as participant consent is usually collected for a specific purpose. This consent might not extend to new studies or alternative purposes, and therefore, security measures may be required to warrant confidentiality. Sharing sensitive data and information within a broader group might entail higher risks for confidentiality breaches (Borgman, 2010). Alternatively, anonymization may reduce that risk, but may also reduce the utility of the data. Finally, it needs to be stressed that knowledge about the data origin and data collection processes is key for the quality and usefulness of stored data, and respective documentation must be available. For example, without knowledge about potential bias throughout the data generating process, it is extremely challenging to merge or combine data from multiple sectors in a OH initiative. The questions in the supplementary online material (ESM-4) derive from a workshop held by NEOH on data and information sharing, in which critical aspects of data sharing were discussed. High scores are recommended for strong facilitation of sharing. The questions focus on the sharing mechanisms, available resources, data quality and accessibility, storage and the resilience of these to change in the system.

3.5.7 One Health index and ratio

Given the lack of current, commonly accepted benchmarks and the fact that OH initiatives are strongly context specific, it is recommended to assess them in relation to a context specific benchmark. Hence, the evaluator should determine what the perfect situation in the given context would look like (using benchmarking data where they exist) and what proportion of this maximum is achieved with the OH initiative.

The aim of the OHI is to combine the assessments conducted in the previous sections of Element 3. To visualise the six assessments, we suggest a spider diagram (Figure 3.6), in which each assessment is represented by a spoke. The diagram depicts the operational aspects 'OH thinking', 'OH planning' and 'OH working' opposed to the infrastructure for 'learning', 'sharing' and 'systemic organisation'. Thus, the operational aspects on the top left of the diagonal are opposed to the infrastructure on the bottom right. Each spoke is scaled to cover a range of values between 0 and 1. Consequently, the plot not only illustrates the degree of integration by the surface, but it also shows the balance between the operation and the supporting means through its symmetry over the diagonal, numerically represented as the OHR.

In Figure 3.6, two exemplary fictive projects are depicted, an example with real data of a comparison of two OH initiatives can be found in the article by Buttigieg and co-workers (Buttigieg *et al.*, 2018). The fictive Project 1 depicted here has a highly developed transdisciplinary team with a very comprehensive multi-dimensional approach. However, it appears to lack learning and sharing infrastructure and has a mismatch between the

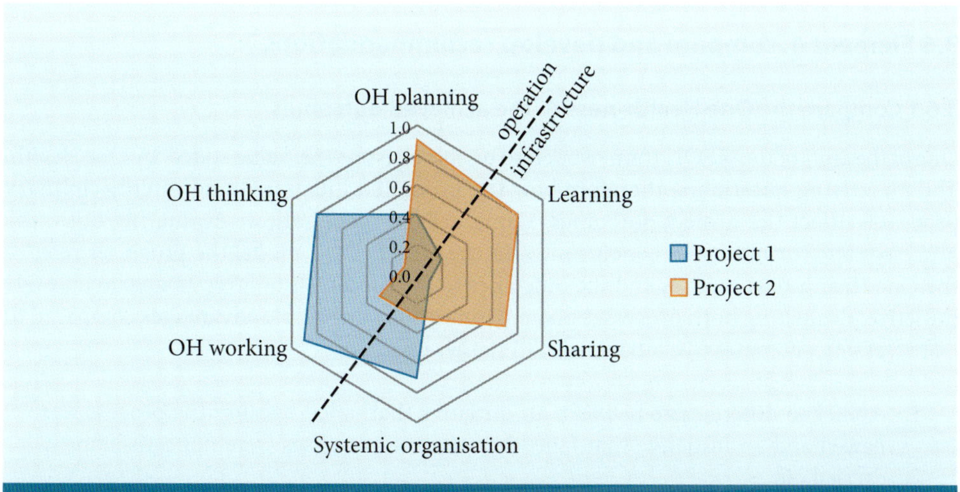

Figure 3.6. Example of the One Health spider diagram for two fictive One Health projects.

responsibilities, authorities and means which affects the transdisciplinary working and hence potentially the OH outcomes. On the other hand, Project 2 has well developed infrastructure and well defined tasks with sufficient funding, but does not explore the inter-disciplinary space nor does it aim at serving multiple species.

The OHI corresponds to the ratio of the surface enclosed by the lines to the surface enclosed if all spokes were equal to 1 (a detailed derivation is provided in the supplementary online material ESM-5). Thus, the OHI is:

$$\text{OHI} = \frac{\{(ScP \times ScT)+(ScL \times ScP)+(ScS \times ScL)+(ScO \times ScS)+(ScW \times ScO)+(ScT \times ScW)\}}{6} \quad (1)$$

where Sc_P is the score obtained in OH planning, Sc_L is the score obtained in learning infrastructure, Sc_S is the score from sharing infrastructure, Sc_O is the score from systemic organisation, Sc_W is the score from OH working, and Sc_T is the score from OH thinking.

The OHR is the relation of the surface covered in the top left of the diagonal to the one in the lower right (a detailed derivation is provided in the supplementary online material ESM-5). To compute the OHR, the surface of the top left surface ($SUR_{operation}$) is calculated:

$$\text{SUR}_{\text{Operations}} = \frac{\sqrt{3}}{4}\left\{\left(\frac{\text{ScO}\times\text{ScW}^2}{\text{ScO+ScW}}\right) + (\text{ScW} \times \text{ScT}) + (\text{ScT} \times \text{ScP}) + \left(\frac{\text{ScP}^2\times\text{ScL}}{\text{ScP+ScL}}\right)\right\} \quad (2)$$

and divided by the surface of the lower right ($SUR_{infrastructure}$)

$$\text{SUR}_{\text{Infrastructure}} = \frac{\sqrt{3}}{4}\left\{\left(\frac{\text{ScP}\times\text{ScL}^2}{\text{ScP+ScL}}\right) + (\text{ScL} \times \text{ScS}) + (\text{ScS} \times \text{ScO}) + \left(\frac{\text{ScO}^2\times\text{ScW}}{\text{ScO+ScW}}\right)\right\} \quad (3)$$

resulting in the following equation:

$$\text{OHR} = \frac{\left(\frac{\text{ScO}\times\text{ScW}^2}{\text{ScO+ScW}}\right)+(\text{ScW}\times\text{ScT})+(\text{ScT}\times\text{ScP})+\left(\frac{\text{ScP}^2\times\text{ScL}}{\text{ScP+ScL}}\right)}{\left(\frac{\text{ScP}\times\text{ScL}^2}{\text{ScP+ScL}}\right)+(\text{ScL}\times\text{ScS})+(\text{ScS}\times\text{ScO})+\left(\frac{\text{ScO}^2\times\text{ScW}}{\text{ScO+ScW}}\right)} \quad (4)$$

3.6 Element 4: compare and develop recommendations

3.6.1 Compare the One Health-ness to the achieved outcomes

One of the aims of the NEOH framework is to be able to assess the 'value added' by One Health. The underlying question is therefore how the promoted integrated and interdisciplinary processes affect the project outcomes. Using a TOC model allows evaluating both, the processes and the outcomes concurrently. In the NEOH TOC model the processes refer to the One Health-ness metrics and the outcomes to the success or failures of a particular OH initiative. Evaluating both processes and outcomes allows multiple advantages compared to just assessing outcomes (De Silva *et al.*, 2014), these are:

1. Ability to differentiate between an initiative that failed because the process was flawed and an initiative that failed because the processes were not satisfactorily carried out.

2. Ability to determine which pathways are most effective at causing the desired outcomes and which are critical for future project success.
3. Ability to identify how the context, i.e. the environment in which the project is being conducted, affects the TOC and thus the outcomes.
4. Ability to identify unexpected outcomes and hypothesise why these occurred.

An additional advantage is the ability to identify and assess intermediate outcomes that can be used as markers of success if the evaluation activities will have ended by the time of the final outcomes (Rogers, 2014). This is particularly important for complex interventions with very long-term goals.

There are a variety of ways to assess the processes and outcomes, which include standard quantitative and qualitative study designs. Used correctly, any study method can be used to measure one or more process(es) or outcome(s). For complete project evaluation every process and outcome should be measured by at least one study; however, if this is not possible the key processes and outcomes must be identified and measured. More advanced statistical modelling methods can also be used and these have the advantage of being able to look at interactions between the metrics. These techniques include: structural equation modelling, discrete simulation models, agent-based modelling, system dynamics modelling, and comparative qualitative analysis. These are described in greater detail in Table 3.4 along with a presentation of their pros and cons.

3.6.2 Develop recommendations

The observations made during the evaluation must now be translated into constructive feedback for the concerned parties. At this science-policy interface careful communication is essential - especially when health is concerned, communication can be a sensitive issue for many involved. Often (external) communication beyond the research or policy institution(s), is merely taken seriously at a later stage, when most developments have taken shape and crucial decisions were made. In the context of OH, we consider communication as a key part of the whole process, from start to finish. We refer to experiences in the field of risk communication, which largely developed around health risks related to environmental issues like nuclear power incidents and the vast diversity of pollutants that we are exposed to. We also propose to frame communication in relation to more than 'just' outcomes, and consider it as decision support: helping receivers of information to make up their own mind about the issue depending on their own stakes, perceptions and preferences, in a well-informed manner, as well as well-argued transparency about key choices involved.

3.6.2.1 A brief history of risk communication

Risk communication has evolved from one-way communication, restricted to the dissemination of information from experts to the public, to two-way risk communication, with a focus on participation and cooperation between scientists, policy-makers and the public (Fischhoff, 1995; Leiss, 1996; McComas, 2006). One-way communication has often been based on the 'deficit model' (Wynne, 1996), i.e. the assumption that clear communication of objective and sound scientific information from experts to the 'ignorant' public is sufficient to make them aware of problems and respond accordingly. However, in most cases, the science is not simple

Table 3.4. A description of some of the advanced statistical modelling techniques available for evaluating processes and outcomes and their pros and cons.

Modelling technique Description	Pros	Cons
Structural equation modelling (SEM) – Nachtigall *et al.*, 2003		
SEM is an umbrella term for multiple modelling techniques. The technique allows the modeller to conduct and combine the techniques of factor analysis, multiple regression analysis, ANOVA, and others. It is able to estimate multiple and interrelated dependence in a single analysis.	• Very flexible as it deals with a system of regression equations (rather than single/multiple linear regression) • Newer software makes this technique now accessible to inexperienced modellers. • Possibility of modelling complex dependencies and latent variables	• Complicated and difficult to understand • Large amount of data required • Sample size requirements often vague • The models are not necessarily assessments of causality • Context can be neglected • The ease of producing a model with new user-friendly interface software means that inexperienced modellers use it but produce statistically flawed models
Discrete event simulation (DES) models – Allen *et al.*, 2015; Caro *et al.*, 2016		
The system is modelled as a series of events that occur over time, individuals can be assigned information and their progress modelled through time. Resources can also be accounted for.	• DES allows for complex decision logic that is not readily available in other modelling techniques • Can be used to test 'what if?' scenarios	• Stochastic approach means that the model output changes slightly each time it is run • Is still a measure of population behaviour not individual but this is often misunderstood as 'entities' represent people with their corresponding attributes
Agent-based modelling – Loomis *et al.*, 2008; Schank, 2010; Siebers, 2013		
A system is modelled as a collection of autonomous decision-making entities ('agents'). Each agent makes a series of decisions based on assigned rules, attributes, and their interactions with their environment and each other. Best for heterogeneous, autonomous, pro-active actors e.g. human-centred systems	• Can allow for complex agent behaviour such as that influenced by memory and motivations. • Can demonstrate individual agents behaviour, not just population behaviour.	• Ability to code is usually needed, languages such as Java are used • Many programmes do not have sufficient power for very complex systems • Models are difficult to validate as the agent-based nature means outputs are not testable with standard statistical techniques

>>>

Table 3.4. Continued.

Modelling technique Description	Pros	Cons
System dynamics modelling – Pitman *et al.*, 2012; Sterman, 2001		
Computer simulation and modelling technique that allows for framing, understanding, and discussing complex issues and problems. Structure is as important as the components of the model.	• Good ability to take into account indirect effects in system • Ability to incorporate time delays, outcomes that are distant in space and time to their cause, and multicausality • Good for identifying causal factors	• Inevitably some components of the complex system will have to be estimated
Qualitative comparative analysis (QCA) – Marshall, 2016		
QCA is a method to analyse the causal contribution of different conditions to an outcome. This method bridges qualitative and quantitative analysis. It is able to handle causal complexity.	• Allows for investigation of multicausality	• Works best on small sample sizes

and consensual, but involves ambiguities and uncertainties. Also the public is more than a mere recipient of information, but consists of actors in the decision process of the strategies to improve and/or preserve situations and in the management of the risks.

3.6.2.2 Communication about complexity

An important challenge in risk communication is how to exchange meaningfully information regarding uncertain, complex and ambiguous knowledge (Renn, 2008). As outlined earlier, framing and dealing with complexity is of crucial importance in OH science, policy and practice (Keune and Assmuth, in press). The number and diversity of factors that may play a role in an OH issue are enormous, and these issues have also a multitude of characteristics and consequences. Framing this complexity is crucial because it sets the boundaries of the system in which the OH initiative is situated in terms of thoughts and actions. This is not merely a technical process of scientific framing, but also a methodological decision-making process with both scientific and societal implications. Mostly the benefits and risks related to such issues cannot be generalized or objectified, and will be distributed unevenly, resulting in health and environmental inequalities. Even more generally, framing is crucial as it reflects cultural factors and historical contingencies, perceptions and mind-sets, political processes, associated values and world-views. Framing is at the core of how we as humans relate to and deal with human, animal and ecosystem health, as scientists, policy makers and practitioners, with models, policies or actions.

The two core issues in risk communication are: 'How can science formulate confident, robust and clear messages when it struggles with uncertainties, unknowns, and ambiguities due to complexity?' and 'How does the traditional scientific evidence base approach live up to expectations of clear communication and of solving problems without pleading for endless ever more detailed research and without too complicated messages due to lack of clear cut scientific understanding?'. The argument that communication should be restricted because of uncertainties is challenged by various authors. Ragas and co-workers (2006) argue that if the information is used by regulators, public managers and risk assessors, then the public equally ought to know. Others dispute the belief that the public is unable to deal with complex issues (e.g. (Marris *et al.*, 2001)), and a third group has shown that withholding data regarding uncertainty often reduces trust (Frewer, 2004; Van Kleef *et al.*, 2007). Hence, as Slovic (1998) has stated, 'The challenge is to communicate the risk estimates so that they are understandable and that the risks and associated uncertainty can be put into a personal perspective'.

3.6.2.3 Communication to support decision making

Framing communication as decision support in a OH context means to inform end-users about relevant elements of complexity in an inclusive, well-structured manner, not as an end-point, but as a basis for end-user decision making about what to do. Decision support methods (Marakas, 1999) can be employed in semi-structured or unstructured decision contexts, can provide support to either an individual or a group, and facilitate learning on the part of the decision maker(s). They are meant to be interactive and user-friendly, and generally are developed in an evolutionary iterative process, using relevant data and models. Multi-criteria decision analysis (MCDA) is a good example of such decision support regarding complex issues: it can simultaneously embrace, combine, and structure various types of often incommensurable diversity: diversity of information (e.g. qualitative and quantitative data, as well as uncertainty), diversity of opinions (among experts), diversity in actor perspectives (stakes) as well as diversity in assessment/decision-making criteria (Keune, 2013). MCDA is not a miracle tool that will objectively solve all problems by unambiguously calculating what is best. It functions more like a 'sounding board': it will structure and visualize the input of actors and factors involved. As such it will offer a basis for well informed and transparent reflection, learning and deliberation. Also, it helps users to be transparent about the decision choices they make, about what they take into account, their preferences and underlying argumentations.

3.6.2.4 Communication is a serious concern

Despite advances in theory and numerous initiatives in practice, the deficit model continues to dominate many attitudes towards the public communication of science (Davies, 2008) as well as practices. Two-way communication is seen as inherently difficult and dangerous. The alternative view – that two-way communication helps to make scientists and policy makers accountable and to empower the public – remains a rarity in many fields of science and policy. Much remains to be done to devise and promote more open, yet workable solution oriented approaches to the communication of science, risk and policy, in the context of complexity.

The epistemological divide between the traditional and alternative approaches largely relies on ambassadors safeguarding their own approach. Without ambassadors of diverse paradigms at the table where crucial methodological choices are being made, especially in practice, under resource constraints and time pressure, the dominant approach will largely steer the process. This also does not imply that traditional experts are not open to alternative approaches or that they do not see the value of it. But in practice, the initial open arms attitude towards two-way communication often is accompanied by closed mind-sets amongst the traditional experts as the process progresses. There may be some exceptional transdisciplinary personalities, but in many settings of real practice the shift to a more collaborative approach often does not easily survive without social scientists being effectively involved. To implement OH it is therefore crucial that the diversity which is considered to be relevant in the process is represented by ambassadors at the epistemological and methodological decision table. This requires including risk communication experts in order to facilitate two-way directional and problem solving collaborations.

Acknowledgements

This chapter is based upon work from COST Action 'Network for Evaluation of One Health' (TD1404), supported by COST (European Cooperation in Science and Technology).

References

Allen, M., Spencer, A., Gibson, A., Matthews, J., Allwood, A., Prosser, S. and Pitt, M., 2015. Right cot, right place, right time: improving the design and organisation of neonatal care networks – a computer simulation study. Health Serv Deliv Res 3(20).

Anderson, V. and Johnson, L., 1997. Systems thinking basics: from concepts to causal loops. Leverage Networks Inc., Acton, MA, USA.

Anonymous, 1999. Interactive terminology for Europe. Available at: https://tinyurl.com/pqlebur.

Anonymous, 2005. In praise of soft science. Nature 435: 1003.

Anonymous, 2009. Developing process evaluation questions. Centers for Disease Control and Prevention Evaluation Briefs. Available at: https://tinyurl.com/y7we4v2p.

Aragrande, M. and Canali, M., 2015. An operational tool to enhance One Health interdisciplinarity. In: Proceedings 3rd GRF One Health Summit. October 4-7, 2015. Davos, Switzerland.

Argyris, C., 1999. On organizational learning, 2nd edition. Wiley-Blackwell, Oxford, UK.

Baum, S.E., Machalaba, C., Daszak, P., Salerno, R.H. and Karesh, W.B., 2016. Evaluating one health: are we demonstrating effectiveness? One Health 3: 5-10.

Befani, B., Ramalingam, B. and Stern, E., 2015. Introduction – towards systemic approaches to evaluation and impact. IDS Bull 46(1): 1-6.

Borgman, C.L., 2010. Scholarship in the digital age: information, infrastructure, and the internet. MIT Press, Cambridge, MA, USA.

Boriani, E., Esposito, R., Frazzoli, C., Fantke, P., Hald, T. and Rüegg, S.R., 2017. Framework to define structure and boundaries of complex health intervention systems: the ALERT project. Front Publ Health 5: 182.

Britt, H., 2016. Discussion note: complexity-aware monitoring. USAID, Washington, DC, USA.

Brown, A.-M., 2016. Differences between the theory of change and the logic model. Available at: https://tinyurl.com/y8tqyx3g.

Bunge, M., 1960. Levels: a semantical preliminary. Rev Metaphys 13: 396-406.

Buttigieg, S.C., Savic, S., Cauchi, D., Lautier, E., Canali, M. and Aragrande, M., 2018. Brucellosis control in Malta and Serbia: a One Health evaluation. Front Vet Sci 5: 147.

Caro, J.J.J., Möller, J. and Moller, J., 2016. Advantages and disadvantages of discrete-event simulation for health economic analyses. Expert Rev Pharmacoecon Outcomes Res 16: 327-329.

Chigas, D., Ehlinger, T., Befani, B., Docherty, J., Michaels, J., Smith, R., Walker, T. and Yoshida, R., 2014. Non-linear impact assessment: challenges, approaches and tools. Dynamical Systems Innovation Lab, Honolulu, Hawai'i.

Chokshi, D.A., Parker, M. and Kwiatkowski, D.P., 2006. Data sharing and intellectual property in a genomic epidemiology network: policies for large-scale research collaboration. Bull World Health Organ 84: 382-387.

Ciborra, C.U. and Hanseth, O., 1998. From tool to Gestell – agendas for managing the information infrastructure. Inf Technol People 11: 305-327.

Commission on Social Determinants of Health of the WHO, 2008. Social determinants of health report: closing the gap in a generation. World Health Organisation, Geneva, Switzerland.

Craig, P., Dieppe, P., Macintyre, S., Michie, S., Nazareth, I. and Petticrew, M., 2013. Developing and evaluating complex interventions: the new Medical Research Council guidance. Int J Nurs Stud 50: 587-592.

Davies, R. and Dart, J., 2005. The Most Significant Change (MSC) technique. CARE International, Genève, Switzerland.

Davies, S.R., 2008. Constructing communication. Sci Commun 29: 413-434.

De Silva, M.J., Breuer, E., Lee, L., Asher, L., Chowdhary, N., Lund, C. and Patel, V., 2014. Theory of change: a theory-driven approach to enhance the Medical Research Council's framework for complex interventions. Trials 15: 267.

Dellinger, M., Tofan, L. and Ehlinger, T., 2012. Predictive analytics and pattern visualisation for human health. Risk Assessm 2197: 2186-2197.

Deprez, S., 2014. The 5 key assumptions of outcome mapping. Available at: https://tinyurl.com/y8bt5f78.

Falzon, L.C., Lechner, I., Chantziaras, I., Collineau, L., Courcoul, A., Filippitzi, M.-E., Laukkanen-Ninios, R., Peroz, C., Pinto Ferreira, J., Postma, M., Prestmo, P.G., Phythian, C.J., Sarno, E., Vanantwerpen, G., Vergne, T., Grindlay, D.J.C. and Brennan, M.L., in press. Quantitative outcomes of a One Health approach to study global health challenges. Ecohealth. https://doi.org/10.1007/s10393-017-1310-5.

Food and Agriculture Organisation (FAO), 2013. The state of food and agriculture. FAO, Rome, Italy.

Fath, B.D., Dean, C.A. and Katzmair, H., 2015. Navigating the adaptive cycle: an approach to managing the resilience of social systems. Ecol Soc 20(2): 24.

Fiol, M.C. and Lyles, M.A., 1985. Organizational learning. Acad Manag Rev 10: 8003-8013.

Fischhoff, B., 1995. Risk perception and communication unplugged: twenty years of process 1. Risk Anal 15: 137-145.

Fonseca, A.G., Torgal, J., De Meneghi, D., Gabriël, S., Coelho, A.C. and Vilhena, M., 2018. One Health-ness evaluation of cysticercosis surveillance design in Portugal. Front Public Health 6: 74.

Frewer, L., 2004. The public and effective risk communication. Toxicol Lett 149: 391-397.

Garcia, J.R. and Zazueta, A., 2015. Going beyond mixed methods to mixed approaches: a systems perspective for asking the right questions. IDS Bull 46: 30-43.

Garvin, D.A., 2000. Learning in action: a guide to putting the learning organization to work. Harward Business Press, Boston, MA, USA.

Gertler, P.J., Martinez, S., Premand, P., Rawlings, L.B. and Vermeersch, C.M.J., 2011. Impact evaluation in practice. World Bank Publications. Available at: https://tinyurl.com/y8c3k6fj.

Giesecke, J. and McNeil, B., 2004. Transitioning to the learning organization. Libr Trends 53: 54.

Gold, J., Wilson-Grau, R., Fisher, S., Roberts, D. and Otoo, S., 2014. Cases in outcome harvesting. Worldbank, Washington, DC, USA.

Gould, N., 2000. Becoming a learning organisation: a social work example. Soc Work Educ 19: 585-596.

Gunderson, L.H., Cosens, B. and Garmestani, A.S., 2016. Adaptive governance of riverine and wetland ecosystem goods and services. J Environ Manag 183: 353-360.

Guns, B., 1998. The faster learning organization, 1st edition. Jossey-Bass, New York, NY, USA.

Hanin, M.C.E., Queenan, K., Savic, S., Karimuribo, E., Rüegg, SR. and Häsler, B., 2018. A One Health evaluation of the Southern African Centre for Infectious Disease Surveillance. Front Vet Sci 5: 33.

Häsler, B., Cornelson, L., Bennani, H. and Rushton, J., 2014a. A review of the metrics for One Health benefits. Rev Sci Tech Off Int Epiz 33: 453-464.

Häsler, B., Gilbert, W., Jones, B.A., Pfeiffer, D.U., Rushton, J. and Otte, M.J., 2012. The economic value of One Health in relation to the mitigation of zoonotic disease risks world organisation for animal health. Curr Top Microbiol Immunol 365: 127-151.

Häsler, B., Hiby, E., Gilbert, W., Obeyesekere, N., Bennani, H. and Rushton, J., 2014b. A One Health framework for the evaluation of rabies control programmes: a case study from Colombo City, Sri Lanka. PLoS Negl Trop Dis 8(10): e3270.

Hirsch Hadorn, G., Hoffmann-Riem, H., Biber-Klemm, D., Grossenbacher-Mansuy, S., Joye, W., Pohl, C., Wiesmann, U. and Zemp, E., 2008. Handbook of transdisciplinary research, 1st edition. Springer International Publishing AG, Cham, Switzerland.

Houe, H., Gardner, I.A. and Nielsen, L.R., 2011. Use of information on disease diagnoses from databases for animal health economic, welfare and food safety purposes: strengths and limitations of recordings. Acta Vet Scand 53, Suppl. 1: S7.

Houghton, J.D., Pearce, C.L., Manz, C.C., Courtright, S. and Stewart, G.L., 2015. Sharing is caring: toward a model of proactive caring through shared leadership. Hum Resour Manag Rev 25: 313-327.

Huysman, M., 1999. Balancing biases: a critical review of the literature on organizational learning. In: Easterby-Smith, M., Araujo, L. and Burgoyne, J.G. (eds.) Organizational learning and the learning organization: developments in theory and practice. SAGE Publications, London, UK, pp. 59-74.

Ifejika Speranza, C., Wiesmann, U. and Rist, S., 2014. An indicator framework for assessing livelihood resilience in the context of social – ecological dynamics. Glob Environ Change 28: 109-119.

Ingram, J. and White, R., 2015. Inaugural lecture – food systems: challenges, concepts and communities. Available at: https://tinyurl.com/yak7cmeu.

INTRAC, 2015. Outputs, outcome and impact. Available at: https://tinyurl.com/y7utlrgh.

Jones, K.E., Patel, N.G., Levy, M.A., Storeygard, A., Balk, D., Gittleman, J.L. and Daszak, P., 2008. Global trends in emerging infectious diseases. Nature 451: 990-994.

Keune, H., 2013. Negotiated complexity: framing multi-criteria decision support in environmental health practice. Am J Oper Res 3: 153-166.

Keune, H., Assmuth, T., in press. Framing complexity in environment and human health. In: Oxford research encyclopedia of environmental science. Oxford University Press, New York, NY, USA. Available at:http://tinyurl.com/yamqsk45.

Khanal, P. and Bhattarai, N., 2016. Health beyond health to bridge the global health gap. Lancet Glob Health 4: e792.

Laing, G., Aragrande, M., Canali, M., Savic, S. and De Meneghi, D., 2018. Control of cattle ticks and tick-borne diseases by acaricide in southern province of Zambia: a retrospective evaluation of animal health measures according to current one health concepts. Front Public Health 6: 45.

Lang, D.J., Wiek, A., Bergmann, M., Stauffacher, M., Martens, P., Moll, P., Swilling, M. and Thomas, C.J., 2012. Transdisciplinary research in sustainability science: practice, principles, and challenges. Sustain Sci 7: 25-43.

Lang, T. and Rayner, G., 2012. Ecological public health: the 21st century's big idea? An essay by Tim Lang and Geof Rayner. BMJ 345: e5466-e5466.

Lattuca, L.R., Knight, D.B. and Bergom, I.M., 2012. Developing a measure of interdisciplinary competence for engineers. In: Conference Proceedings 2012 of the American Society for Engineering Education, pp. 1-19. Available at: https://tinyurl.com/yc9k6beb.

Ledford, H., 2015a. How to solve the world's biggest problems. Nature 525: 308-311.

Léger, A., Stärk, K., Rushton, J. and Nielsen, L.R., 2018. A One Health evaluation of the University of Copenhagen research centre for control of antibiotic resistance. Front Vet Sci, https://doi.org/10.3389/fvets.2018.00194.

Leiss, W., 1996. Three phases in the evolution of risk communication practice. Ann Am Acad Pol Soc Sci 545: 85-94.

Lélé, S. and Norgaard, R.B., 2005. Practicing interdisciplinarity. Bioscience 55: 967.

Lerner, H. and Berg, C., 2015. The concept of health in One Health and some practical implications for research and education: what is One Health? Infect Ecol Epidemiol 5: 25300.

Levitt, B. and March, J.G., 1988. Organizational learning. Ann Rev Sociol 14: 319-340.

Loomis, J., Bond, C. and Harpman, D., 2008. The potential of Agent-Based modelling for performing economic analysis of adaptive natural resource management. J Nat Resour Policy Res 1: 35-48.

Manolakos, E., Virani, H. and Novotny, V., 2007. Extracting knowledge on the links between the water body stressors and biotic integrity. Water Res 41: 4041-4050.

Marakas, G.M., 1999. Decision Support Systems in the 21st century. Pearson Education, Hoboken, NJ, USA.

Marris, C., Wynne, B., Simmons, P. and Weldon, S., 2001. Public perceptions of agricultural biotechnologies in Europe. PABE, Brussels, Belgium. Available at: http://csec.lancs.ac.uk/archive/pabe/.

Marshall, G., 2016. Qualitative comparative analysis. Better Evaluation. Available at: https://tinyurl.com/y9h6xhu4.

McComas, K.A., 2006. Defining moments in risk communication research: 1996-2005. J Health Commun 11: 75-91.

Meadows, D.H., 2008. Thinking in systems – a primer. Chelsea Green Publishing Co, White River Junction, VT, USA.

Moore, G., Audrey, S., Barker, M. and Bond, L., 2014. Process evaluation of complex interventions. UK Medical Research Council (MRC) guidance, Southampton, UK.

Moore, G.F., Audrey, S., Barker, M., Bond, L., Bonell, C., Hardeman, W., Moore, L., O'Cathain, A., Tinati, T., Wight, D. and Baird, J., 2015. Process evaluation of complex interventions: Medical Research Council guidance. BMJ 350: h1258.

Mowles, C., 2014. Complex, but not quite complex enough: the turn to the complexity sciences in evaluation scholarship. Evaluation 20: 160-175.

Muñoz-Prieto, A., Nielsen, L.R., Martinez-Subiela, S., Mazeikiene, J., Lopez Jornet, P.L., Savic, S. and Tvarijonaviciute, A., 2018. Evaluation of One Health operations and supporting infrastructure in a questionnaire-based study of obesity among dogs and their owners in 11 European countries. Front Vet Sci 5: 163.

Nachtigall, C., Kroehne, U., Funke, F. and Steyer, R., 2003. (Why) should we use SEM? Pros and cons of structural equation modeling. Methods Psychol Res 8: 1-22.

Nancarrow, S.A., Booth, A., Ariss, S., Smith, T., Enderby, P. and Roots, A., 2013. Ten principles of good interdisciplinary team work. Hum Resour Health 11: 19.

Nikitina, S., 2005. Pathways of interdisciplinary cognition. Cogn Instr 23: 389-425.

Novotny, V., Bartošová, A., O'Reilly, N. and Ehlinger, T., 2005. Unlocking the relationship of biotic integrity of impaired waters to anthropogenic stresses. Water Res 39: 184-198.

Ostrom, E., 2009. A general framework for analyzing sustainability of social-ecological systems. Science 325: 419-422.

Paternoster, G., Tomassone, L., Tamba, M., Chiari, M., Lavazza, A., Piazzi, M., Favretto, A.R., Balduzzi, G., Pautasso, A. and Vogler, B.R., 2017. The degree of One Health implementation in the West Nile virus integrated surveillance in northern Italy, 2016. Front Publ Health 5: 1-10.

Pfeiffer, D.U., 2014. From risk analysis to risk governance – adapting to an ever more complex future. Vet Ital 50: 169-176.

Pitman, R., Fisman, D., Zaric, G.S., Postma, M., Kretzschmar, M., Edmunds, J. and Brisson, M., 2012. Dynamic transmission modeling: a report of the ISPOR-SMDM modeling good research practices task force-5. Val Health 15: 828-834.

Piwowar, H.A., Becich, M.J., Bilofsky, H. and Crowley, R.S., 2008. Towards a data sharing culture: recommendations for leadership from academic health centers. PLoS Med 5: 1315-1319.

Piwowar, H.A. and Chapman, W.W., 2010. Public sharing of research datasets: a pilot study of associations. J Informetr 4: 148-156.

Piwowar, H.A., Day, R.S. and Fridsma, D.B., 2007. Sharing detailed research data is associated with increased citation rate. PLoS ONE 2(3): e308.

Pumain, D., Pavé, A., Venant, Fa., Verdier, N., Victorri, B. and West, G.B., 2006. Hierarchy in natural and social sciences, 1st edition. Springer, Dordrecht, the Netherlands.

Rabinowitz, P.M., Kock, R., Kachani, M., Kunkel, R., Thomas, J., Gilbert, J., Wallace, R., Blackmore, C., Wong, D., Karesh, W., Natterson, B., Dugas, R. and Rubin, C., 2013. Toward proof of concept of a one health approach to disease prediction and control. Emerg Infect Dis 19: e130265.

Radeski, M., O'Shea, H., De Meneghi, D. and Ilieski, V., 2018. Positioning animal welfare in the One Health concept through evaluation of an animal welfare center in Skopje, Macedonia. Front Vet Sci 4: 1-11.

Ragas, A.M.J., Huijbregts, À.M.A.J., Van Kaathoven, À.E.H., Wolsink, J.H. and Wemmenhove, J., 2006. Development and implementation of a right-to-know web site that presents estimated cancer risks for air emissions of large industrial facilities. Integr Environ Assess Manag 2(4): 365-374.

Rayner, G. and Lang, T., 2012. Ecological public health. Reshaping the conditions for good health. Routledge, Abingdon, UK.

Redding, J.C. and Catalanello, R.F., 1994. Strategic readiness: the making of the learning organization. Jossey-Bass, San Francisco, CA, USA.

Renn, O., 2008. Risk governance: coping with uncertainty in a complex world. Earthscan, London, UK.

Retief, F., Bond, A., Pope, J., Morrison-Saunders, A. and King, N., 2016. Global megatrends and their implications for environmental assessment practice. Environ Impact Assess Rev 61: 52-60.

Reynolds, M., 2015. (Breaking) the iron triangle of evaluation. IDS Bull 46: 71-86.

Rogers, P.J., 2014. Theory of change (Unicef), methodological briefs: impact evaluation. Unicef, Firenze, Florence, Italy.

Romanelli, C., Cooper, D., Campbell-Ledrum, D., Maiero, M., Karesh, W.B., Danny, H. and Golden, C.D., 2015. Connecting global priorities : biodiversity and human health - a state of knowledge review. WHO, Geneva, Switzerland.

Rooke, D. and Torbert, W.R., 2005. Seven transformations of leadership. Harv Bus Rev. Available at: https://tinyurl.com/p5vmhhh.

Rüegg, S., Rosenbaum Nielsen, L., Buttigieg, S., Santa, M., Aragrande, M., Canali, M., Ehlinger, T., Chantziaras, I., Boriani, E., Radeski, M., Bruce, M. and Häsler, B., 2018. A systems approach to evaluate One Health initiatives. Front Vet Sci 5: 23.

Rüegg, S.R., McMahon, B.J., Häsler, B., Esposito, R., Nielsen, L.R., Ifejika Speranza, C., Ehlinger, T., Peyre, M., Aragrande, M., Zinsstag, J., Davies, P., Mihalca, A.D.A., Buttigieg, S.C., Rushton, J., Carmo, L.P., De Meneghi, D., Canali, M., Filippitzi, M.-E.M.E., Goutard, F.L., Ilieski, V., Milićević, D., O'Shea, H., Radeski, M., Kock, R., Staines, A., Lindberg, A., Rosenbaum Nielsen, L., Ifejika Speranza, C., Ehlinger, T., Peyre, M., 2017. A blueprint to evaluate One Health. Front Public Health 5: 1-5.

Rushton, J., Häsler, B., De Haan, N. and Rushton, R., 2012. Economic benefits or drivers of a 'One Health' approach: why should anyone invest? Onderstepoort J Vet Res 79: 1-5.

Santa, M., 2014a. Framework for multivariate continuous transformation towards learning organization. University Paris 1 Pantheon, Sorbonne, Paris, France.

Santa, M., 2014b. The learning organization atlas framework. Available at: http://tinyurl.com/yah5glgu.

Santa, M., 2015. Learning organisation review – a 'good' theory perspective. Learn Organ 22: 242-270.

Saunders, R.P., 2005. Developing a process-evaluation plan for assessing health promotion program implementation: a how-to guide. Health Promot Pract 6: 134-147.

Schank, J., 2010. Agent-based modeling. Available at: https://tinyurl.com/zvwp2gb.

Schelling, E. and Zinsstag, J., 2015. Transdisciplinary research and One Health. In: Zinsstag, J., Schelling, E., Waltner-Toews, D., Whittaker, M. and Tanner, M. (eds.) One Health: the theory and practice of integrated health approaches. CABI, Wallingford, UK, pp. 366-373.

Scott, L. and Caress, A.-L., 2005. Shared governance and shared leadership: meeting the challenges of implementation. J Nurs Manag 13: 4-12.

Siebers, P.-O., 2013. Simulation in a nutshell. Available at: https://tinyurl.com/yapycd3j.

Slovic, P., 1998. The risk game. Science 59: 73-77.

Sterman, J.D., 2001. System dynamics modeling: tools for learning in a complex world. Calif Manage Rev 43: 8-25.

Stern, E., 2015. Impact evaluation. a guide for commissioners and managers. Bond, London, UK, 35 pp. Available at: http://tinyurl.com/yby7kcvy.

Stokols, D., Fuqua, J., Gress, J., Harvey, R., Phillips, K., Baezconde-Garbanati, L., Unger, J., Palmer, P., Clark, M.A., Colby, S.M., Morgan, G. and Trochim, W., 2003. Evaluating transdisciplinary science. Nicotine Tob Res 5: 21-39.

Stokols, D., Hall, K.L. and Vogel, A.L., 2013. Transdisciplinary public health: definitions, core characteristics and strategies for success. In: Haire-Joshu, D. and McBride, T.D. (eds.) Transdisciplinary public health: reserach, methods and practice. Jossey-bass Publishers, San Francisco, CA, USA, pp. 3-30.

Strang, V. and McLeish, T., 2015a. Evaluating interdisciplinary research: a practical guide. Durham University, Durham, UK, 20 pp.

Taplin, D.H., Clark, H., Collins, E. and Colby, D.C., 2013. Theory of change. Technical papers: a series of papers to support development of theories of change based on practice in the field. ActKnowledge, New York, NY, USA, 23 pp.

Tenopir, C., Allard, S., Douglass, K., Aydinoglu, A.U., Wu, L., Read, E., Manoff, M. and Frame, M., 2011. Data sharing by scientists: practices and perceptions. PLoS ONE 6: 1-21.

Thygeson, M., Morrissey, L. and Ulstad, V., 2010. Adaptive leadership and the practice of medicine: a complexity-based approach to reframing the doctor-patient relationship. J Eval Clin Pract 16: 1009-1015.

Trochim, W.M., Cabrera, D.A., Milstein, B., Gallagher, R.S. and Leischow, S.J., 2006. Practical challenges of systems thinking and modeling in public health. Am J Public Health 96: 538-546.

Tsang, E.W.K., 1997. Organizational learning and the learning organization: a dichotomy between descriptive and prescriptive research. Hum Relations 50: 73-89.

Van Kleef, E., Houghton, J.R., Krystallis, A., Pfenning, U., Rowe, G., Van Dijk, H., Van der Lans, I.A. and Frewer, L.J., 2007. Consumer evaluations of food risk management quality in Europe. Risk Anal 27: 1565-1580.

Wallace, R.G. and Wallace, R., 2016. Neoliberal Ebola. Springer International Publishing, Cham, Switzerland.

Walter, A.I., Helgenberger, S., Wiek, A. and Scholz, R.W., 2007. Measuring societal effects of transdisciplinary research projects: design and application of an evaluation method. Eval Program Plann 30: 325-338.

Ward, M.P. and Lewis, F.I., 2013. Bayesian graphical modelling: applications in veterinary epidemiology. Prev Vet Med 110: 1-3.

Watkins, K.E. and Marsick, V.J., 1993. Sculpting the learning organization: lessons in the art and science of systemic change, 1st edition. Jossey-Bass, San Francisco, CA, USA.

Whitehead, N.P. and Scherer, W.T., 2015. Quantifying the quality of a systems approach. In: 2015 Annual IEEE Systems Conference (SysCon) Proceedings. April 13-16, 2015. Vancouver, BC, Canada, pp. 44-49.

Whitehead, N.P., Scherer, W.T. and Smith, M.C., 2015. Systems thinking about systems thinking. IEEE Syst J 9: 1117-1128.

Whitmee, S., Haines, A., Beyrer, C., Boltz, F., Capon, A.G., De Souza Dias, B.F., Ezeh, A., Frumkin, H., Gong, P., Head, P., Horton, R., Mace, G.M., Marten, R., Myers, S.S., Nishtar, S., Osofsky, S.A., Pattanayak, S.K., Pongsiri, M.J., Romanelli, C., Soucat, A., Vega, J. and Yach, D., 2015. Safeguarding human health in the Anthropocene epoch: report of The Rockefeller Foundation – Lancet commission on planetary health. Lancet 6736: 1-56.

World Health Organisation (WHO), 2009. Systems thinking for health systems strengthening. WHO, Geneva, Switzerland.

Williams, B., 2016. Using systems concepts in evaluation design: a workbook, 1st edition.

Wilson-Grau, R. and Britt, H., 2012. Outcome harvesting. Ford Foundation, Cairo, Egypt. Available at: https://tinyurl.com/y9rpdrjr.

Wolf, M., 2015. Is there really such a thing as 'One Health'? Thinking about a more than human world from the perspective of cultural anthropology. Soc Sci Med 129: 5-11.

Woodward, D., Drager, N., Beaglehole, R. and Lipson, D., 2001. Globalization and health: a framework for analysis and action. Bull World Health Organ 79: 875-881.

Wynne, B., 1996. Misunderstood misunderstandings: social identities and public uptake of science. In: Irwin, A. and Wynne, B. (eds.) Misunderstanding science? The public reconstruction of science and technology. Cambridge University Press, Cambridge, UK, pp. 19-46.

Yukl, G., 2012. Effective leadership behavior : what we know and what questions need more attention. Acad Manag Perspect 26(4): 66-85.

Zinsstag, J., Meisser, A., Schelling, E., Bonfoh, B. and Tanner, M., 2012. From 'two medicines' to 'One Health' and beyond. Onderstepoort J Vet Res 79(2): 492.

Zinsstag, J., Schelling, E., Waltner-Toews, D. and Tanner, M., 2011. From 'one medicine' to 'One Health' and systemic approaches to health and well-being. Prev Vet Med 101: 148-156.

Chapter 4

Evaluating the contributions of One Health initiatives to social sustainability

Photo: Chinwe Ifejika Speranza

Chinwe Ifejika Speranza[1,2*], Tamara Wüthrich[1], Simon Rüegg[3], Jakob Zinsstag[4], Hans Keune[5,6], Sébastien Boillat[1], Lauren Blake[7], Susan Thieme[1], Chris Degeling[8] and Stephan Rist[1,2]

[1]Institute of Geography, University of Bern, Hallerstrasse 12, 3012 Bern, Switzerland; [2]Centre for Development and Environment, University of Bern, Mittelstrasse 43, 3012 Bern, Switzerland; [3]Section of Epidemiology, Vetsuisse Faculty, University of Zürich, Winterthurerstrasse 270, 8057 Zürich, Switzerland; [4]Department of Epidemiology and Public Health, Swiss Tropical and Public Health Institute, University of Basel, P.O. Box, 4002 Basel, Switzerland; [5]Belgian Biodiversity Platform, Research Institute Nature & Forest (INBO), Herman Teirlinckgebouw, Havenlaan 88 bus 73, 1000 Brussels, Belgium; [6]University of Antwerp, Campus Drie Eiken, gebouw R R.3.07. Universiteitsplein 1, 2610 Wilrijk, Belgium; [7]Department of Pathobiology and Population Sciences, Royal Veterinary College, University of London, Hawkshead Lane, North Mymms, Hatfield, Hertfordshire, AL9 7TA, United Kingdom; [8]Research for Social Change, Faculty of Social Science, University of Wollongong, Wollongong, NSW, Australia; chinwe.ifejika.speranza@giub.unibe.ch

Abstract

One Health is an approach that integrates perspectives from human, animal and environmental health to address health challenges. As the idea of One Health is grounded in achieving sustainable outcomes, an important aspect is the contribution of One Health to social sustainability. In this chapter we ask, what social sustainability is, what the indicators of social sustainability related to One Health are, and, through what measures we can evaluate the contributions of One Health to social sustainability, in terms of its operations, its supporting infrastructures and outcomes. We adopt a wider conceptualization of social sustainability and propose an approach based on basic needs, capabilities and emancipation, environmental justice, solidarity and social cohesion. First, we identify indicators used in literature to capture social sustainability in human, animal and environmental health and propose ways to integrate them into a framework for the evaluation of One Health initiatives. Second, we formulate questions that can be used to evaluate the social sustainability of One Health initiatives. Third, we discuss the viability of operationalising the indicators, the trade-offs that might arise and identify how they can be minimised. We then discuss methodological issues and highlight the importance of transdisciplinary deliberative approaches for adapting the framework to specific contexts.

Keywords: One Health, social sustainability, capabilities, emancipation, environmental justice, solidarity, social cohesion

4.1 Introduction

Interacting agro-ecological, physical, economic, socio-cultural and political conditions (commonly understood as social-ecological systems) can contribute in various ways to human, animal and ecosystem health. For instance, human health benefits from contact with nature through improved mental and physical well-being, and human interactions with nature can improve pro-nature attitudes and behaviours (Frumkin *et al.*, 2017; Hofmann *et al.*, 2017; Richardson *et al.*, 2016).

Yet stressors on resources and the environment increase wider health risks, including and beyond disease. Human-induced environmental impacts such as greenhouse gas emissions, deforestation, and land degradation, or natural processes such as volcanic eruptions or pest infestations can drive environmental change and make environments unconducive for animal and human health. This includes the increase of climatic hazards such as floods and storms (Zinsstag *et al.*, 2018), respiratory diseases due to air pollution (Thurston *et al.*, 2017), bioaccumulation of pollutants and endocrine disrupters in the food chain (Frazzoli and Mantovani, 2010; Frazzoli *et al.*, 2009), water pollution by pharmaceuticals and plastics (cf. Caliman and Gavrilescu, 2009), and the development of antibiotic resistance and subsequent adverse impacts on human and animal health. Pollution also affects animal health and this can compromise the functioning of ecosystems, such as the reduction of crop pollination due to bee colonies affected by pesticides and herbicides (Henry *et al.*, 2012).

Impaired health often results from complex interactions between different components of a social-ecological system; hence, there is a need to address the interdependencies between humans, animals and the environment, and the social and environmental determinants of health. This requires a consideration of the distal and proximate drivers of disease and health, as well as disease detection, prevention and control (WHO *et al.*, 2015, p. 41; Zinsstag *et al.*, 2011). The One Health (OH) approach that integrates societal and scientific perspectives to address the sectoral interlinkages between human, animal and environmental health (Rüegg *et al.*, 2017) can meet this challenge. An integrative OH-approach promises to be more effective in reducing losses (e.g. economic loss from production and trade; human and animal mortality from disease, habitat loss) that would have occurred if single sector approaches were followed (Berthe *et al.*, 2018; Zinsstag *et al.*, 2015a). Moreover, it is also expected that OH leads to more sustainable outcomes for humans, animals and the environment.

So far, no frameworks or methods exist for evaluating *how* OH, through its operations and outcomes contributes to social sustainability. This chapter thus aims to develop a framework for defining what social sustainability is about and presents a methodological framework for evaluating the contributions of OH to social sustainability.

4.2 Understanding social sustainability, its dimensions and indicators

Contemporary ideas of social sustainability primarily build on the Brundtland Report that defines sustainable development as 'development which meets the needs of the present without compromising the ability for future generations to meet their own needs' (WCED, 1987, p. 43). Subsequently, various attempts highlight the social in sustainable development; addressing system characteristics and properties such as welfare in the present and future, and the interdependence between society and the environment (cf. Garcés *et al.*, 2003; Hodge and Hardi, 1997).

A first distinction in framing social sustainability is the focus on: (1) the capability of institutions to address societal concerns; and (2) the ability to maintain a dynamic balance between social agents and social structure.

In the social quality/capability perspective, social sustainability is understood as a 'quality of societies' that encompasses not only basic needs but also the ability to address societal concerns in the face of risks, such as coping with climate change (Eizenberg and Jabareen, 2017) or being resilient to health challenges (Obrist *et al.*, 2010). It thus focuses on whether institutional configurations are able to satisfy both human needs and preserve the social and ecological capabilities required to fulfil these needs by including criteria of social justice, human dignity and meaningful participation (Littig and Griessler, 2005).

Thus, the above-mentioned focus on capabilities and needs is part of a social structure-agent perspective. It focuses on and unpacks the relations between agents (individual(s)) and social structure (society). It builds on the assumption that agents and social structures are constitutive of each other, with individual and collective perceptions shaping historical and contemporary social developments (Giddens, 1984). In this sense, Empacher and Wehling (1999) argue that the social is innately bipolar, with tensions between the individual social actor who strives for autonomy and achievement of own goals; and the society (social system) within which the individual actor is situated, which strives for conformity, cohesion and stability. Social sustainability thus entails securing individual and social stability and securing the capacity of society to develop and function (Empacher and Wehling, 1999). For the individual, this relates to physical and material well-being (income, employment), social recognition and social integrity, and opportunities for self-development and autonomy. Securing social stability concerns peaceful coexistence, distributive justice, and participation.

According to Empacher and Wehling (1999), such development and functional capacity can best be achieved through maintaining cultural diversity, diversity of social structures, social cohesion (inter-generational, solidarity principle) and availability of education and learning facilities. They thus identify five key elements of social sustainability: (1) livelihood security for all; (2) development capacity of social subsystems and structures; (3) maintenance and further development of social norms and values; (4) equal access to resources; and (5) participation in decision-making.

A second distinction can be made between analytical (what are the relations between society and nature?), normative ('what kind of social values are needed?') and political framings of social sustainability ('what practical strategies should be adopted to achieve social sustainability') (Littig and Griessler, 2005). Analytical, normative and political aspects of sustainability can also be interpreted as system-, target- and transformation knowledge (Pohl and Hirsch Hadorn, 2007). While social sustainability and its different theoretical, political and practical framings are inherently normative (Littig and Griessler, 2005; Pareja-Eastaway, 2012; Vallance *et al.*, 2011), different perspectives have primarily focused on either the analytical or normative aspects.

The analytical perspective departs from theories concerning the relationship between nature and society in terms of the social values to be attained through sustainable development (Littig and Griessler, 2005). This perspective is both descriptive and prescriptive, with a focus on describing the social processes that shape society's interrelations with nature, and inquiring about how processes and structures can be transformed to ensure development chances of future generations (Littig and Griessler, 2005).

In contrast, a normative perspective to social sustainability is concerned with what kind of social values are needed. It captures a set of social principles (Box 4.1) as reflected in the contents of the Brundtland report. This perspective seeks to set value standards such as participation, equal opportunities, justice, etc., which are considered inherently legitimate and define social development ideals for present and future generations (Becker *et al.*, 1999, p. 5; in Littig and Griessler, 2005, p. 70). In addressing the contributions of OH initiatives to social sustainability, various principles drawn from across United Nations charters may be assumed, and are made explicit in Box 4.1. Grounded in human rights principles, these tenets can help to guide transdisciplinary deliberations on the social sustainability of OH initiatives.

Box 4.1. Underlying principles for assessing social sustainability.

- Human rights principles: Based on the United Nations, Universal Declaration on Human Rights in 1948, all humans have equal rights and freedoms, which are protected by law, irrespective of race, gender, nationality and other differences. In 30 articles, various indispensable rights for human dignity and free development of human personality are specified. Article 25 highlights the 'right to a standard of living adequate for the health and well-being..., including food, clothing, housing, medical care and necessary social services and the right to security...' (United Nations, 1948, p. 76). Article 29.2 highlights the limitations of individual rights for 'securing due recognition and respect for the rights and freedoms of others ..., public order and the general welfare in a democratic society' (p. 77).
- The principle of intragenerational equity proposes that social impacts of interventions should not fall disproportionately on certain groups, in particular, children and women, the disabled and socially excluded, certain generations or certain regions (op. cit. Vanclay, 2006, p. 5). A critical aspect of intragenerational equity is gender equity.
- The principle of intergenerational equity proposes to manage interventions in ways that allow meeting the needs of the present generation without jeopardising the ability of future generations to meet their own needs (op. cit. Vanclay, 2006, p. 5). This relates to discounting in economic evaluation (see Chapter 6, Section 6.2.2.5).
- The uncertainty principle acknowledges that our knowledge of the natural and social world and of social processes is incomplete as the social environment and the processes affecting it are changing constantly and vary from place to place and over time (op. cit. Vanclay, 2006, p. 5).
- The precautionary principle states that strategies of precaution must be prioritized against strategies of reaction, especially when there are serious or irreversible threats to the health of humans or ecosystems, and even when there is acknowledged scientific uncertainty. In this sense, precaution should guide 'public health decisions under conditions of uncertainty with an appropriate consideration of power, ownership, equity and dignity' (cf. Martuzzi and Tickner, 2004, pp. 3; 7).
- The prevention principle states that it is generally preferable and cheaper in the long-term to prevent negative social impacts and ecological damage than having to restore or rectify damage after the event (op. cit. Vanclay, 2006, p. 6).
- The recognition and preservation of diversity states that planned interventions should not lead to the loss of social diversity (age, gender, value systems and different skills) in a community or diminish social capital (op. cit. Vanclay, 2006, p. 5). Social-ecological systems with diverse resources are likely to be more inclusive and more resilient to stress and shocks (Ifejika Speranza *et al.*, 2014).
- The polluter pays principle proposes that the full costs of avoiding or compensating social impacts should be borne by the proponent of the planned intervention. (op. cit. Vanclay, 2006, p. 6). This includes the internalisation of costs, so that the full social and ecological costs of a planned intervention should be included into the cost of the intervention using economic and other instruments. Thus, no intervention can be cost-effective if they create hidden costs to current or future generations or to the environment (op. cit. Vanclay, 2006, p. 6). (For details, see Chapter 6).
- The protection and promotion of health and safety proposes that all interventions should be assessed for their health impacts and accident risks, paying particular attention to those groups that are more vulnerable and more likely to be harmed. This generally includes the economically deprived, indigenous groups, children and women, the elderly, and the disabled as well as the population most exposed to risks arising from the planned intervention (op. cit. Vanclay, 2006, p. 6)
- The principle of multi-sectoral integration argues that social development needs and social issues should be properly integrated into all interventions (op. cit. Vanclay, 2006, p. 6). >>>

Box 4.1. Continued.

- The principle of subsidiarity proposes that decision-making power should be decentralised, with accountable decision-making as close to an individual citizen as possible, with local people having an input into the approval and management process (op. cit. Vanclay, 2006, p. 6).
- The principle of emancipation means setting people free from the coercive control or constraint of more powerful or dominant other people or social groups, and from subjection to them. It emphasises altering the relationship between dominant and subordinate social groups, and lessening the opportunities for the one to harm the interests of the other (Williamson, 2010, p. 2).
- The principle of Common But Differentiated Responsibilities and Respective Capabilities acknowledges that in view of the different contributions of countries to global environmental degradation, countries have common but differentiated responsibilities and respective capabilities in view of their level of economic development to address this global environmental challenge, in particular, climate change (UNCED, 1992, p. 2 Principle 7; United Nations, 1992).

Considering social sustainability from a transformative perspective requires an additional conceptual step as made explicit by Opielka (2017, pp. 10-11), who identifies three inter-related discourses of social sustainability: (1) a narrow framing; (2) internal conceptualization (differentiated into conservative and liberal perspectives); and (3) a wider conception. The narrow framing captures social sustainability as one of the three pillars of sustainability, as 'conflict reduction and redistribution of resources' that allies with ecological sustainability but opposes the dominance of economic sustainability. For Opielka, a conservative perspective of the internal conceptualisation addresses 'social sustainability as the sustainability of the social', thereby maintaining the core values of a society while avoiding institutional transformation and social redistribution. The bridge between this internal conception and the human responsibility for nature and the environment is made through public debates on the commons: e.g. air, biodiversity, water, as they relate to local communities and the world society. A liberal perspective emphasises the sustainability of economic functioning as captured by 'intergenerational justice concerning the distribution of resources' such as old age allowances or financial debt. Opielka (2017) identifies the wider conception of social sustainability in cases, whereby social sustainability becomes a goal in societal transformation towards post-growth, green growth, de-growth or as captured by the sustainable development goals (Box 4.2), thus opening up the concept of social sustainability towards the type of economic system to which social sustainability should be related.

We argue that the above-mentioned 'wider conceptualization' of social sustainability is adequate for the OH-context, as it leads more concretely, to the consideration of a broader set of interests and actors than standard public health approaches.

First, it means integrating human, animal and ecological health (Hinchliffe, 2015; Rock *et al.*, 2014). Thus, even when public health interventions are humanist in orientation, efforts to sustain the health of our ecological communities might require the prioritisation of non-human interests (Capps and Lederman, 2015; Degeling *et al.*, 2016).

Box 4.2. Social sustainability in the sustainable development goals (adapted from United Nations, 2015).

The challenge of achieving social sustainability in health interventions is reflected in the global Sustainable Development Goals (SDGs). Maintaining a healthy world population remains a challenge that is being addressed through SDG 3 that aims to 'ensure healthy lives and promote well-being for all at all ages' (United Nations, 2015, p. 20ff.). Within this goal, health equity is reflected in the Target 3.8., which aims to 'achieve universal health coverage, including financial risk protection, access to quality essential health-care services and access to safe, effective, quality and affordable essential medicines and vaccines for all.' Target 3.9 foresees 'by 2030, [to] substantially reduce the number of deaths and illnesses from hazardous chemicals and air, water and soil pollution and contamination.' Target 3b plans among others to 'provide access to medicines for all' and Target 3d aims to 'strengthen the capacity of all countries, in particular developing countries, for early warning, risk reduction and management of national and global health risks.'

Building on a pledge to leave no one behind – the SDGs strive to ensure the social foundations of society. SDG 1 aims to end poverty in all its forms (p. 19), targeting 'nationally appropriate social protection systems and measures for all, including floors,...(SDG 1.3)', ensuring that people 'have equal rights to economic resources, as well as access to basic services, ownership and control over land and other forms of property,... (SDG 1.4)', SDG 1.5 supports building the 'resilience of the poor and those in vulnerable situations...'. It also aims to secure 'significant mobilization of resources from a variety of sources...' (1.A), and 'create sound policy frameworks at the national, regional and international levels, based on pro-poor and gender-sensitive development strategies, to support accelerated investment in poverty eradication actions' (1.B).

Life in dignity and equality are important social conditions. Ensuring equality is a focus in various SDGs (e.g. SDG 1: no poverty; SDG 5: 'achieve gender equality and the empowerment of all women and girls' (p. 22); SDG 10: 'reduce inequality within and among countries', p. 25).

The SDGs also aim to foster social cohesion through a focus on peace – 'to foster peaceful, just and inclusive societies which are free from fear and violence. There can be no sustainable development without peace and no peace without sustainable development'. Further, the United Nations regards partnership (also in SDG 17) as a basis for Sustainable Development, 'based on a spirit of strengthened global solidarity, focussed in particular on the needs of the poorest and most vulnerable and with the participation of all countries, all stakeholders and all people' (p. 6).

Second, it means focussing also on the implications that social sustainability has for the currently dominant capitalist economic systems. Although there are varying forms of capitalist economic systems – they are commonly understood as the results of the specific interactions between marketization (driven by private or public interests) and socio-environmental protection (driven by protecting basic rights of people, other living entities and the environment) from the liabilities of marketization.

It was the merit of Polanyi (2001) to show that modern market-based economic systems result from, and are reproduced through, a progressive dis-embedding of the economic system from the related social systems. He also showed that this process was- and still is – only possible to the degree that labour, nature and money (as a means of exchange) are stripped of their use value[1] and are turned through this into eventually factious commodities as the only way these human and natural elements can circulate in a predominantly market-based economy.

These considerations from a historical perspective of political economy are important for the discussion on social sustainability because they allow integrating a main root cause of social unsustainability (i.e. exclusive commodification of human, animal or ecosystem health) that affects the present and future generations. Hence, the definition of social sustainability must also consider to what degree it is able to contribute to the re-embedding of economic relations into the realm of wider society. This is of course not equal with reinstituting pre-capitalist societies. As Fraser (2011), based on feminist theories and practices shows, it is possible to broaden Polanyi's notion of a 'double' to a 'triple' movement, adding the notion of emancipation to the processes of marketization and state-based protection.[2]

Fraser (2013, p. 129) claims that the triple movement serves as an analytical lens that – unlike the double movement of Polanyi – '... delineates a three-sided conflict among proponents of marketization, adherents of social protection and partisans of emancipation. However, the aim here is not simply greater inclusiveness. It is rather to capture the shifting relations among those three sets of political forces, whose projects intersect and collide. The triple movement foregrounds the fact that each can ally, in principle, with either of the other two poles against the third.' The triple movement approach means therefore connecting the critique of commodification to the critique of domination, implying to understand social sustainability as also related to the transformation of the economic system. Such framing implies taking into account the opportunities and constraints offered by political collective action not only in view of a marketization vs state (regulation or protection), but in the

[1] According to economic theory and the political economy used by Polanyi (2001) and many others, all these items have use value i.e. they can be used for other purposes than engaging in market relations.

[2] In his seminal work, the 'great transformation' Karl Polanyi (2001) uncovers a double movement that acts as a major driver of economic processes in modern history. A first movement of 'marketization' refers to the establishment of hegemonic discourses and related institutions through which the economic elites are praising market utopia as the best way of organizing modern societies. Accordingly, the economic realms are progressively dis-embedded from the social and cultural ties used by society for gaining control over economic institutions and eventually determine the scope of the market. Polanyi argues that free-market utopians and related liberals are pushing towards a situation in which societies are increasingly subject to the rules of the market. This creates 'modern' capitalist economic systems that are powered by transforming nature, humans and means of exchange (money) into 'fictitious commodities' which are bought and sold in the market just like any other commodity.
The second movement is the reaction against the social, cultural, economic, health and environment related costs of the first movement. This reaction was not foreseen by the promoters of the first movement and was the result of the manifold protests against the ravages of the forces of free markets. Actors of the second movement are civil society organizations, trade unions, progressive and social-democratic political parties, social and liberation movements, organizations fighting for human, labour, ethnic, political, social, cultural – and more recently – also for environmental rights of people and ecosystems that were coming under pressure through the expansion of the forces of 'free markets'. The common ground of these movements was the establishment of sometimes powerful discourses and institutions (laws, rules and regulations) aiming at protecting social and environmental realms of societal life, from the negative influences of the 'free markets'. The actors of this second movement are mainly operating through governments and states that have the legitimacy to define, enforce and sanction the economic actors, based on socially and culturally defined rights that have to be respected, even if they contradict purely economic interests.

wider more complex and dynamic interplays between marketization, protection (state) and emancipation (social movements).

Considering the above discussions on the principles underlying social actions and relations (Box 4.1), the analytical and normative dimensions[3] of social sustainability and the need to emphasise the social, there is a need for a framework that integrates these dimensions. The suggestion by Littig and Grieβler (2005) to track progress towards social sustainability, using the following three core indicators, namely: (1) the satisfaction of basic needs and quality of life; (2) social justice and equal opportunities; and (3) social coherence, follows such an integrative approach. While building on this approach, we propose to extend considerations on capabilities to emancipation processes, and extend notions of social justice to an approach based on environmental justice that encompasses both human and non-human dimensions (cf. Fraser, 2009).

We thus define social sustainability as a condition, process or outcome whereby the needs and capabilities of current generations are secured, environmental justice, solidarity, social cohesion, as well as emancipation and self-determination thrive in a context of ecological sustainability, while ensuring to the extent possible the capacity of future generations to meet their own capabilities (Figure 4.1). This definition builds on the capability approach as proposed by Sen and Nussbaum (Nussbaum, 2011, 2000; Nussbaum and Sen, 2002; Sen, 2009, 2000, 1993, 1992, 1985), while integrating the notion of basic human needs and emancipation. This wider approach also emphasises the non-human interests and needs central to the health and sustainability of our ecological basis (Figure 4.1).

Figure 4.1 illustrates the biophysical environment comprising soil, water, animals, plants, other biodiversity, physical and built resources as the context within which underlying values and principles (Box 4.1) are negotiated, socio-economic conditions thrive and institutional arrangements are deliberated. It shows that social sustainability builds on a biophysical context and can be realized in terms of conditions, processes and outcomes. The three dimensions of human well-being as captured by: (1) basic needs, capabilities and emancipation; (2) environmental justice; and (3) solidarity and social cohesion influence one another and are reflected in the sustainable development goals (Box 4.2). Well-being, which refers to quality of life, is thus likely to be high in the face of achieved functionings and capabilities and in a context of environmental justice, solidarity and social cohesion. Evaluating social sustainability thus means analysing the extent to which processes are socially sustainable and are likely to lead to socially and ecologically sustainable outcomes. In the following, we discuss the three overlapping dimensions.

4.2.1 Basic needs, capabilities and emancipation

In this section, we propose a 'basic needs, capabilities and emancipation' dimension of social sustainability. We consider capability as a broader conception of needs that goes beyond the basic needs ensuring the material basis of life, to providing people scope for action

[3] Dimension as used in this chapter refers to a component, an aspect, a feature, or a facet.

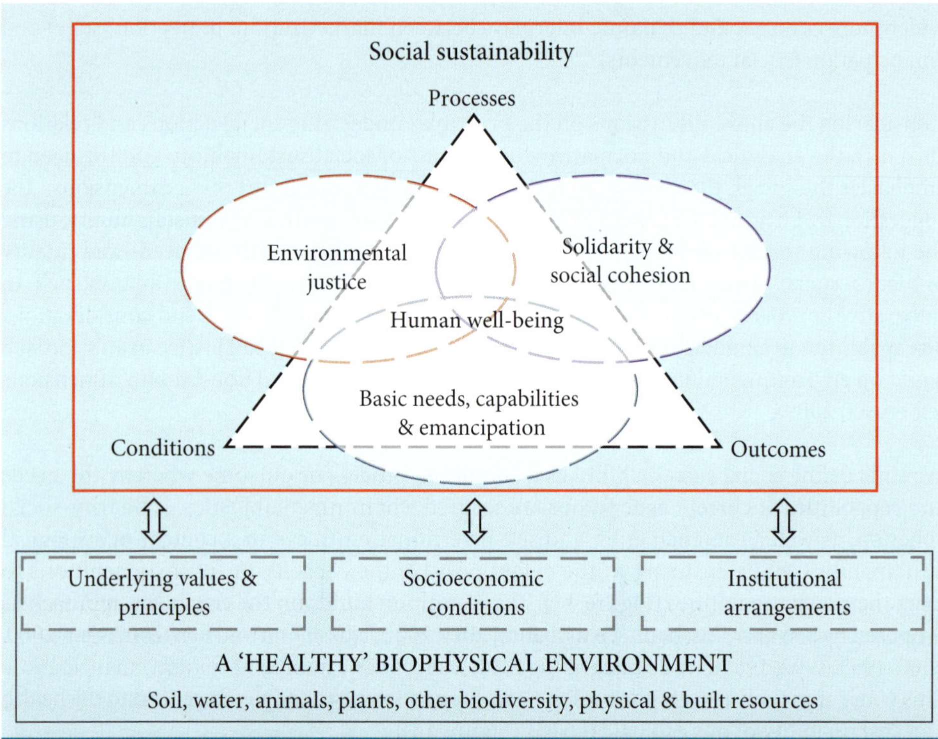

Figure 4.1. Conceptualising social sustainability.

and opportunity that allows choice, with emancipation playing a key role in creating the conditions to enable choice.

4.2.1.1 Basic needs

The concept of basic needs captures the universal need of humanity for food, shelter, clothing, bodily integrity, health, healthy environment, and access to clean drinking water and sanitation infrastructure, and security during illness, childhood and old age, and social crises (Empacher and Wehling, 1999; United Nations, 1948). The satisfaction of basic needs and quality of life can be extended to encompass education, employment, health security as well as subjective satisfaction with social processes and conditions (Littig and Griessler, 2005).

It can also be extended to non-material (psychological, spiritual, mental) and cultural needs that include integration in cultural and social networks, and free time and leisure (Empacher and Wehling, 1999). Such an extension means that action opportunities experienced by individuals must be expanded to make agency[4] and enable them to cater for their basic needs (Empacher and Wehling, 1999).

According to Vallance *et al.* (2011), basic needs are also relevant to contexts of high economic development, because access to necessary goods and services are subject to change and are the foundations of the 'so-called 'higher-order' needs. We argue therefore that a capabilities lens to social sustainability that captures both basic- and higher-order needs is applicable to contexts with different (whether high or low) levels of economic development.

4.2.1.2 The capability approach

The capability approach is an evaluative normative and theoretical framework that asserts that the freedom to achieve well-being is critical for human development and justice. It frames this freedom in terms of capabilities, that is, people's opportunities 'to achieve outcomes that they value and have reason to value', that is, to do and be what they have reason to value (Sen, 1999, p. 291). It evaluates the extent to which a person is able to be (has capability) or to do something (function) with or without having chosen to be or do something in a particular way (Coast *et al.*, 2008; Sen, 1993). For example, starving and fasting are similar functionings but fasting is dependent on the person haven chosen to fast (Sen, 1993). Because it focuses on capacity and opportunity, applying the capability approach can help understand and address conditions, processes and well-being outcomes of people. The capability approach has been widely applied in the social sciences and has been conceptualised to comprise the following dimensions: resources, conversion factors, capabilities (opportunities to achieve beings and doings), choice and functionings (beings and doings) (Figure 4.2).

1. Resources (goods and services; Figure 4.2) can be categorised into human capital – e.g. knowledge and skills; ability to work/labour; physical and cognitive limitations (e.g. Stafford *et al.*, 2017), social capital – e.g. family and friends, financial capital – e.g. incomes and savings, natural capital – e.g. personal relationship with environment/animals (species), and physical capital – e.g. housing (cf. Ifejika Speranza *et al.*, 2014).
2. Conversion factors capture the extent to which a functioning (e.g. being healthy) can be derived out of resources. Conversion factors refer to the ability to convert resources (means) into opportunities (capabilities) or outcomes (functionings) (Sen, 1992). This ability is often an interplay of three types of conversion factors: (1) Internal/individual conversion factors refer to individual abilities, which are internal characteristics of an individual such as sociodemographic and socioeconomic characteristics. External conversion factors comprise: (2) social conversion factors, which capture the social context within which an individual lives – the formal and informal norms, policy landscape, levels of social cohesion, power dynamics, impacts of class, gender and other intersectionality such as race, ethnicity, and caste; and (3) environmental conversion factors that depict the biophysical environment of a person (Sen, 1992). In ideal cases, the combination of

[4] Agency refers to the possibility of people to shape actions and societal structures in which they are embedded in such a way that the members of a society have equal chances to bring their views to social, economic and material expressions.

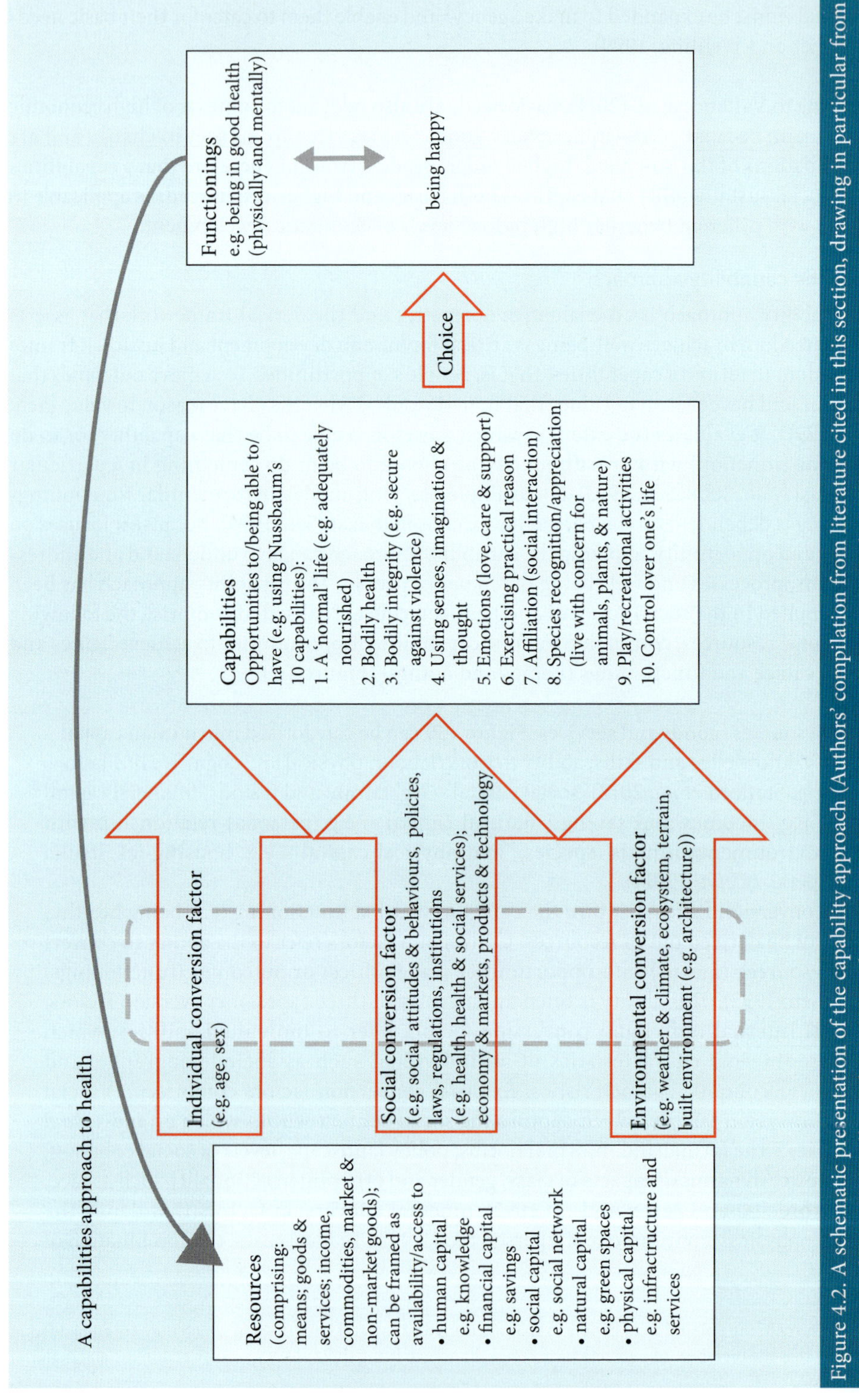

Figure 4.2. A schematic presentation of the capability approach (Authors' compilation from literature cited in this section, drawing in particular from Bussière *et al.*, 2016; Lorgelly *et al.*, 2015; Nussbaum, 2006; Sen, 1992, 1993, 1999; Simon *et al.*, 2013).

these conversion factors would foster the ability of individuals to use opportunities or combinations of opportunities available/accessible to them. Thus, human agency, that is, the ability to pursue valued objectives (e.g. aspirations to next generation's better health), to act and bring about change (Sen, 1992, p. 19) depends on individual and contextual factors.

3. Capabilities refer to (sets of) opportunities for achieving functionings and well-being (Figure 4.2). As levels of capabilities per individual are different, Sen (1992, p. 45 n. 19) captures this notion in *basic capabilities*, that is, 'the ability to satisfy certain elementary and crucially important functionings up to certain levels'. This can be related to poverty thresholds and issues related to human survival such as food and basic needs as discussed in the section on basic needs. Sen's conception of capabilities builds on the idea of social justice, which Nussbaum further concretised using the idea of a life in human dignity. Nussbaum (2006, pp. 76-78) (cf. Nussbaum, 2011) identified ten central capabilities as a minimum standard for a dignified and just life: life; bodily health; bodily integrity; senses, imagination and thought; emotions; practical reason; affiliation; other species; play; and control over one's environment (Figure 4.2). Nussbaum (2006) has also proposed a parallel set of species-specific capabilities to guide our treatment of non-human animals.
4. Functionings refer to 'beings and doings' (well-being outcomes) in the sense of what a person can be (e.g. being malnourished, unhealthy, wealthy, poor, excluded) or do (e.g. working, participating in a meeting) (Sen, 1992). Achieved functionings depends on the interactions of resources available to an individual, the individual, social and environmental conversion factors as well as the choice an individual makes out of the opportunities (capabilities) available/accessible (Figure 4.2).

4.2.1.3 Emancipation

Expanding the capability framing with emancipation (Figure 4.2) allows capturing broader processes of structural change that people can strive for in order to improve capabilities and functionings. According to Fraser (2013), emancipation implies being part of a society of autonomous subjectivities that have equal possibilities of taking part in the configuration of socio-cultural, political and economic structures defining the choices that a society offers to its members. Emancipation has thus the potential to transform the conversion factors that enable capabilities. Hence, emancipatory processes are more likely to be successful if they (1) open new spaces for communicative action, allowing for an intersubjective re-definition of the present situation, (2) contributed to rebalance the relationships between social capital and social, emotional and cognitive competencies within and between local and external actors (Rist *et al.*, 2006).

4.2.1.4 Evaluating the contributions of One Health to basic needs, capabilities and emancipation

To evaluate the impacts of OH initiatives on social sustainability, we propose to assess the four different dimensions – resources, conversion factors, capabilities (opportunities) and achieved functioning (achievements/outcomes) as these, while interacting, capture different dimensions of the capability approach. These dimensions are also active at different levels/ scales: e.g. resources may be at the scale of an individual or a community (society). In line with

OH-principles, there is a need for a transdisciplinary and participatory process in defining what resources, conversion factors, capabilities and functionings are to be achieved.

A departure point would be to ask whether people have the resources (means; goods and services: e.g. health services) to make choices (e.g. use health services; select safe consumer products and foods) to achieve functioning (e.g. being healthy). A strength of the capabilities approach is that it can highlight how the impacts and outcomes of interventions – such as the provision of a good or service – varies across and between settings because people live under different conditions, and/or have different types and levels of capabilities.

The importance of resources that people value vary from person to person, so also the focus of OH initiatives. Applied to OH, the question then is what health resources are at the disposal of all people, how each individual has access to the resources and opportunities, how the social-ecological environment influences each individual's opportunities from which s/he can make choices of which actions to implement.

Considering that individuals differ in their abilities to convert resources into outcomes, the extent to which the social and environmental conditions empower people to achieve functionings (outcomes) becomes critical.

Individual conversion factors include physical body conditions, education, knowledge and skills, age, sex and health conditions, cognitive ability, coping styles, social background, profession, past and current experiences, attitude, behaviour, character, and other factors that influence individual experiences of health and well-being (United Nations, 2008, p. 24).

Socio-economic factors affect health operations and outcomes (Braveman and Gottlieb, 2014; CSDH, 2008; Marmot *et al.*, 2012; WHO, 2011), and policies and regulatory frameworks as well as community social capital can enhance health functionings. Socio-economic and political stability and relevant regulations are needed to sustain health programmes (Gruen *et al.*, 2008). Community involvement and participation can also improve the social sustainability of a OH-initiative (Pareja-Eastaway, 2012). Compensations, such as social support services (WHO, 2013), and social security payment systems can address gaps in social conversion factors for those unable to participate. Moreover, preventive and precautionary strategies such as taxes on unhealthy food (Roberto *et al.*, 2015) have potentials to reduce disease.

Emancipation can be captured by evaluating the degrees of self-determination in health-related aspects (access to different health traditions, treatments, institutional equity independently from class, gender or race categories). It thus reflects the intersection of individual and socio-economic conversion factors.

Bussière *et al.* (2016) categorise environmental factors into barriers and facilitators. Facilitators include assistive technology and access to built-environment, such as curb ramps or to transportation, or provisions in law or social policy, family and community support. Barriers are unaccommodating physical or built environments, as well as stereotypical and stigmatizing attitudes. Favourable social and/or economic environment can compensate for the negative effects of cognitive and physical limitations (Bussière *et al.*, 2016). Urban planning can be relevant to prevention, hence involving different user groups in the planning and decision process is important (Kabisch and Haase, 2014). Spaces for green infrastructure and physical activity have been found to reduce cardiovascular risks, obesity and diabetes and reduce health costs in developed countries (Carter and Horwitz, 2014; Grabow *et al.*, 2012; Jarrett *et al.*, 2012; Pucher *et al.*, 2010). Green spaces also provide habitat for wild animals.

Capabilities and functionings have been assessed through different methods and measures. Ruger (2012a, p. 79) proposes examining health functionings (achievements – e.g. being healthy) and a person's health agency (e.g. the capability of an individual to pursue healthy behaviour) as indicators of health capabilities since health capabilities are not directly observable. In operationalising these measures, a focus should be on assessing whether each individual has the same opportunity (capability) as outcomes (functionings) may vary depending on the choices people make[5].

Health capabilities represent 'the ability of individuals to achieve certain health functionings and the freedom to achieve those functionings' (Ruger, 2012b, p. 81). Socioeconomic capabilities can be in the form of health insurance, education level, and income. In basic terms, the question here is whether people have the freedom (choice) to undertake the relevant basic actions for them to avoid exposure to mortalities or fatalities arising from diseases. Mitchell *et al.*, (2017) conducted a review of the applications of the capability approach and the measurement of capability in the health field. The authors found that most studies focussed on the 'sufficiency of capabilities' whereby health status is one out of the many indicators evaluated. However, as health is an outcome of One Health (cf. Rüegg *et al.*, 2017), health status can be omitted as an indicator of social sustainability (another outcome of One Health) in order to avoid double counting[6]. Various authors have applied the capability approach in health (Al-Janabi *et al.*, 2012; Callander *et al.*, 2013a,b; Mitra *et al.*, 2013; Netten *et al.*, 2012; Simon *et al.*, 2013). Gender aspects have also been considered (Mabsout, 2011; Nikiema *et al.*, 2012). Lorgelly *et al.* (2015) operationalised the capability approach for public health (Box 4.1) using Nussbaum's 10 capabilities.

The WHO (2001, 2010, 2013) proposed the International Classification of Functioning, Disability and Health (ICF) framework for measuring health and disability, whereby functioning is conceptualised as a 'dynamic interaction between a person's health condition, environmental factors and personal factors' (WHO, 2013, p. 5).

Capabilities and well-being can be evaluated in terms of identifying the resources people value in terms of 'agency goals' (being able to do/capability) (Coast *et al.*, 2008) that enables them to achieve the functioning of being healthy and having well-being. Thus, the aim of an evaluation of the contribution of OH-initiatives to capabilities is to measure the capability set of people to be healthy and to achieve well-being. A first step in this analysis would be to identify from people the aspects of health and well-being capabilities they value and in a

[5] This means measuring an intermediate outcome or output that – according to a theory of change – may lead to a final outcome or impact.
[6] In analyses outside a one health context, health can be incorporated as an indicator of social sustainability.

next step to assess the extent to which they have options to achieve these values, the enabling factors and whether they succeed in achieving them (see for example Figure 4.2).

To operationalise such a framework, health and well-being values can be identified through literature review, surveys and expert assessments. These can then be developed into a questionnaire to collect data that is differentiated according to social categories (age, gender, class, ethnicity, etc.) as well as self-rated health condition (scale 1-5: poor, fair, good, very good, excellent). By not pre-defining values or resources important to people to achieve their capability well-being, context specific values can be captured and adapted for analysis (See for example Bussière *et al.*, 2016; Lorgelly *et al.*, 2015; Ruger, 2012b; Stafford *et al.*, 2017; Üstün *et al.*, 2010).

Table 4.1 illustrates the different ways the capability approach has been applied to analyse health issues. It shows that not all the 10 capabilities identified by Nussbaum are applicable to all cases and that authors have adapted them to fit their purpose.

Following Abma *et al.* (2016), questions can be asked about: (1) the aspects of health and well-being that are important/valuable (captures resources) to a person; (2) Whether the person has sufficient opportunities to realise the aspects of health important/valuable to him/her (captures capabilities); and (3) whether the person realises/achieves such identified aspects

Table 4.1. Illustration of the uses of the capability approach in health.

Illustration 1	**Illustration 2**	**Illustration 3**
Lorgelly *et al.*'s (2015, p. 80) operationalisation of the capability approach (using Nussbaum's 10 Central Human Capabilities) for public health.	Bussière *et al.*'s (2016, p. 72) operationalisation of health capabilities (disabilities) based on the International Classification of Functioning, Disability and Health (ICF) framework, using 5 latent constructs.	Üstün *et al.*'s (2010, p. 816) WHO Disability Assessment Schedule 2.0 based on the ICF (using 6 constructs).
1. LIFE: 'Given my family history, dietary habits, lifestyle and health status', I expect to live up to… 2. BODILY HEALTH. My health limits my daily activities, compared to most people my age.	HEALTH CONDITION CAPABILITIES: Number and presence of diseases: diseases, impairments, perceived health status and symptoms (e.g. sleep disorders, tiredness, stress, palpitations, discomfort) PHYSICAL CAPABILITIES: (Scoring scale): 'physical activity limitations (e.g. walking, raising arms, seeing, hearing) and resulting activity restrictions, primarily in terms of activities of daily living (e.g. washing, using the toilet, dressing)'.	SELF-CARE: Ability to attend to personal hygiene, dressing and eating, and to live/stay alone
3. BODILY INTEGRITY: I feel safe walking alone in the area near my home.		MOBILITY: Ability to move and get around
		>>>

Table 4.1. Continued.

Illustration 1	**Illustration 2**	**Illustration 3**
4. SENSES, IMAGINATION AND THOUGHT: 'I am able to express my views, including political and religious views.'	COGNITIVE CAPABILITIES: (Scoring scale): 'Cognitive activity limitations (e.g. understanding what people say, concentrating, remembering, being aggressive) and the resulting activity restrictions (e.g. establishing relationships, being disturbed in daily life because of a psychological problem)'.	COGNITION: Understanding and communicating
5. EMOTIONS: At present, I enjoy the love, care and support of my family and friends; In the past 4 weeks, I have lost sleep over worry. 6. PRACTICAL REASON: 'I am free to decide for myself how to live my life.'	SOCIETAL CAPABILITIES: (Scoring scale): 'participation restrictions of an individual in society (e.g. instrumental activities of daily living, leisure, employment, living as a couple), including environmental barriers (e.g. negative attitudes, inaccessible transportation and public buildings, limited social support, and the need for human/technical assistance)'.	
7. AFFILIATION: I am able to 'meet socially with friends, relatives or work colleagues'. 8. SPECIES: 'I am able to appreciate and value plants, animals and the world of nature' 9. PLAY: In the past 4 weeks, I have been able to enjoy recreational activities. 10. CONTROL OVER ONE'S LIFE: 'I am able to influence decisions affecting' my health and well-being; In the past 4 weeks I have experienced discrimination.	SOCIOECONOMIC CAPABILITIES: (Scoring scale): 'personal factors, specifically socioeconomic factors (educational level, insurance, income, home ownership, savings)'.	GETTING ALONG: Ability to interact with other people PARTICIPATION: Ability to engage in community, civil and recreational activities LIFE ACTIVITIES: Ability to carry out domestic responsibilities, leisure, work and school

of health and well-being. To capture the influence of the conversion factors, questions can be asked on what and to what extent self-reported individual (4), social environmental (5) and natural environmental (6) factors enable the person to achieve the aspects of health and well-being he/she values. Following Abma *et al.* (2016, p. 36), all selected items can be ordered from (1) to (5) and scored: 1='not at all'; 2='not', 3='neutral', 4='yes', and 5='very much'.

Box 4.3. An example of questions for capturing health and well-being capabilities.

Assuming for example that 'capability to be healthy and to achieve well-being' is being evaluated, the following questions can be asked to capture this capability:

1. Resources: To what extent does the initiative foster/facilitate/has positive influence on the resource you have at your disposal?
2. Capabilities: Depending on focus, questions can be chosen/adapted from Table 4.1.
3. Functioning: Depending on focus, questions can be chosen/adapted from Table 4.1.
4. Personal conversion factors:
   a. Under the circumstances, what is the most important personal (individual) factor that influences your ability to be healthy/achieve well-being?
   b. To what extent does the initiative foster/facilitate/has positive influence on this personal factor?
5. Social-political environments:
   a. How much does your social environment (family and friends, community, society) support you in your activities to be healthy/achieve well-being?
   b. How much do health policies, regulations and procedures support you in your activities to be healthy/achieve well-being?
6. Natural and built environments:
   a. How much does your natural environment support you in your activities to be healthy/achieve well-being?
   b. How much does your built environment support you in your activities to be healthy/achieve well-being?

To add depth to the collected information, 'explanations' (e.g. please explain why you assigned this score) can be requested from the respondents for the scores they assign to (1) to (6).

In the foregoing, the needs and capability dimensions of social sustainability have been elaborated. In the next section, the environmental justice dimension of social sustainability is discussed.

4.2.2 Environmental justice

Considering human, animal and environmental interconnections, justice needs to be expanded to encompass human species and ecosystems. Environmental justice focuses on the right of all humans to a healthy environment irrespective of their social positions and wealth status (Griffiths, 2006; Schlosberg, 2007), thus extending the concept of social justice to account for human and non-human dimensions of justice. Environmental justice refers to three interrelated dimensions of justice that include recognition (mutual respect), procedural justice including participation and self-determination in decision-making, and distributional justice in terms of equitable access to resources, benefits and burdens (Fraser, 2009; Schlosberg, 2007).

Furthermore, it has been proposed to extend subjects of justice beyond the human individual to include human communities, non-human animals and environmental elements (Schlosberg, 2013; Sikor *et al.*, 2014). Applied to non-human animals, environmental justice can be interpreted in terms of animal welfare (Carrel *et al.*, 2016). Thus, evaluating the contributions of OH to social sustainability in the dimension of environmental justice entails examining the extent to which human individuals and communities are recognised, can participate and equitably share in resources and burdens, and the extent to which animal welfare and environmental health are concerns. In the following, we discuss these dimensions and their relations to OH initiatives.

4.2.2.1 Recognition

Recognition is about respecting identities and cultural differences (Fraser and Honneth, 2003). It is about the 'extent to which different agents, ideas and cultures are respected and valued in interpersonal encounters and in public discourse and practice' (Martin *et al.*, 2016, p. 255). Recognition means 'acknowledging that individuals in groups construct different cases about what is right and wrong based on a complex assemblage of ideas and circumstances, which shape the way they experience a particular problem or issue' (Martin, 2017, p. 14).

Since contexts are different and people's reactions to circumstances are often mediated by their ideas, beliefs and interpretations that are locally and historically situated and less homogenous, rational and predictable (Parsons, 2007), social justice is understood/perceived differently. Recognition means therefore opening up equity concerns to the plurality of contextual and cultural framings of justice. Thus, Fraser (2000) proposes to address cultural inequalities in addition to economic and political inequality, with a focus on the complementarities between redistribution and social recognition.

Recognition therefore entails that all actors (privileged, disadvantaged or vulnerable), are recognized in terms of having an appropriate share of burdens, benefits and opportunities, voice, and their identities respected (Eizenberg and Jabareen, 2017). Through accounting for social categories such as gender, race, age, class and their intersections, recognition and guaranteeing diversity can ensure that the different needs of members of society are not addressed in a one-size fits all approach but in diverse ways relevant to the social context (Borgonovi and Compagni, 2013). Recognition thus provides some clues to navigate through the delicate terrain of universally accepted capabilities and the plurality of framings, aspiration and values. In other words, it means to acknowledge and account for plurality of justice values without abandoning the attachment to a general basic normative principle (Martin, 2017). Recognition also refers to Fraser's (2013) notion of emancipation pointing to the right of self-determination of societal actors in a context of equality to reduce asymmetric power relations.

4.2.2.2 Procedural justice and participation

In social justice theory, distributive justice is intimately tied to procedural justice (Rawls, 1971), which focuses on whether procedures of decision-making ensure equity (e.g. who was involved in the decision-making process? How were the persons involved? Was due diligence followed? Was the process transparent?). Procedural justice has been conceptualized as meaningful participation. Fraser (1996, pp. 30-31), proposes the concept of parity of participation, which focuses on the extent to which social arrangements allow all '(adult) members of society to interact with one another as peers'. Parity of participation thus depends on legal and political factors but also on the equal distribution of material resources and economic independence, that allows a person's independent participation and voice.

Participation as a measure of procedural justice assumes that people are likely to perceive a decision as just if they participated in making that decision, and is thus often associated with democratic decision-making (Barnes and Coelho, 2009; Iroz-Elardo, 2015). Arnstein (1969) proposed a hierarchical ladder of participation reflecting different levels of engagement with the higher levels reflecting the highest levels of participation and expected effectiveness. Participation is expected to improve ownership and the tailoring of interventions so they are appropriate and relevant for people (cf. Gruen *et al.*, 2008). However, Hurlbert and Gupta (2015) have highlighted shortcomings of participation, especially when it is deemed as inherently good without examining whether it is implemented with appropriate mechanisms or addressed in a technocratic manner. Participation is the most fundamental element that links the dialectic relationship between agency and social structure. This means that participation must allow creating governance conditions that aim at social learning processes that involve all relevant actors. Such social learning processes also aim at creating spaces for transforming strategic action (oriented towards optimizing ego-centric individual or collective interests) into communicative action, oriented in collective efforts, based on a common understanding about what problems, conflicts and solutions are (Rist *et al.*, 2007).

Colquitt and Rodell (2015, p. 189), propose to evaluate procedures based on: (1) 'Process control: procedures provide opportunities' for influencing/controlling a process – voice; (2) 'Decision control: influence over outcomes'; (3) 'Consistency: procedures are consistent across person and time'; (4) 'Bias suppression: procedures are neutral and unbiased'; (5) 'Accuracy: procedure is based on accurate information'; (6) 'Correctability: procedures offer opportunities to correct an outcome'; (7) 'Representativeness: procedures take into account concerns of subgroups'; and (8) 'Ethicality: procedures uphold standards of morality'.

4.2.2.3 Distributive justice

In establishing processes to pursue health equity[7], Litman (2015, p. 3) considers equity as closely related to the social distributive justice, whereas equity is 'also called justice and fairness and refers to the distribution of impacts (benefits and costs) and whether that distribution is considered fair and appropriate'. The notion of distributive justice goes back to Rawls (1971), who considers the greatest benefits of the least advantaged as an outcome to attain through rational impartiality in procedures. Social distributive justice thus ensures that people have (equal) rights (Eizenberg and Jabareen, 2017), and comprises two dimensions: intergenerational (between present and future generations) and intra-generational (between

[7] 'Health inequality is the generic term used to designate differences, variations, and disparities in the health achievements of individuals and groups' (Kawachi *et al.*, 2002, p. 647). 'Health inequity refers to those inequalities in health that are deemed to be unfair or stemming from some form of injustice' (Kawachi *et al.*, 2002, pp. 647-648). The crux of distinguishing 'between equality and equity is that the identification of health inequities entails normative judgment premised upon (1) one's theories of justice; (2) one's theories of society; and (3) one's reasoning underlying the genesis of health inequalities. Because identifying health inequities involves normative judgment, science alone cannot determine which inequalities are also inequitable, nor what proportion of an observed inequality is unjust or unfair' (Kawachi *et al.*, 2002, p. 648).

different societal categories using an intersectional lens (e.g. race, ethnic groups, gender, age, class, etc.) in allocating resources, burdens, benefits and opportunities.

Distributive justice can be evaluated based on (1) 'Equity: outcomes are allocated according to contributions'; (2) 'Equality: outcomes are allocated equally'; (3) 'Need: outcomes are allocated according to need' (Colquitt and Rodell, 2015, p. 189).

4.2.2.4 Justice for animals and other non-human entities

Incorporating environmental justice into health concerns makes the link between humans, animals and ecosystems, and is thus particularly relevant to the OH approach. Griffiths (2006, p. 582) proposes promoting environmental justice as a way towards reducing health inequalities as the concept includes the right of all to a healthy environment. Through an environmental justice approach, health inequalities associated with environmental inequalities can be reduced. Applying the concept of 'environmental justice' helps identify whether exposure to health risks is 'socially patterned' and/or due to the impact of reduced or uneven availability of health facilities, and reduced access or 'access deprivation' (op. cit. Smith, 2016).

Because OH explicitly prioritises the health of non-human animals and ecological systems, there are emerging questions as to whether non-human entities (organisms and ecologies) should also be subjects of distributive justice, rather than this social good being the strict preserve of human interests and human benefits (Capps and Lederman, 2014; Rock and Degeling, 2015). A capability based approach to justice – with its commitment to flourishing – also seeks to promote conditions for health and a good life and could be meaningfully extended to more-than-human concerns (Haraway, 2008; Nussbaum, 2006). Broadening the scope of environmental health justice to include non-humans will require us to share the risks, burdens and goods of OH interventions across species boundaries. Such a move would be both politically and ethically controversial because traditional public health approaches to disease risks are steadfastly humanist in orientation and distribute the costs of control (where possible) onto the environment and other species (Verweij and Bovenkerk, 2016). Nevertheless, consistent with the social sustainability agenda, OH could be a vehicle to prioritise approaches that seek to share both risks and benefits of interventions, where humans and non-humans are considered to be prone to much of the same environmental risks, and have a converging set of interests to their integrity (Capps and Lederman, 2015; Degeling *et al.*, 2016).

Dealing with human and animal health as OH inevitably sheds light on the human-animal relationship and bond. Animals such as dogs contribute to human health, biologically (e.g. reduce cardiovascular health risks), psychologically (e.g. reduce depression and loneliness), socially (e.g. more positive perception of people and the environment), and have educational effects on children (Hediger and Beetz, 2015). Domestication of wild animals has been one of the fundamental cultural achievements of humans and the use of animals for hunting and as livestock was critical for human development and culture.

OH, even in a more restricted definition, faces challenging questions regarding cultural differences in view of what animals are and how they are valued. According to Zinsstag *et al.* (2015b, op. cit. p. 19), culture and religion determine the norms and values governing human-animal relationship. As intimate companions, animals have high emotional value for humans but also have financial and consumption values as many humans consume their meat. The authors argue that despite the general protective attitude in most cultures and religions, domestic animals are still massively handled and slaughtered in terrible conditions, hence the need for an urgent and much stronger engagement for animal protection and welfare. They explain that under given circumstances, humans are prey for animals and this is one of the reasons for deep-seated fears against wildlife, which have led to the extinction or threat of extinction of predators in large parts of the world and one of the reasons for the current ecological crisis. Culture, religion and economic considerations thus influence the human-animal relationship and by extension the potential of OH within the dilemma of aspirations of a globalized economy, social development and animal welfare (op.cit. Zinsstag *et al.*, 2015b).

Thus, OH initiatives need to account for the normative aspects (values) of the human-animal relationship with emphasis on improving animal protection and welfare. Acknowledging animals' rights implies considering their well-being such as through animal welfare regulations (Wettlaufer *et al.*, 2015). As OH outcomes should be socially sustainable from the perspective of the user (human and animals, plants, microbiota ecosystems) a non-speciesist, or a 'less speciesist' position with all its dilemmas, need to be taken. The consideration that non-human animals have direct entitlements to justice remains nevertheless a debated issue (Berkey, 2017; Liberto, 2017; Plunkett, 2016).

4.2.2.5 Evaluating the contributions of One Health to environmental justice

Equity and health equity

The emergence and perpetuation of health problems is often related to multiple causal pathways, which make it difficult to assess which health problems manifest in social injustices, and which constitute human rights deficits or violations (Pogge, 2015, 2016). Such factors include exclusive and discriminatory barriers to health system access, lack of enforcement of legal restrictions, and the agency and environment of people with avoidable health problems. Tanner (2005) and Zinsstag *et al.* (2011) have proposed to assess the effectiveness of health interventions and policies in terms of social equity, through an integrative analysis of social, economic and cultural, as well as biological and environmental determinants of health and well-being.

Ruger (2012b) links societal health strategies with the notion of equity through attainment equality, which focuses on absolute levels of achievement, and shortfall equality focusing on shortfalls of actual achievement from the optimal average (such as longevity or physical performance) for individuals. Attainment equality highlights social variables such as education, gender, social class and location, whereas a shortfall equality draws attention to the reasons for the deficit from the optimum (Ruger, 2012b).

Equity and right issues in health have often been restricted to the access to universal health care issues than for equal health or the equal right to health (Ruger, 2012b, p. 120). This is visible in the WHO definition of universal health coverage (UHC), whereby, 'all people can use the promotive, preventive, curative, rehabilitative and palliative health services they need, of sufficient quality to be effective, while also ensuring that the use of these services does not expose the user to financial hardship' (WHO, 2016). However, Ruger proposes to rethink equal access as ensuring 'the social conditions in which all individuals have the capability to be healthy' [...] thereby emphasising 'effective access so that all have the ability to achieve health functionings and health agency' (Ruger, 2012b, p. 134).

Equity and rights issues have also been expanded to intergenerational equity. In such framing, the question would be whether OH initiatives that aim to preserve current capabilities increase the burden for future generations (e.g. disease burden, tax burden) and constrain their capabilities (e.g. in terms of increasing expenditure on health, increasing pensioners purchasing power and creating imbalance with the working population; (cf. Garcés *et al.*, 2003)). This concern is also visible in the recent report from the Lancet Commission on planetary health (Whitmee *et al.*, 2015), which emphasizes the preservation of the health of future generations through maintaining the integrity of biophysical systems, while addressing health inequities in the present generation.

Health issues and the dimensions of environmental justice

In relation with health issues, recognition has two main implications. First, it means to acknowledge better the contribution of individuals and communities to health, such as for example, volunteer care of sick or geriatric family members that needs to be legally and economically recognised (Garcés *et al.*, 2003). This also implies to avoid presenting people in a negative light related with their health condition, such as dismissing the poor for poor sanitation behaviour. Second, it also implies recognizing how people manage to improve their subjective welfare and optimize their accessibility to services (Garcés *et al.*, 2003). This requires acknowledging the plurality of views on notions of health, disease and treatment, thus opening up to the recognition of culturally appropriate, complementary and alternative medical care options.

In health issues, participation is usually understood as the extent to which individual actors and communities engage with and are committed to health policies and activities through collaborative partnerships. The WHO (2018) acknowledges that community engagement is key to successfully controlling disease outbreaks. In relation with OH initiatives, the question then is whether OH initiatives have procedural ways to ensure the participation that is representative enough to capture existing diversity that creates ownership and belonging (Borgonovi and Compagni, 2013).

Participation is therefore about how democratic a health care system is, and to what extent citizens can be involved in defining priorities and criteria, and shape a health care system that responds to collective expectations (Borgonovi and Compagni, 2013, pp. 36-37). Collaborative partnerships, active health-system policy-making, incentive structures, and population-based performance measures are some forms (Plochg *et al.*, 2006; in Gruen *et al.*, 2008, p. 1583). Indicators of participation in vital spaces to empower individuals who suffer disabilities have also been formulated in the ICF framework (WHO, 2013, p. 114).

Social distributive justice has been broadly related to health through *the causes for the persistence of health* inequalities, such as limited resources for public health systems, larger burden of disease in poorer countries, limited means to purchase health insurance, inadequate allocation of resources, unethical trade (e.g. dumping of banned products and waste), to legal entitlement to health services (Pogge, 2015, 2016). However, Ruger (2010) highlights that social distributive justice is often reduced to the right to health care services, thereby neglecting philosophical reflections on a right to health. Benatar *et al.* (2016, p. 323) make a more direct link to distribution, highlighting the 'inequalities in the global distribution of conditions necessary for human health and well-being'. Among them is the structure of the global economy that advantages the wealthy, hence is deemed unjust (Kochhar, 2015; Benatar, 2003; in Benatar *et al.*, 2016, p. 325; cf. Schrecker, 2017; Pew Research Center, 2018).

A focus on environmental justice (as a wider conception of social justice) highlights the processes that underpin the achievement of capabilities and functionings. This implies that capabilities can be regarded as a metric of environmental justice in that it specifies thresholds (foundations) of a set of capabilities that all persons need to achieve and which governments need to ensure for their citizens (Nussbaum, 2006, p. 71). Yet a focus on environmental justice exposes the interlinkages (including the tensions) between addressing individual needs and collective needs as well as individual responsibilities and collective responsibilities. For example, public health concerns in a context of the spread of infectious diseases: How can the health care needs of the individual be ensured and at the same time ensure that public health is secured? What roles must the individual play and what the collective and how can a just division of responsibilities (Robeyns, 2017, p. 157) be ensured? Justice is thus about rights and responsibilities (duties), which need to be negotiated in a political and social process. Box 4.4 displays possible question for evaluating the contributions of OH to environmental justice. To facilitate the move from theory to implementation/application, example questions and references to literature are provided in Box 4.4.

4.2.3 Solidarity and social cohesion

In this section, we discuss solidarity and social cohesion, why they are important, what their constituent parts are and in what ways a OH-initiative positively/negatively affects these dimensions.

4.2.3.1 Solidarity

Solidarity is closely related to recognition, in that recognition can occur without solidarity, but recognition is a precondition for solidarity. Solidarity is therefore 'a measure of relatedness toward the achievement of mutual interests and goals and emerges between people who share common interests and perceive the advantages of pursuing them collectively (Goffee and Jones, 1998; in Pinto *et al.*, 2011, p. 379).

Forst (2002 in; Juul, 2010) identifies four normative contexts in which people are situated as ethical persons, legal persons, political citizens and moral persons. 'Ethical norms' is about sharing values as a member of an ethnic group or a local community. 'Legal norms' captures situations in which individuals are equal bearers of rights, binding for all members of a legal community. 'Political-democratic norms' captures civic solidarity characterised by tolerance and respect for different ways of life, protection through a system of equal rights, recognition of people as equal participants in public life and the avoidance of exclusion for ethical, social or political reasons. 'Moral norms' protect people in situations where ethical, legal or political

Box 4.4. Potential points for evaluating the contributions of One Health to environmental justice.

Prior to an evaluation, an agreement has to be reached in a deliberative process among those involved about which social determinants will be considered (e.g. age, wealth and income, gender) in the local social-ecological context. The following examples serve to illustrate potential focus or questions and are not comprehensive. A yes/no data can be generated or 'a five point scale where 1 = To a Very Small Extent; 2 = To a Small Extent; 3 = To a Moderate Extent; 4 = To a Large Extent; 5 = To a Very Large Extent' adopted (Colquitt and Rodell, 2015, p. 191).

- Equity – The opportunities and outcomes are equitable for all involved.
  Does the OH-initiative 'describe how equity issues will be addressed' (period, mechanisms, outcome targets) (Association of Local Public Health Agencies/Ontario Public Health Association: ALPHA/OPHA, 2013, p. 10)?
- Recognition – The OH-initiative
  1. acknowledges diverse social categories; ensures right to information in a form and language that is accessible to all actors.
  2. does not discriminate against people.
  3. considers the diverse viewpoints/perspectives.
  4. communicates information in a language that all understand.
  5. incorporates the identification and planning for priority populations[1]
     5.1. 'Identification of priority populations': 'Standardized and explicit process (e.g. specified in a policy and procedure for operational planning)' (ALPHA/OPHA, 2013, p. 5)
     5.2. 'Identification of priority populations': 'Standardized and explicit template (e.g. separate column for priority population)' (ALPHA/OPHA, 2013, p. 5)
     5.3. 'Process for identification of priority populations': 'has a comprehensive list of possible priority populations (e.g. list of 10 subgroups) for consideration' (ALPHA/OPHA, 2013, p. 5)
- Procedural justice
  The questions can be applied to the procedures of a OH-initiative (*adapted from Colquitt and Rodell, (2015, p. 191)).
  1. Process control: '*Are you able to express your views during those procedures?' How satisfied are you with the procedure used to determine health insurance premiums? (Lind and Tyler, 1988, p. n.p.); How much opportunity did you have to participate in the decision-process to grant/not grant you health benefits? (Lind and Tyler, 1988, p. n.p.); Affected actors are involved in decisions relating to their own welfare.
  2. Decision control: '*Can you influence the decisions arrived at by those procedures?'; 'How much control did you have over decisions that were made in your case' (Lind and Tyler, 1988, p. n.p.)
  3. Consistency: '*Are those procedures applied consistently?'
  4. Bias suppression: '*Are those procedures free of bias?*
  5. Accuracy: '*Are those procedures based on accurate information?'
  6. Correctability: '*Are you able to appeal the decisions arrived at by those procedures?'
  7. Representativeness: 'Were the concerns of other groups considered in the procedures?' The initiative has procedures/mechanisms that ensure the participation that is representative enough to capture existing diversity, respecting equality, autonomy and self-determination as the maximum expression of emancipation.
  8. Ethicality: 'Do those procedures uphold ethical and moral standards?'

>>>

Box 4.4. Continued.

- Distributive justice

  The questions below could refer to the outcomes of a OH-initiative such as health benefits/subsidies, insurance premiums, etc. (*adapted from Colquitt and Rodell (2015, p. 191))

  The initiative:

  1. Equity: ensures that resources and the various costs (financial, social and cultural e.g. eroding cultural identity) to address health issues are allocated in an equitable manner; prioritises actions that promote benefits across species boundaries; maintains intergenerational solidarity – by not impairing the decision-making capacity of future generations; '*Do those outcomes reflect the effort you have [e.g. made to keep healthy]?'
  2. Fairness: assigns obligations according to the dispositions (resources) of the actors. '*Are those outcomes appropriate [e.g. for the efforts you have made to keep healthy]?'
  3. Equality: '*Do those outcomes reflect what you have contributed [e.g. in terms of health insurance paid; efforts to keep healthy]?'
  4. Needs: '*Are those outcomes justified, [e.g. given your health needs]?'
  5. Maintains or at least does not worsen the rights of various actor categories (intra/inter-generational) – youths, adults, the aged, and vulnerable (chronically ill; disabled), gender (men/women), ethnic groups.
  6. Ensures patients right to access second expert opinions

- Animal welfare

  The OH-initiative

  1. Ensures that non-human beings including animals are not adversely affected.

[1] Priority populations refer to 'those populations that are at risk [whether due to socially-produced factors, e.g. low income, or due to biological or physiological reasons [e.g. age], and for which public health interventions may be reasonably considered to have a substantial impact at the population level' (ALPHA/OPHA, 2013, p. 6).

norms are not sufficient, focusing on how one ought to treat all people as fellow human beings (Forst, 2002, p. 268).

Based on these considerations, Juul (2010, p. 266) formulates a concept of solidarity in inter-human relations as the recognition of a person as an equal and worthy partner of interaction and a just distribution of possibilities for recognition. Contemporary solidarity is thus about recognition and a fair distribution of chances for recognition (Juul, 2010, p. 235). In a pluralistic and individualised society, a multidimensional concept of solidarity is required that captures its affective (based on emotions), conventional (interests), and reflective (based on individual moral choice) dimensions (Juul, 2010). Yet as a frame of reference for ethical considerations, solidarity needs to be interpreted in context.

In practice, solidarity can also be extended to non-human others (Rock *et al.*, 2014). Given that caring relationships are foundational to health, people's diverse connections with places and non-human forms of life can also be relevant to social cohesion and to public health (Burgess *et al.*, 2005; Johnston *et al.*, 2007; Rock *et al.*, 2007). Amiot and Bastian (2017) developed constructs to assess solidarity with animals, moral concern toward animals, solidarity with humans, and identification with nature. In terms of solidarity with animals questions can be asked about strength of bonds, concern, closeness, connection to other animals/humans and commitment towards animals/humans. Bréchon (2014) applied identity altruism (compassion for family, neighbourhood, region, country) and social altruism (compassion for underprivileged other persons) to capture solidarity. Bonnie *et al.* (2010) assessed the degree of solidarity by asking respondent opinions whether persons that can cause high health costs should pay higher, equal or lower health insurance premiums, assessing expectations of 'lower, equal and higher costs' as high to low solidarity respectively.

4.2.3.2 Social cohesion

Solidarity is often seen as a prerequisite for social cohesion (Juul, 2010). Social cohesion refers to the process of development of shared values, shared challenges and equal opportunity within a community, which relies on trust, hope and reciprocity, and which fosters a sense of belonging and recognition among all its members (adapted from Jenson (1998, p. 4)). Carron and Hausenblas (1998) define social cohesion as 'a dynamic process that reflects a group's tendency to stick together and remain united in satisfying member needs' (in Bruhn, 2009, p. 34).

Having harmonious relations between people is likely to produce conducive conditions for societal development and sustaining beneficial social life than without. Social cohesion is thus critical for creating and maintaining social order, that is, the norms, rules and laws that define 'living together' as a group, community or society (Staerklé, 2013, p. 49). In other words, 'the structuring and structured processes of social reality' that is 'constantly generated by the interplay of worldviews and institutions' (Mielke *et al.*, 2011, p. 1). As the persistence of social life (social order) depends on different factors and perspectives, 'various social orders may exist at the same time' with the dominant social order being the preferred order of society by its constituent (and often powerful) members (Mielke *et al.*, 2011, p. 3).

However, the understanding of social cohesion on the basis of shared values (Jenson, 1998, p. v) has been critiqued to overlook conflicts and political action (Jenson, 1998). These different ways to conceptualise social cohesion raises a question about cohesion of what and for whom and whether social cohesion enhances or hinders social equity (Jenson, 1998). Hence, integrating institutions into the definition of social cohesion can address this equity question.

Building on previous studies (e.g. Chan *et al.*, 2006; Jenson, 1998) Bottoni (2018) highlights social cohesion can be studied in three domains – individual, groups and institutions (captured as state government and its constituents e.g. regions). The author identified at least six levels of social cohesion – (1) relationships between individuals; (2) between individuals and groups; (3) individuals and institutions; (4) between (within) groups as a whole; (5) between groups and institutions; and (6) between and within institutions (understood as state, regions). The author highlights the 'within dimensions' (e.g. within a group, connections of different parts of an institution) as well as the horizontal (e.g. peers) or vertical relations (e.g. European Union and member states) can also be analysed.

Yet, what constitutes social cohesion may differ from context to context. In traditional societies, social cohesion may be through similarity between people through sharing values, having a sense of belonging and feeling a part of the community (collective identity) (Jenson, 2002). However, in multicultural societies, which many societies are increasingly likely to become, tolerance and openness to diversity are critical for maintaining social order (Bottoni, 2018; Jenson, 2002). We find that sharing values (Table 4.2) still remains critical for social cohesion and does not negate tolerance and openness to diversity or sharing interests as in bonding and bridging networks, and have thus included it in the constituents of social cohesion. Jenson (2002, p. 5) highlights the need for mechanisms and institutions to balance social justice and social cohesion by simultaneously valuing and promoting 'equality of opportunity and fairness across all dimensions of diversity', while 'fostering the capacity to act together, collectively and democratically'.

Table 4.2. Constituents of social cohesion from a perspective of outcomes for individuals (adapted from Bottoni, 2018; Chan *et al.*, 2006; Jenson, 1998; Littig and Griessler, 2005).

To what extent does the initiative contribute to...

Levels	**Subjective perspective (attitudinal)**	**Objective perspective (behavioural)**
Micro level – relationships among individuals (informal connections: interpersonal relations, family, primary groups)	1. Interpersonal trust 2. Social support (giving and receiving help and support)	3. Density of social relations (number and frequency of social relations, and compared to age-peers)
Meso level – relationships among individuals and groups (formal connections: neighbourhood, secondary groups, working groups)	4. Belonging: sharing values – (e.g. incorporating/respecting the norms/unwritten rules of the community in OH activities), collective identity 5. Openness: 'acceptance and openness toward diversity' (recognition/respect of differences and equality in possibilities of self-determined participation in the definition of agency and structures of a society)	6. Participation and emancipation: social and political participation. 7. Bridging (inter-group) and bonding (intra-group) ties 8. Inclusion: equality of opportunity
Macro level – relationships among individuals and society (institutions)	9. Institutional trust – (e.g. trust in parliament, legal system, police, health system, and other organs relevant to a OH issue at hand)	10. Legitimacy of institutions: Quality/ conditions of various social services – health, education; Satisfaction with government and its policies) 11. Partnerships/collaboration

4.2.3.3 Evaluating the contributions of One Health to solidarity and social cohesion

The principle of solidarity in health issues is visible in the international arena, such as the need for solidarity among all member states required to achieve the SDG goal of health and well-being at all ages (Hill *et al.*, 2014, p. 3). Working on a sustainable health care system for European countries, Garcés *et al.* (2003) emphasise that intergenerational solidarity (a time dimension) should underline every decision in a sustainable health care system. To achieve this, they propose changes in the legal, care, economic, administrative and cultural dimensions (Garcés *et al.*, 2003, pp. 210-212). Social and financial co-responsibility are demanded and welfare gained as a right depends on the citizens' ability to invest during working life in order to cope with a possible future state of dependency and care need (Garcés *et al.*, 2003, p. 210). For Borgonovi and Compagni (2013), solidarity and interconnectedness are essential, for pooling resources to ensure an adequate level of cover for those in need. The question then is to what extent does a OH initiative advocate for social and financial co-responsibility for taking on the burdens and receiving the benefits of a specific intervention? And, following on from this, to what extent does the OH initiative ensure that people pay for the costs and receive the benefits according to their social and financial capabilities? Box 4.5 provides some examples of how to apply solidarity to a OH-initiative.

As OH-issues are complex, interacting across scales and factors, partnerships between health and other organisations are critical for effective OH initiatives (Degeling *et al.*, 2015). Such partnerships could be through a national coordinating body, the consideration of health aspects by industry, public education and social mobilisation (cf. Gruen *et al.*, 2008). Table 4.2 displays the constituents of social cohesion from the perspective of the relations between individuals and other social units (e.g. groups; institutions). Thus from the perspective of outcomes for individuals, OH initiatives can be evaluated for their contributions to the 11 constituents of social cohesion (Table 4.2).

Box 4.5. Evaluating the contributions of One Health to solidarity.

Solidarity – The OH-initiative

- supports cost-sharing for medical expenses /insurance premiums
- motivates people to engage in volunteer care of sick or geriatric persons
- fosters the maintenance or improvement (or does not adversely affect) ties with family/friends/the community
- makes people feel concerned/more concerned about the health situation of other people, non-human animals, our environment, interlinkages between environment, human and animal health; E.g. 'I feel a strong bond toward animals/other humans'; 'I think of myself as part of nature, not separate from it' (Amiot and Bastian, 2017, p. 4).

These questions can be assessed on 'a 1 (strongly disagree) to 5 (strongly agree) Likert-type scale' (adapted from Amiot and Bastian, 2017, p. 4).

4.3 Operationalising the social sustainability framework for evaluating OH Initiatives

Based on the foregoing, a socially sustainable OH initiative at the minimum, does not undermine individual needs and capabilities, fosters emancipation, environmental justice, solidarity and social cohesion, and thereby improves human well-being (Figure 4.1).

If social sustainability is regarded as an outcome of OH interventions, there is a need to outline how the process that is likely to lead to the achievement of this outcome as well as the outcome can be evaluated. As health cases that require a OH-approach have the potential to become fatal and pandemic, it becomes ethically questionable to apply a case control or a before and after research approach by analysing cases where a OH-approach has been adopted and cases where they have not been adopted. Yet retrospective analysis of health crises can provide insights on the added values of a OH-approach. As social sustainability is not only determined by an initiative, whether OH or other initiatives, there is a need for adopting an analytical approach that accounts for the contributions of contextual factors to social sustainability in terms of using control groups. An alternative could be simulations of integrative versus single sector approaches which was for example applied to the control of human rabies by post-exposure prevention (PEP) in humans alone versus the mass vaccination of dogs and PEP (Zinsstag *et al.*, 2009).

The limitations in defining social sustainability also affects its current measurements – whether in terms of indicators, which are incomplete and differ across time, cultures and places hence posing difficulties for measurements and comparisons (cf. Pareja-Eastaway, 2012; Popovic *et al.*, 2014). To resolve the different societal perspectives of what social sustainability ought to be requires discursive processes, social learning and deliberative negotiations in transdisciplinary processes.

A baseline and a follow-up where the various domains that constitute social sustainability are analysed at least between two time points is proposed (OH-operations and OH-outcomes). We thus conceptualise two approaches to evaluate the added value of a OH initiative from a social sustainability perspective. First, a process-based approach focuses on OH-operations and asks to what extent these operations are socially sustainable. Second, an outcome-based approach asks to what extent OH-outcomes are socially sustainable from the perspective of humans, animals, plants, microbiota, and ecosystems.

With Table 4.3, we provide a summary of an analytical framework comprising key questions (for details see Figure 4.2, Box 4.3, 4.4 and 4.5, Table 4.1 and 4.2), whose answers can provide insights on the social sustainability of OH operations and how OH operations contribute to social sustainability.

It is important to note that the boundaries between some concepts remain fuzzy, for example social justice (a narrow dimension of environmental justice) and social cohesion overlap to a considerable extent. Participation for instance fits well under both dimensions. Participation can either be placed in 'social justice' or 'social cohesion' according to the context, and should be only counted once in the analysis. Ideally, specific questions should be adapted to the contexts being analysed through a transdisciplinary process of co-producing the research

Table 4.3. A framework for evaluating the contributions of One Health initiatives to social sustainability.

Dimensions	Indicators
Resources	Goods and services (livelihood assets – human, natural, social, financial, physical assets)
Conversion factors	Individual, social, environmental
Basic needs	Shelter, food, income
The 10 central capabilities	Life; bodily health; bodily integrity; senses, imagination and thought; emotions; practical reason; affiliation; other species; play; and control over one's environment
Emancipation	Degrees of self-determination in health-related aspects (access to different health traditions, treatments, institutional equity independently from class, gender or race categories).
Environmental justice	Recognition; Procedural justice; Distributional justice; Animal welfare
Solidarity and social cohesion	Interpersonal trust; Social support; Density of social relations; Belonging; Openness; Participation; Bridging (inter-group) and bonding (intra-group) ties; Inclusion; Institutional trust; Legitimacy of institutions; Partnerships/ collaboration

questions with stakeholders. Collected data can then be analysed using regular qualitative and quantitative measures (cf. Bussière *et al.*, 2016) and discussed with stakeholders.

4.4 Conclusions

The goal of this chapter was to show how the contributions of OH to social sustainability can be evaluated. This is an important topic considering that many health challenges result from the interaction of humans, animals and the environment, and are thus interconnected in their drivers, impacts and outcomes. Yet social sustainability has received relatively little attention and there are few comprehensive frameworks for its evaluation, especially in relation with health issues. We first analysed social sustainability and identified its key dimensions as well as the associated indicators. We then explored how each of the dimensions and indicators have been applied to health or health related issues, and adapted them for our purpose. The obtained framework can be used for evaluating the contributions of OH to social sustainability. However, considering multiple social perspectives and values, such indicators need to be adapted to contexts and concretised through transdisciplinary deliberative processes in order to operationalise it for use in evaluating actual OH interventions. Finally, the developed framework on social sustainability can be applied to other contexts beyond

OH initiatives. With little adaptation to the specific research objectives, questions can be formulated for evaluating the contributions of other initiatives/projects/programmes to social sustainability. As this is an initial attempt at a comprehensive framework for evaluating social sustainability, we expect that future work can improve on this basis.

Acknowledgements

This chapter is based upon work from COST Action 'Network for Evaluation of One Health' (TD1404), supported by COST (European Cooperation in Science and Technology).

The authors would like to thank Barbara Häsler and Chiara Frazzoli who reviewed this chapter and provided useful further suggestions for the work.

References

Abma, F.I., Brouwer, S., De Vries, H.J., Arends, I., Robroek, S.J., Cuijpers, M.P.J., Van der Wilt, G.J., Bültmann, U. and Van der Klink, J.J.L., 2016. The capability set for work: development and validation of a new questionnaire. Scand J Work Environ Health 42: 34-42.

Al-Janabi, H., N. Flynn, T. and Coast, J., 2012. Development of a self-report measure of capability wellbeing for adults: the ICECAP-A. Qual Life Res 21: 167-176.

Association of Local Public Health Agencies / Ontario Public Health Association (ALPHA/OPHA), 2013. Health equity indicators. Health Equity Workgroup, Ontario, Canada.

Amiot, C.E. and Bastian, B., 2017. Solidarity with animals: assessing a relevant dimension of social identification with animals. PLoS ONE 12: e0168184.

Arnstein, S.R., 1969. A ladder of citizen participation. J Am Inst Plann 35: 216-224.

Barnes, M. and Coelho, V.S., 2009. Social participation in health in Brazil and England: inclusion, representation and authority. Health Expect 12: 226-236.

Becker, E., Jahn, T. and Stieβ, I., 1999. Exploring uncommon ground: sustainability and the social sciences. In: Becker, E. and Jahn, T. (eds.) Sustainability and the social sciences. a cross-disciplinary approach integrating environmental considerations into theoretical reorientation. Zed Books, London, UK, pp. 1-22.

Benatar, S., Daibes, I. and Tomsons, S., 2016. Inter-philosophies dialogue: creating a paradigm for global health ethics. Kennedy Inst Ethics J 26: 323-346.

Benatar, S.R., 2003. Bioethics: power and injustice: IAB Presidential Address. Bioethics 17: 387-399.

Berkey, B., 2017. Prospects for an inclusive theory of justice: the case of non-human animals. J Appl Philos 34: 679-695.

Berthe, F., Bouley, T., Karesh, W.B., LeGall, F., Planté, C. and Seifman, R., 2018. Operational framework for strengthening human, animal and environmental public health systems at their interface. World Bank Group, Washington, DC, USA.

Bonnie, L.H.A., Van den Akker, M., Van Steenkiste, B. and Vos, R., 2010. Degree of solidarity with lifestyle and old age among citizens in the Netherlands: cross-sectional results from the longitudinal SMILE study. J Med Ethics 36: 784-790.

Borgonovi, E. and Compagni, A., 2013. Sustaining universal health coverage: the interaction of social, political, and economic sustainability. Value Health J Int Soc Pharmacoeconomics Outcomes Res 16: 34-38.

Bottoni, G., 2018. Validation of a social cohesion theoretical framework: a multiple group SEM strategy. Qual Quant 52: 1081-1102.

Braveman, P. and Gottlieb, L., 2014. The social determinants of health: it's time to consider the causes of the causes. Public Health Rep Nurs 3D Divers Disparit Soc Determinant 129: 19-31.

Bréchon, P., 2014. Measuring solidarity values: not that easy. In: EVS Meeting, Bilbao, Spain, pp. 10.

Bruhn, J., 2009. The group effect. Social cohesion and health outcomes. Springer US, Boston, MA, USA.

Burgess, C.P., Johnston, F.H., Bowman, D.M.J.S. and Whitehead, P.J., 2005. Healthy country: healthy people? Exploring the health benefits of indigenous natural resource management. Aust NZ J Public Health 29: 117-122.

Bussière, C., Sicsic, J. and Pelletier-Fleury, N., 2016. Simultaneous effect of disabling conditions on primary health care use through a capability approach. Soc Sci Med 154: 70-84.

Caliman, F.A. and Gavrilescu, M., 2009. Pharmaceuticals, personal care products and endocrine disrupting agents in the environment – a review. CLEAN – Soil Air Water 37: 277-303.

Callander, E., Schofield, D. and Shrestha, R., 2013a. Chronic health conditions and poverty: a cross-sectional study using a multidimensional poverty measure. BMJ Open 3: e003397.

Callander, E., Schofield, D.J. and Shrestha, R.N., 2013b. Freedom poverty: a new tool to identify the multiple disadvantages affecting those with CVD. Int J Cardiol 166: 321-326.

Capps, B. and Lederman, Z., 2015. One Health, vaccines and ebola: the opportunities for shared benefits. J Agric Environ Ethics 28: 1011-1032.

Capps, B. and Lederman, Z., 2014. One Health and paradigms of public biobanking. J Med Ethics 41.

Carrel, M., Young, S.G. and Tate, E., 2016. Pigs in space: determining the environmental justice landscape of swine concentrated animal feeding operations (CAFOs) in Iowa. Int J Environ Res Public Health 13: 849.

Carron, A.V. and Hausenblas, H.A., 1998. Group dynamics in sport. Fitness Information Technology.

Carter, M., Horwitz, P., 2014. Beyond proximity: the importance of green space useability to self-reported health. EcoHealth 11: 322-332.

Chan, J., To, H.-P. and Chan, E., 2006. Reconsidering social cohesion: developing a definition and analytical framework for empirical research. Soc Indic Res 75: 273-302.

Coast, J., Smith, R. and Lorgelly, P., 2008. Should the capability approach be applied in Health Economics? Health Econ 17: 667-670.

Colquitt, J.A. and Rodell, J.B., 2015. Measuring justice and fairness. In: Cropanzano, R. and Ambrose, M.L. (eds.) The Oxford Handbook of Justice in the Workplace, Oxford Library of Psychology. Oxford University Press, Oxford/New York, UK, USA, pp. 187-202.

Commission on Social Determinants of Health (CSDH), 2008. Closing the gap in a generation: health equity through action on the social determinants of health: Final Report of the Commission on Social Determinants of Health. World Health Organization, Geneva, Switzerland.

Degeling, C., Johnson, J., Kerridge, I., Wilson, A., Ward, M., Stewart, C. and Gilbert, G., 2015. Implementing a One Health approach to emerging infectious disease: reflections on the socio-political, ethical and legal dimensions. BMC Public Health 15: 1-11.

Degeling, C., Lederman, Z. and Rock, M., 2016. Culling and the common good: re-evaluating harms and benefits under the One Health paradigm. Public Health Ethics 9: 244-254.

Eizenberg, E. and Jabareen, Y., 2017. Social sustainability: a new conceptual framework. Sustainability 9: 68.

Empacher, C. and Wehling, P., 1999. Indikatoren sozialer Nachhaltigkeit. Grundlagen und Konkretisierungen. ISOE-Diskussionspapiere.

Forst, R., 2002. Contexts of justice: political philosophy beyond liberalism and communitarianism, Philosophy, social theory, and the rule of law. University of California Press, Berkeley, CA, USA.

Fraser, N., 2013. A triple movement? Parsing the politics of crisis after Polanyi. New Left Rev 81: 119-132.

Fraser, N., 2011. Marketization, social protection, emancipation: toward a neo-Polanyian conception of capitalist crisis. In: Calhoun, C. and Derluguian, G. (eds.) Business as usual: the roots of the global financial meltdown. NYU Press, New York, NY, USA.

Fraser, N., 2009. Scales of justice: reimagining political space in a globalizing world, New directions in critical theory. Columbia University Press, New York, NY, USA.

Fraser, N., 2000. Rethinking recognition. New Left Rev: 107-120.

Fraser, N., 1996. Social justice in the age of identity politics: redistribution, recognition, and participation. The Tanner Lectures on Human Values. Stanford University.

Fraser, N. and Honneth, A., 2003. Redistribution or recognition? A political-philosophical exchange. Verso, London/New York, UK/USA.

Frazzoli, C. and Mantovani, A., 2010. Toxicants exposures as novel zoonoses: reflections on sustainable development, food safety and veterinary public health: toxicants exposures as novel zoonoses. Zoonoses Public Health 57: e136-e142.

Frazzoli, C., Petrini, C. and Mantovani, A., 2009. Sustainable development and next generation's health: a long-term perspective about the consequences of today's activities for food safety. Ann Ist Super Sanita 45: 65-75.

Frumkin, H., Bratman, G.N., Breslow, S.J., Cochran, B., Kahn Jr, P.H., Lawler, J.J., Levin, P.S., Tandon, P.S., Varanasi, U., Wolf, K.L. and Wood, S.A., 2017. Nature contact and human health: a research agenda. Environ Health Perspect: 125.

Garcés, J., Rodenas, F. and Sanjosé, V., 2003. Towards a new welfare state: the social sustainability principle and health care strategies. Health Policy 65: 201-215.

Giddens, A., 1984. The constitution of society: outline of the structuration theory. Berkeley University, CA, USA.

Goffee, R. and Jones, G., 1998. The character of a corporation: how your company's culture can make or break your business, 1st edition. HarperBusiness, New York, NY, USA.

Grabow, M.L., Spak, S.N., Holloway, T., Stone, B., Mednick, A.C. and Patz, J.A., 2012. Air quality and exercise-related health benefits from reduced car travel in the Midwestern United States. Environ Health Perspect 120: 68-76.

Griffiths, J., 2006. Mini-Symposium: health and environmental sustainability. Public Health 120: 581-584.

Gruen, R.L., Elliott, J.H., Nolan, M.L., Lawton, P.D., Parkhill, A., McLaren, C.J. and Lavis, J.N., 2008. Sustainability science: an integrated approach for health-programme planning. Lancet 372: 1579-1589.

Haraway, D.J., 2008. When species meet, Posthumanities. University of Minnesota Press, Minneapolis, MN, USA.

Hediger, K. and Beetz, A., 2015. The role of human-animal interactions in education. In: Zinsstag, J., Schelling, E., Waltner-Toews, D., Whittaker, M. and Tanner, M. (eds.) One Health: the theory and practice of integrated health approaches. CABI, Wallingford, UK, pp. 73-84.

Henry, M., Beguin, M., Requier, F., Rollin, O., Odoux, J.-F., Aupinel, P., Aptel, J., Tchamitchian, S. and Decourtye, A., 2012. A common pesticide decreases foraging success and survival in honey bees. Science 336: 348-350.

Hill, P.S., Buse, K., Brolan, C.E. and Ooms, G., 2014. How can health remain central post-2015 in a sustainable development paradigm? Glob Health 10: 18.

Hinchliffe, S., 2015. More than one world, more than one health: re-configuring interspecies health. Soc Sci Med 129: 28-35.

Hodge, R.A. and Hardi, P., 1997. The need for guidelines: the rationale underlying the Bellagio principles for assessment. In: Assessing sustainable development. principles in practice. International Institute for Sustainable Development. Winnipeg, Manitoba, Canada, pp. 7-20.

Hofmann, M., Young, C., Binz, T., Baumgartner, M. and Bauer, N., 2017. Contact to nature benefits health: mixed effectiveness of different mechanisms. Int J Environ Res Public Health 15: 31.

Hurlbert, M. and Gupta, J., 2015. The split ladder of participation: a diagnostic, strategic, and evaluation tool to assess when participation is necessary. Environ Sci Policy 50: 100-113.

Ifejika Speranza, C., Wiesmann, U. and Rist, S., 2014. An indicator framework for assessing livelihood resilience in the context of social-ecological dynamics. Glob Environ Change 28: 109-119.

Iroz-Elardo, N., 2015. Health impact assessment as community participation. Community Dev J 50: 280-295.

Jarrett, J., Woodcock, J., Griffiths, U.K., Chalabi, Z., Edwards, P., Roberts, I. and Haines, A., 2012. Effect of increasing active travel in urban England and Wales on costs to the National Health Service. Lancet 379: 2198-2205.

Jenson, J., 2002. Identifying the links: social cohesion and culture. Can J Commun: 27.

Jenson, J., 1998. Mapping social cohesion: the state of Canadian research. Family Network, CPRN, Ottawa, Canada.

Johnston, F.H., Jacups, S.P., Vickery, A.J. and Bowman, D.M.J.S., 2007. Ecohealth and Aboriginal testimony of the nexus between human health and place. EcoHealth 4: 489-499.

Juul, S., 2010. Solidarity and social cohesion in late modernity: a question of recognition, justice and judgement in situation. Eur J Soc Theory 13: 253-269.

Kabisch, N. and Haase, D., 2014. Green justice or just green? Provision of urban green spaces in Berlin, Germany. Landsc Urban Plan 122: 129-139.

Kawachi, I., Subramanian, S.V. and Almeida-Filho, N., 2002. A glossary for health inequalities. J. Epidemiol. Community Health 56: 647-652.

Kochhar, R., 2015. A global middle class is more promise than reality: from 2001 to 2011, nearly 700 million step out of poverty, but most only barely. Pew Research Center, Washington, DC, USA.

Liberto, H., 2017. Species membership and the veil of ignorance: what principles of justice would the representatives of all animals choose? Utilitas 29: 299-320.

Lind, E.A. and Tyler, T.R., 1988. Appendix. Measurement of procedural justice beliefs. In: The social psychology of procedural justice. Critical Issues in Social Justice. Springer Science & Business Media, New York, NY, USA, pp. 243-248.

Litman, T., 2015. Evaluating transportation equity: guidance for incorporating distributional impacts in transportation planning. Vic Transp Policy Inst: 65.

Littig, B. and Griessler, E., 2005. Social sustainability: a catchword between political pragmatism and social theory. Int J Sustain Dev 8: 65.

Lorgelly, P.K., Lorimer, K., Fenwick, E.A.L., Briggs, A.H. and Anand, P., 2015. Operationalising the capability approach as an outcome measure in public health: the development of the OCAP-18. Soc Sci Med 142: 68-81.

Mabsout, R., 2011. Capability and health functioning in Ethiopian households. Soc Indic Res 101: 359-389.

Marmot, M., Allen, J., Bell, R., Bloomer, E. and Goldblatt, P., 2012. WHO European review of social determinants of health and the health divide. Lancet 380: 1011-1029.

Martin, A., 2017. Just conservation: biodiversity, wellbeing and sustainability. Taylor & Francis, London, UK.

Martin, A., Coolsaet, B., Corbera, E., Dawson, N.M., Fraser, J.A., Lehmann, I. and Rodriguez, I., 2016. Justice and conservation: the need to incorporate recognition. Biol Conserv 197: 254-261.

Martuzzi, M. and Tickner, J., 2004. Introduction – the precautionary principle: protecting public health, the environment and the future of our children. In: WHO Europe: the precautionary principle: protecting public health, the environment and the future of our children. WHO Regional Office for Europe, Copenhagen, Denmark, pp. 7-14.

Mielke, K., Schetter, C. and Wilde, A., 2011. Dimensions of social order: empirical fact, analytical framework and boundary concept. Dep Polit Cult Change Cent Dev Res Univ Bonn, ZEF Working Paper Series, Bonn, Germany.

Mitchell, P.M., Roberts, T.E., Barton, P.M. and Coast, J., 2017. Applications of the capability approach in the health field: a literature review. Soc Indic Res 133: 345-371.

Mitra, S., Jones, K., Vick, B., Brown, D., McGinn, E. and Alexander, M.J., 2013. Implementing a multidimensional poverty measure using mixed methods and a participatory framework. Soc Indic Res 110: 1061-1081.

Netten, A., Burge, P., Malley, J., Potoglou, D., Towers, A.-M., Brazier, J., Flynn, T., Forder, J. and Wall, B., 2012. Outcomes of social care for adults: developing a preference-weighted measure. Health Technol Assess: 16: 1-166.

Nikiema, B., Haddad, S. and Potvin, L., 2012. Measuring women's perceived ability to overcome barriers to healthcare seeking in Burkina Faso. BMC Public Health 12: 147.

Nussbaum, M. and Sen, A., 2002. The quality of life, 1st ed. Clarendon Press, Oxford, New York and Aukland, USA.

Nussbaum, M.C., 2011. Creating capabilities: the human development approach. First Harvard University Press paperback edition. The Belknap Press of Harvard University Press, Cambridge, MA, USA.

Nussbaum, M.C., 2006. Frontiers of justice: disability, nationality, species membership. Harvard University Press, Cambridge, MA, USA.

Nussbaum, M.C., 2000. Women and human development: the capabilities approach, 13. print. ed, John Robert Seeley lectures. Cambridge University Press, Cambridge, MA, USA.

Obrist, B., Henley, R. and Pfeiffer, C., 2010. Multi-layered social resilience: a new approach in mitigation research. Prog Dev Stud 10: 283-293.

Opielka, M., 2017. Soziologie Sozialer Nachhaltigkeit – Zur Idee der Internalisierungsgesellschaft 16.

Pareja-Eastaway, M., 2012. Social sustainability. Int Encycl Hous Home: 502-505.

Parsons, C., 2007. How to map arguments in political science. Oxford University Press, New York, NY, USA.

Pew Research Center, 2018. Majorities say government does too little for older people, the poor and the middle class. Partisan, age gaps in views of government help for younger people 11.

Pinto, L.H., Cabral-Cardoso, C. and Werther, W.B., 2011. Why solidarity matters (and sociability doesn't): the effects of perceived organizational culture on expatriation adjustment. Thunderbird Int Bus Rev 53: 377-389.

Plochg, T., Delnoij, D.M., Hoogedoorn, N.P. and Klazinga, N.S., 2006. Collaborating while competing? The sustainability of community-based integrated care initiatives through a health partnership. BMC Health Serv Res 6: 37.

Plunkett, D., 2016. Justice, non-human animals, and the methodology of political philosophy. Jurisprudence 7: 1-29.

Pogge, T., 2016. Are we violating the human rights of the world's poor? In: Gaisbauer, H.P., Schweiger, G. and Sedmak, C. (eds.) Ethical issues in poverty alleviation, studies in global justice. Springer International Publishing, Cham, Switzerland, pp. 17-42.

Pogge, T., 2015. Health impact fund: aligning incencitives. In: Karan, A. and Sodhi, G. (eds.) Protecting the health of the poor: social movements in the South. CROP International Studies in Poverty Research. Zed Books, London, UK.

Pohl, C. and Hirsch Hadorn, G., 2007. Principles for designing transdisciplinary research. Propos Swiss Acad Arts Sci: 35-40.

Polanyi, K., 2001. The great transformation: the political and economic origins of our time, 2nd Beacon Paperback edition. Beacon Press, Boston, MA, USA.

Popovic, T., Kraslawski, A., Heiduschke, R. and Repke, J.-U., 2014. Indicators of social sustainability for wastewater treatment processes. Compu Aid Chem Engin: 723-728.

Pucher, J., Buehler, R., Bassett, D.R. and Dannenberg, A.L., 2010. Walking and cycling to health: a comparative analysis of city, state, and international data. Am J Public Health 100: 1986-1992.

Rawls, J., 1971. A theory of social justice. Cambridge, MA, USA.

Richardson, M., Cormack, A., McRobert, L. and Underhill, R., 2016. 30 days wild: development and evaluation of a large-scale nature engagement campaign to improve well-being. PLoS ONE 11: e0149777.

Rist, S., Chidambaranathan, M., Escobar, C., Wiesmann, U. and Zimmermann, A., 2007. Moving from sustainable management to sustainable governance of natural resources: the role of social learning processes in rural India, Bolivia and Mali. J Rural Stud 23: 23-37.

Rist, S., Chiddambaranathan, M., Escobar, C. and Wiesmann, U., 2006. 'It was hard to come to mutual understanding…' – the multidimensionality of social learning processes concerned with sustainable natural resource use in India, Africa and Latin America. Syst Pract Action Res 19: 219-237.

Roberto, C.A., Swinburn, B., Hawkes, C., Huang, T.T.-K., Costa, S.A., Ashe, M., Zwicker, L., Cawley, J.H. and Brownell, K.D., 2015. Patchy progress on obesity prevention: emerging examples, entrenched barriers, and new thinking. Lancet 385: 2400-2409.

Robeyns, I., 2017. Wellbeing, freedom and social justice: the capability approach re-examined. Open Book Publishers, Cambridge, UK.

Rock, M., Mykhalovskiy, E. and Schlich, T., 2007. People, other animals and health knowledges: towards a research agenda. Soc Sci Med 64: 1970-1976.

Rock, M.J. and Degeling, C., 2015. Public health ethics and more than human solidarity. Soc Sci Med 129: 61-67.

Rock, M.J., Degeling, C. and Blue, G., 2014. Toward stronger theory in critical public health: insights from debates surrounding posthumanism. Crit Public Health 24(3): 337-348.

Rüegg, S.R., McMahon, B.J., Häsler, B., Esposito, R., Nielsen, L.R., Ifejika Speranza, C., Ehlinger, T., Peyre, M., Aragrande, M., Zinsstag, J., Davies, P., Mihalca, A.D., Buttigieg, S.C., Rushton, J., Carmo, L.P., De Meneghi, D., Canali, M., Filippitzi, M.E., Goutard, F.L., Ilieski, V., Milićević, D., O'Shea, H., Radeski, M., Kock, R., Staines, A. and Lindberg, A., 2017. A blueprint to evaluate One Health. Front Public Health 5: 20.

Ruger, J.P., 2012a. Health and social justice, 1st edition. Oxford University Press, Oxford, UK.

Ruger, J.P., 2012b. Global health justice and governance. Am J Bioeth 12: 35-54.

Ruger, J.P., 2010. Health and social justice. Oxford University Press, Oxford, UK.

Schlosberg, D., 2013. Theorising environmental justice: the expanding sphere of a discourse. Environ Polit 22: 37-55.

Schlosberg, D., 2007. Defining environmental justice. Oxford University Press, Oxford, UK.

Schrecker, T., 2017. Was Mackenbach right? Towards a practical political science of redistribution and health inequalities. Health Place 46: 293-299.

Sen, A., 2009. The idea of justice. Belknap Press of Harvard University Press, Cambridge, MA, USA.

Sen, A., 2000. Development as freedom. Oxford University Press, New Delhi, India.

Sen, A., 1999. Development as freedom, 6th print. edition. Oxford University Press, Oxford, UK.

Sen, A., 1993. Capability and well-being. In: Nussbaum, M. and Sen, A. (eds.) The quality of life. Oxford University Press, Oxford, UK, pp. 30-53.

Sen, A., 1992. Inequality reexamined. Clarendon Press, Oxford, UK.

Sen, A., 1985. Commodities and capabilities. Oxford University Press, Oxford, UK.

Sikor, T., Martin, A., Fisher, J. and He, J., 2014. Toward an empirical analysis of justice in ecosystem governance: justice in ecosystem governance. Conserv Lett 7: 524-532.

Simon, J., Anand, P., Gray, A., Rugkåsa, J., Yeeles, K. and Burns, T., 2013. Operationalising the capability approach for outcome measurement in mental health research. Soc Sci Med 98: 187-196.

Smith, M., 2016. Cycling on the verge: the discursive marginalisation of cycling in contemporary New Zealand transport policy. Energy Res Soc Sci 18: 151-161.

Staerklé, C., 2013. Othering in political lay thinking: a social representational approach to social order. In: Magioglou, T. (ed.) Culture and political psychology: a societal perspective. Advances in cultural psychology: constructing human development. Information Age Publishing, INC, Charlotte, NC, USA, pp. 49-74.

Stafford, M., Cooper, R., Cadar, D., Carr, E., Murray, E., Richards, M., Stansfeld, S., Zaninotto, P., Head, J. and Kuh, D., 2017. Physical and cognitive capability in mid-adulthood as determinants of retirement and extended working life in a British cohort study. Scand J Work Environ Health 43: 15-23.

Tanner M., 2005. Better health for the poor: a systems approach. The right for health: a duty for whom? Symposium Report 2004. Novartis Foundation for Sustainable Development, Basel, Switzerland.

Thurston, G.D., Kipen, H., Annesi-Maesano, I., Balmes, J., Brook, R.D., Cromar, K., De Matteis, S., Forastiere, F., Forsberg, B., Frampton, M.W., Grigg, J., Heederik, D., Kelly, F.J., Kuenzli, N., Laumbach, R., Peters, A., Rajagopalan, S.T., Rich, D., Ritz, B., Samet, J.M., Sandstrom, T., Sigsgaard, T., Sunyer, J. and Brunekreef, B., 2017. A joint ERS/ATS policy statement: what constitutes an adverse health effect of air pollution? An analytical framework. Eur Respir J 49: 1600419.

United Nations Conference on Environment and Development (UNCED), 1992. The Rio Declaration on Environment and Development. UNCED, Rio de Janeiro, Brazil.

United Nations, 2015. Transforming our world: the 2030 agenda for sustainable development. Available at: sustainabledevelopment.un.org.

United Nations, 2008. Training manual on disability statistics. World Health Organization/United Nations Economic and Social Commission for Asia and the Pacific. United Nations, Bangkok, Thailand.

United Nations, 1992. United Nations framework convention on climate change. UN, New York, NY, USA.

United Nations, 1948. The United Nations' universal declaration of human rights. UN, New York, NY, USA.

Üstün, T.B., Chatterji, S., Kostanjsek, N., Rehm, J., Kennedy, C., Epping-Jordan, J., Saxena, S., von Korff, M. and Pull, C., 2010. Developing the World Health Organization Disability Assessment Schedule 2.0. Bull World Health Organ 88: 815-823.

Vallance, S., Perkins, H.C. and Dixon, J.E., 2011. What is social sustainability? A clarification of concepts. Geoforum 42: 342-348.

Vanclay, F., 2006. Conceptual and methodological advances in social impact assessment. In: Becker, H.A. and Vanclay, F. (eds.) The international handbook of social impact assessment: conceptual and methodological advances. Elgar, Cheltenham, UK, pp. 1-9.

Verweij, M. and Bovenkerk, B., 2016. Ethical promises and pitfalls of OneHealth. Publ Health Ethic 9: 1-4.

Wettlaufer, L., Hafner, F. and Zinsstag, J., 2015. The human-animal relationship in the law. In: Zinsstag, J., Schelling, E., Waltner-Toews, D., Whittaker, M. and Tanner, M. (eds.) One Health: the theory and practice of integrated health approaches. CABI, Wallingford, UK, pp. 26-37.

Whitmee, S., Haines, A., Beyrer, C., Boltz, F., Capon, A.G., de Souza Dias, B.F., Ezeh, A., Frumkin, H., Gong, P., Head, P., Horton, R., Mace, G.M., Marten, R., Myers, S.S., Nishtar, S., Osofsky, S.A., Pattanayak, S.K., Pongsiri, M.J., Romanelli, C., Soucat, A., Vega, J. and Yach, D., 2015. Safeguarding human health in the Anthropocene epoch: report of The Rockefeller Foundation-Lancet Commission on planetary health. Lancet 386: 1973-2028.

World Commission on Environment and Development (WCED), 1987. Report of the WCED: our common future. Oxford University Press, New York, NY, USA.

World Health Organisation (WHO), 2018. Ebola virus disease. World Health Organ. News Fact Sheets. Available at: Available at: http://www.who.int/news-room/fact-sheets/detail/ebola-virus-disease.

World Health Organisation (WHO), 2016. Director-General summarizes the outcome of the Emergency Committee regarding clusters of microcephaly and Guillain-Barré syndrome. News Detail Statement. Available at: Available at: http://tinyurl.com/y8u5b5ss.

World Health Organisation (WHO), 2013. How to use the ICF: a practical manual for using the International Classification of Functioning, Disability and Health (ICF). Exposure draft for comment. October 2013.

World Health Organisation (WHO), 2011. Rio Political Declaration on Social Determinants of Health. Presented at the World Conference on social determinants of health, Rio de Janeiro, Brazil.

World Health Organisation (WHO), 2010. Measuring health and disability: manual for WHO Disability Assessment Schedule (WHODAS 2.0). WHO, Geneva, Switzerland.

World Health Organisation (WHO), 2001. International classification of functioning, disability and health: ICF. WHO, Geneva, Switzerland.

World Health Organisation (WHO), Convention on Biological Diversity (CBD), United Nations Environment Programme (UNEP), 2015. Connecting global priorities: biodiversity and human health. A state of knowledge review. WHO, Geneva, Switzerland.

Williamson, C., 2010. Towards the emancipation of patients: patients' experiences and the patient movement. Policy Press, Bristol, UK / Portland, OR, USA.

Zinsstag, J., Bonfoh, B., Cissé, G., Nguyen Viet, H., Silué, B., N'Guessan, T.S., Weibel, D., Schertenleib, R., Obrist, B. and Tanner, M., 2011. Towards equity effectiveness in health interventions. In: Wiesmann, U. and Hurni, H. (eds.) Research for sustainable development: foundations, experiences, and perspectives. Perspectives of the Swiss National Centre of Competence in Research (NCCR) North-South, University of Bern. NCCR North-South, Centre for Development and Environment (CDE) and Instiute of Geography, University of Bern, Bern, Switzerland, pp. 623-640.

Zinsstag, J., Crump, L., Schelling, E., Hattendorf, J., Maidane, Y.O., Ali, K.O., Muhummed, A., Umer, A.A., Aliyi, F., Nooh, F., Abdikadir, M.I., Ali, S.M., Hartinger, S., Mäusezahl, D., De White, M.B.G., Cordon-Rosales, C., Castillo, D.A., McCracken, J., Abakar, F., Cercamondi, C., Emmenegger, S., Maier, E., Karanja, S., Bolon, I., De Castañeda, R.R., Bonfoh, B., Tschopp, R., Probst-Hensch, N. and Cissé, G., 2018. Climate change and One Health. FEMS Microbiol Lett 365(11).

Zinsstag, J., Durr, S., Penny, M.A., Mindekem, R., Roth, F., Gonzalez, S.M., Naissengar, S. and Hattendorf, J., 2009. Transmission dynamics and economics of rabies control in dogs and humans in an African city. Proc Natl Acad Sci 106: 14996-15001.

Zinsstag, J., Schelling, E., Waltner-Toews, D., Whittaker, M. and Tanner, M., 2015a. One Health: the theory and practice of integrated health approaches. CAB International, Wallingford, UK.

Zinsstag, J., Waltner-Toews, D. and Tanner, M., 2015b. Theoretical issues of One Health. In: Zinsstag, J., Schelling, E., Waltner-Toews, D., Whittaker, M. and Tanner, M. (eds.) One Health: the theory and practice of integrated health approaches. CABI, Wallingford, UK, pp. 16-25.

Chapter 5
Assessing the ecological dimension of One Health

Photo: Tomas Hulik/Shutterstock.com

K. Marie McIntyre[1*], Barbara Vogler[2], Ilias Chantziaras[3], Ana Cláudia Coelho[4], Juan García Díez[5], Alexandra Esteves[6], Gilberto Igrejas[7,8,9], Peter Lurz[10], Sara Babo Martins[11,12], Alexander Mastin[13], Alexander Murray[14], Stephen R. Parnell[13], Patricia Poeta[6,15], Sónia Ramos[6], Cristina Saraiva[6], Chinwe Ifejika Speranza[16,17] and Barry J. McMahon[18*]

[1]Department of Epidemiology & Population Health, Institute of Infection & Global Health, University of Liverpool, Leahurst Campus, Neston, Cheshire CH64 7TE, United Kingdom; [2]Department of Poultry and Rabbit Diseases, Institute of Veterinary Bacteriology, Vetsuisse Faculty Zurich, University of Zurich, Switzerland; [3]Department of Reproduction, Obstetrics and Herd health, Faculty of Veterinary Medicine, Ghent University, 9820, Merelbeke, Belgium; [4]Animal and Veterinary Research Centre (CECAV), University of Tras-os-Montes and Alto Douro, 5000-801 Vila Real, Portugal; [5]CECAV – Animal and Veterinary Research Centre, University of Trás-os-Montes e Alto Douro, Quinta de Prados, 5001-801 Vila Real, Portugal; [6]Department of Veterinary Sciences, School of Agrarian and Veterinary Sciences, Animal and Veterinary Research Centre, University of Trás-os-Montes e Alto Douro, 5000-801 Vila Real, Portugal; [7]Department of Genetics and Biotechnology, University of Trás-os-Montes and Alto Douro, Vila Real, Portugal; [8]Functional Genomics and Proteomics Unit, University of Trás-os-Montes and Alto Douro, Vila Real, Portugal; [9]Associated Laboratory for Green Chemistry (LAQV-REQUIMTE), University NOVA of Lisbon, Caparica, Portugal; [10]Honorary Fellow, The Royal (Dick) School of Veterinary Studies, University of Edinburgh, Easter Bush Campus, Midlothian, EH25 9RG, Scotland, United Kingdom; [11]Department of Production and Population Health, Royal Veterinary College, Hawkshead Lane, North Mymms, Hatfield, Hertfordshire, AL9 7TA, United Kingdom; [12]SAFOSO AG, Bern-Liebefeld, Waldeggstrasse 1, 3097 Liebefeld, Switzerland; [13]Ecosystems and Environment Research Centre, School of Environment and Life Sciences, University of Salford, Peel Building, Salford, Greater Manchester, M5 4WT, United Kingdom; [14]Marine Laboratory, Marine Scotland Science, 375, Victoria Road, Aberdeen, AB11 9DB, United Kingdom; [15]Associated Laboratory for Green Chemistry (LAQV-REQUIMTE), University NOVA of Lisbon, Caparica, Portugal; [16]Institute of Geography, University of Bern, Hallerstrasse 12, 3012 Bern, Switzerland; [17]Centre for Development and Environment, University of Bern, Mittelstrasse 43, 3012 Bern, Switzerland; [18]UCD School of Agriculture & Food Science, University College Dublin, Belfield, Dublin 4, Ireland; k.m.mcintyre@liverpool.ac.uk; barry.mcmahon@ucd.ie

Abstract

This chapter provides a conceptual framework describing the main ecological components of the global ecosystem, which need to be considered when using a One Health approach, including incorporating examples of metrics which both reflect the connectedness of different environments and quantify the complex interactions between humans, domesticated and non-domesticated animals and the environment in which they live, and the direct and indirect drivers which impact them. The set of ecological components described should be used to inform the audience on how to quantify the sustainability and thus the 'One Health-ness' of any environment. It is the quantification of an array of these components which demonstrate the 'added value' of One Health, through the savings of lives, improvements in life-lived (quality of life), qualitative gains and financial savings. One Medicine recognises that there is virtually no difference in the paradigm between human and veterinary medicine and both disciplines can contribute to the development of each other; animals should thus be positioned in the social and not the environmental realm, taking a 'less speciecist' stance. It should be understood that when quantifying the health of anything, be it an organism or an ecosystem, the variables measured are all context-dependent, particularly for ecosystem and environmental health. The interpretation of resulting measurements will differ dependent upon a human, an animal and ecosystem perspective, and each of these perspectives has its own value, when thinking about 'One Health'. The environment is a major determinant of health with an estimated 25-33% of the global burden of disease attributed to environmental risk factors. Accordingly, when measuring the ecological dimension of One Health, account needs to be taken of the fitness and sustainability (including integrity) of the ecosystem and environment. This is not easy to quantify, as it results in the creation of indexes of heterogeneous variables, which do not provide an easily interpretable output of resilience. Various indices have been developed which aim to quantify environmental health, including: the long-term sustainability of different ecosystems; the state of the world's biological diversity; describing the status of ecosystem services. Metrics to measure the health status of the world include the 'One Health-ness' of water, air, soil, biodiversity, and ecosystems. They require ecosystem approaches to health to factor in ecosystem interactions in health research. Methods to quantify the health status of populations under closer management of humans also need to be mentioned including those of humans, domestic animals, plants, and aquaculture. Finally, antimicrobial resistance issues across the ecological dimension should be considered. Many of these metrics are very 'human-centric' and should therefore be interpreted with caution. A major challenge for mankind to achieving a One Health in the future is to examine the trade-off from producing food and look for synergies with food quality for both animals and humans but also related to zoonoses emergence. Contaminants, including biological such as pathogens, chemical elements or compounds need to be identified and acted upon. Balancing all ecological components (water, air, soil, biodiversity) is critically important for food security, and both food safety and food security should be interlinked for a One Health approach to sustained global food security. It is critical that a focus on food quality is not just on outcomes such as food nutrients or food safety (e.g. maximum residue levels) but also on how one-health-ness interacts with the food system at different stages of the value chain (production, processing, transport and consumption of food) to affect food quality. Given the growing human population, a set of One Health indicators which capture the link between human health, animal health and ecological health is to inform future global developments, particularly as we enter a period of unprecedented anthropogenic influence on global ecosystems.

Keywords: One Health, metric, sustainability, ecosystem, biodiversity, water quality, ecosystem services, health impact assessment, quantitative microbial risk assessment, material flow analysis

5.1 Introduction

'Ecology is the scientific study of interactions that determine the distribution and abundance of organisms' (Krebs, 1972). Health in an ecological context, encompasses the health of all organisms, including humans, connected to the ecosystems in which they live (Tabor, 2002).

This chapter aims to provide a conceptual framework describing the main ecological components of the global ecosystem, which need to be considered when using a One Health approach, including incorporating examples of metrics which both reflect the connectedness of different environments and quantify the complex interactions between humans, domesticated and non-domesticated animals and the environment in which they live, and the direct and indirect drivers which impact them (Woolhouse and Gowtage-Sequeria, 2005). The set of ecological components which are described should be used to inform the audience of the methods to quantify the sustainability and thus the 'One Health-ness' of any environment, (not just those within the sphere of natural sciences), with a suite of relevant components incorporated, dependent upon the context of the problem. It is the quantification of an array of these components which demonstrate the 'added value' of One Health, through the savings of lives, improvements in life-lived (quality of life), qualitative gains and financial savings (Zinsstag *et al.*, 2015b). Environmental sustainability refers to the maintenance of natural capital in ways that facilitate meeting the needs of current and future generations while neither exceeding the capacity of supporting ecosystems to continue to regenerate ecosystem services to meet those needs, nor by diminishing biological diversity (Morelli, 2011). For guidance on integration of human and animal health to social-ecological systems, see Ostrom (2007) and Cumming and Cumming (2015) and text on health in social-ecological systems (HSES) (Chapter 2; Kock *et al.*, 2018). One Medicine recognises that there is virtually no difference in the paradigm between human and veterinary medicine and both disciplines can contribute to the development of each other (Schwabe, 1984); animals should thus be positioned in the social and not the environmental realm, taking a 'less speciecist' stance (Zinsstag *et al.*, 2011).

It should be understood that when quantifying the health of anything, be it an organism or an ecosystem, the variables measured are all context-dependent, particularly for ecosystem and environmental health; see discussion on definition and measurement in Cumming and Cumming (2015). The interpretation of resulting measurements will differ dependent upon a human, animal and ecosystem perspective, and each of these perspectives has its own value, when thinking about 'One Health'. Whilst describing available metrics to quantify the ecological dimension of One Health, we do not aim to define in detail, the quantitative modelling techniques which are utilised to examine the often non-linear, unidirectional or reciprocal drivers of health or causes of disease transmission (Cumming and Cumming, 2015); we instead highlight where the reader can find further information.

5.2 Specific ecological components, an introduction

The environment is a major determinant of health (World Health Organization, 2016a) with an estimated 25-33% of the global burden of disease attributed to environmental risk factors (Smith *et al.*, 1999). Accordingly, when measuring the ecological dimension of One Health, account must be taken of the fitness and sustainability (including integrity) of the ecosystem and environment. This is not easy to quantify, as it results in the creation of indexes of heterogeneous variables, which do not provide an easily interpretable output of resilience (Suter, 1993). Various indices have been developed to quantify environmental health, including:

- the long-term sustainability of different ecosystems;
- the state of the world's biological diversity;
- describing the status of ecosystem services.

Metrics to measure the health status of the world include the 'One Health-ness' of water, air, soil, biodiversity, and ecosystems. They require ecosystem approaches to health to factor in ecosystem interactions in health research (Charron, 2012a,b). Methods to quantify the health status of populations under closer management of humans also need to be mentioned including those of humans, domestic animals, plants, and aquaculture. Finally, antimicrobial resistance issues across the ecological dimension should be considered. Many of these metrics are very 'human-centric' and should therefore be interpreted with caution. Issues associated with the measurement of health status are also discussed within chapter conclusions, to give the audience an awareness of biases associated with data collection and interpretation.

5.2.1 Water

5.2.1.1 Fresh-water quality

Brief description and outputs

Spring, summer and autumn sampling of macro-invertebrates of river systems in conjunction with environmental data (e.g. pH, O_2 concentration) as outlined by (Wright *et al.*, 1984) provides an index of water quality and resource protection.

Critique

The approach relies on the standardisation of methods across sampling sites and has been aided by the implementation of the EU Water Framework Directive (2000/60/EC) (The European Parliament and the Council of the European Union, 2000). There are limitations in sampling of macro-invertebrates in river systems in conjunction with the collection of environmental data (e.g. pH, O_2 concentration) in relation to the direct detection of bacterial and other disease-causing organisms. However, detection of bacterial or other diseases may indicate the resilience of benthic-macro invertebrate communities.

How would it impact the assessment of One Health initiatives?

The measurement has the potential to indicate wider environmental health while also potentially reflecting the impact that human systems (e.g. agricultural or otherwise) may have on water quality. Human sustainability is dependent on the availability of appropriate water resources.

5.2.1.2 Ocean health index

Brief description and outputs

The ocean health index comprises ten diverse public goals (food provision, fishing opportunity, natural products, carbon storage, coastal protection, tourism and recreation, coastal livelihoods and economies, sense of place, clean waters and biodiversity) important for a healthy ocean (Halpern *et al.*, 2012). The index quantifies the health of the human-ocean system for each coastal country.

Critique

The index provides a collection of methods for on-going assessment of ocean health related to well-accepted societal goals. However, as with other global-scale analysis tools, the local precision and application are questionable, because the usefulness of the ocean health index is reliant on the data available, which reflects the public goals at the scale at which data is collected (Halpern *et al.*, 2012).

How would it impact the assessment of One Health initiatives?

This index has potential to quantify the impact that human activities have on oceans while also providing information on how the health of the oceans could impact human sustainability.

5.2.2 Air

Brief description and outputs

The air quality health index (AQHI) is a quantitative measure which has been applied to different geographical locations including Ontario (Ontario Ministry of the Environment and Climate Change, 2010) and Tehran (Ahmadi *et al.*, 2015) to communicate to the public how polluted the air currently is or how polluted it is forecasted to become in urban environments. AQHI measures and informs on three key urban air pollutants: ozone (O_3), particulate matter (PM2.5), and nitrogen dioxide (NO_2) (Ontario Ministry of the Environment and Climate Change, 2010). These pollutants are weighted and the calculated sum provides the numerical index. The air quality index is assumed to be related to the proportion of the human population likely to experience severe adverse health effects, however the direct relationship between the two has not been quantified.

Critique

This index provides a framework for comparable air quality across global urban environments. However, the lack of standardised data collection and agreement on the weighting of pollutants means that the applications of the index across geographical and political boundaries, are therefore limited (Ahmadi *et al.*, 2015). In addition, the consequences of a change in the index are not captured using this method; this plausibly explains why this index has not been widely applied.

How would it impact the assessment of One Health initiatives?

The AQHI can be applied to define the actual condition or conditions of urban air quality and to achieve sustainability and resilience (Ahmadi *et al.*, 2015). There may be scope to apply the AQHI to the health of companion animals and to food producing animals within urbanised and surrounding areas.

5.2.3 Soil

Brief description and outputs

It has been suggested that a key number of soil indicator functions could inform best practices in relation to the management of soils (Andrews *et al.*, 2002). Considering the complex nutrient and element flow through the food web it becomes clear that plants and animals do not stand alone, but are part of an ecosystem. As mentioned by Oliver and Gregory (2015), the soil is at the core of (terrestrial) food production and it mechanically filters, absorbs and transforms substances, thus acting as a buffer that controls the transport of substances to the atmosphere, aquatic ecosystems, and plants (Dudka and Miller, 1999).

Critique

Although there is a comprehensive appreciation for the importance of soils to animal and human health as well as food production (McBratney *et al.*, 2014; Oliver and Gregory, 2015, Brevik and Sauer, 2015), there is no single indicator that encapsulates all these components.

How would it impact the assessment of One Health initiatives?

Protecting the quality of soils is important to the environment and associated species, and links directly to water and food quality. Therefore, soil quality is core for One Health and methods for understanding and capturing this information have been outlined (Keith *et al.*, 2016; Oliver and Gregory, 2015).

5.2.4 Biodiversity

5.2.4.1 Biodiversity-living plant index

Brief description and outputs

Biodiversity is a complex term, referring to the sum total of all biological variation from genes to ecosystems (Groombridge, 1992). The Living Planet Index is a well-recognised method to describe an aggregated population trend among terrestrial and freshwater vertebrate species (i.e. mammals, birds, reptiles, amphibians and fish) that shows the rate of change in the status of biodiversity over time (Collen *et al.*, 2009).

Critique

The index relates to vertebrates, therefore there are limitations in how it can be applied to overall biodiversity, as taxonomic groups such a vascular plants and invertebrates are not represented. In addition, there are considerable gaps and heterogeneity in geographic, taxonomic, and temporal coverage of existing indicators, with fewer data for developing countries (Butchart *et al.*, 2010), although these data gaps could perhaps be supplemented by information provided in the IUCN red list (International Union for Conservation of Nature and Natural Resources, 2017).

How would it impact the assessment of One Health initiatives?

While literature exists describing how biodiversity tends to decrease human disease risk through a dilution effect on the reservoir host (Keesing *et al.*, 2010), the presentation of empirical data is limited and the applicability of this appears to be context dependent (Salkeld *et al.*, 2013). This index has potential to evolve into a method that can quantify the impact that human activities have on biodiversity while also providing information on how biodiversity could impact human economic and environmental sustainability.

5.2.4.2 IUCN ecosystem red list

Brief description and outputs

This provides a consistent, practical and theoretically grounded framework designed by the International Union for Conservation of Nature and Natural Resources (IUCN) for establishing a systematic red list of the world's ecosystems. The output categorises ecosystems as: least concern, vulnerable, endangered, critically endangered, etc. (Keith *et al.*, 2013).

Critique

This could be used in conjunction with the IUCN red list for species. However, there are reservations regarding the validity and applications of this framework, such as there being no consistent means to classify ecosystems for assessing conservation status, and the framework not considering global drivers e.g. climate change (Boitani *et al.*, 2015). This debate will hopefully help develop a better discussion and communication of the key components of sustainable ecosystem utilisation.

How would it impact the assessment of One Health initiatives?

The IUCN Ecosystem red list could inform how a One Health initiative is impacting on the status of the habitats within an area. The measurement would have to be at the appropriate temporal and spatial scale to capture the effect of One Health Initiatives.

5.2.4.3 The 'omics' as a tool to quantify biodiversity

Brief description and outputs

Evaluation of health at a genomic scale can increasingly be undertaken using modern genomic, transcriptomic and proteomic and metabolomic methods. For example, combinatorial chemistry and structural biology are being applied to rapidly explore and optimise the interactions between lead compounds and their biological targets. Searching for similarities between biological sequences, using genomics as a tool to quantify abundance of genomes as opposed to informatics looking for functionality e.g. using basic local alignment search tools such as BLAST (Pertsemlidis and Fondon, 2001), is the major way by which bioinformatics can contribute to understanding One Health.

Critique

Bioinformatics can be used to quantify genomic diversity and versatility in a range of species from pathogens to environmental DNA (eDNA) (Lodge *et al.*, 2012), however, how these outputs can be combined with other indices of One Health is a field still in its infancy.

How would it impact the assessment of One Health initiatives?

The 'Omics' provide an insight into the diversity, stability and resilience of communities at a different biological scales with which to inform on One Health.

5.2.5 Ecosystem services

Brief description and outputs

Ecosystems services refer to the direct and indirect contributions of ecosystems to human well-being. There are three categories:

- Provisioning ecosystem services, which refers to all nutritional, material and energetic outputs from living systems.
- Regulating and maintenance ecosystem services, referring to all the ways in which living organisms can mediate or moderate the ambient environment that affects human performance.
- Cultural ecosystem services, which refers to all the non-material, and normally non-consumptive, outputs of ecosystems that affect people's physical and mental states (TEEB, 2015).

The trade-offs between differences in land-use activities have been conceptually discussed by Foley *et al.* (2005) (Figure 5.1). In addition, there are a number of ecological and environmental evaluation techniques developed across Europe to access the effectiveness of agri-environmental schemes at farm level (Purvis *et al.*, 2009).

Critique

Ecosystem services is a useful means of communication relating to the importance of ecosystems for a range of different human requirements. However, it should be emphasised that the perspective is human although the concept of ecosystem functioning, which should benefit many species, is core. Sustained or restored environmental services can create important added value for One Health but their assessment requires advanced study design capable of measuring a causal relationship between health in humans and animals and ecosystem services (Zinsstag *et al.*, 2015c).

How would it impact the assessment of One Health initiatives?

The ecosystems services framework has many potential overlaps with One Health e.g. soils (Keith *et al.*, 2016). Concepts like sustainability and resilience are common to both. In addition, the ecosystem framework could be used as a communication tool for integration of One Health in environmental research. One Health assessment could also adopt some of the methods used with ecosystems services to translate to economic and human sustainability (TEEB, 2015).

Holistic ecosystem health indicator

It should be noted the holistic ecosystem health indicator (HEHI) is a framework for evaluating the outputs from collaborative processes, which uses ecological, social, and interactive indicators to monitor conditions through time (Munoz-Erickson *et al.*, 2007). The HEHI involves the building and managing a human designed ecosystem or a social-ecological system (Cumming and Cumming, 2015).

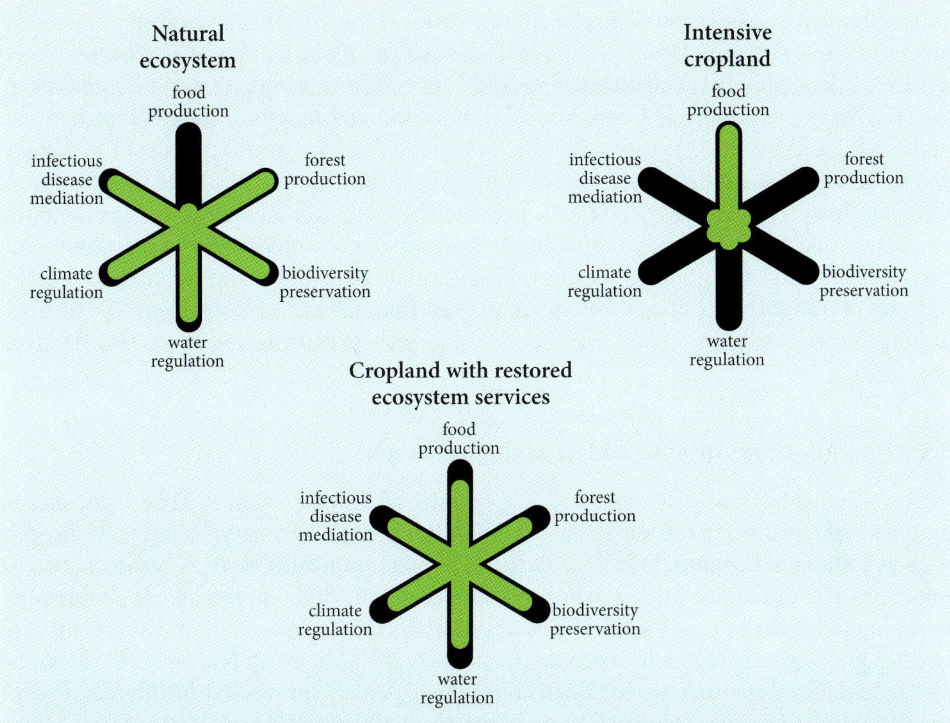

Figure 5.1. Conceptual framework for comparing land use and trade-offs of ecosystem services. The provisioning of multiple ecosystem services under different land-use regimes can be illustrated with these simple 'flower' diagrams, in which the condition of each ecosystem service is indicated along each axis. (In this qualitative illustration, the axes are not labeled or normalized with common units.) For purposes of illustration, we compare three hypothetical landscapes: a natural ecosystem (left), an intensively managed cropland (middle), and a cropland with restored ecosystem services (right). The natural ecosystems are able to support many ecosystem services at high levels, but not food production. The intensively managed cropland, however, is able to produce food in abundance (at least in the short term), at the cost of diminishing other ecosystem services. However, a middle ground – a cropland that is explicitly managed to maintain other ecosystem services – may be able to support a broader portfolio of ecosystem services (adaopted from Foley *et al.*, 2005).

5.3 Health status of vertebrate populations

5.3.1 Disease as a modulating factor in vertebrate populations

Diseases can cause host populations to crash, for example Rinderpest epidemics in Africa in the early twentieth century or the Spanish flu epidemic after the First World War. Although not a metric to be considered in quantifying health status, population biology describes how the disease processes between hosts and disease agents (infectious pathogens or parasites) are played out with both groups fighting for their lives and their fitness in a trade-off, at multiple scales, between transmission of disease, the virulence of the infection and density-dependent processes impacting population structures (Hudson *et al.*, 2002). Lambin *et al.* (2010) explored dynamic interactions between disease and landscape elements using case studies, emphasising the importance of adopting a dynamic view of spatial and temporal interactions between both scales and infectious agents, vectors, infected organisms and the range of biotic and abiotic factors that influence disease (Cumming and Cumming, 2015). The modulation of host populations by disease is likely to be impacted by biotic factors such as: the age structure and sexual bias in the host population (immunocompetence differs between age groups and between the sexes), the populations' overall health status, the behaviour populations exhibit in terms of preventive health measures such as vaccinations and preventables (i.e. using bed nets against vector-biting), and any underlying genetic predisposition to disease (Hudson *et al.*, 2002).

5.3.2 Non-infectious disease and its ecological drivers

Chronic non-infectious diseases, for example, rates of illnesses such as Type 1 diabetes in humans and lameness in cattle associated with milking parlour floor-type, and acute diseases such as asthma in humans, affect the status of population health. Both non-infectious and infectious diseases can be impacted by social and economic differences between populations, for example, higher rates of disease are observed with a low socio-economic status compared to when populations are richer in people (Hajat *et al.*, 2010, Taylor-Robinson *et al.*, 2015), and also in livestock, because poor livestock holders may perceive preventive health measures or the upholding of welfare standards as unaffordable (Gilbert and Rushton, 2016).

5.3.3 Infectious disease and its ecological drivers

The ecology of infectious diseases is important. Examples include: food systems which propagate infection by acting as reservoirs for pathogen infectivity; environmental drivers such as temperature impacting the likelihood of pathogens persisting but also of them mutating and further evolving; and emerging antibiotic resistance issues being influenced by spill-over points at the human, domesticated animal, and wildlife interface (Karesh *et al.*, 2012). The health of populations is also likely to be impacted by demographic change, because the spatial distribution of individuals influences the likelihood of disease transmission, as a consequence of contact patterns.

5.3.4 Metrics to quantify disease in populations

Various methods can be used to quantify the impact of infectious diseases using qualitative, semi-quantitative, and quantitative approaches including those used to rank pathogens within risk assessment. Some examples and assessment of basic techniques are provided below, however best practice has recently been reviewed at EU level (O'Brien *et al.*, 2016). The main reason to effectively quantify disease is to work out whether the population affected is increasing, in order to reduce the time to detection, facilitate earlier intervention at source and reduce the overall impact of infection (Zinsstag *et al.*, 2015b).

5.3.4.1 Quantitative methods

Brief description and outputs

Quantitative methods aim to quantify the impact of diseases/pathogens, but may use a mixture of numerical estimates and ranking where data is missing. In some quantitative studies, the impact of a single infection is calculated in detail, for example (Fosse *et al.*, 2008), where risk scores were used and considered as cross-functions of the incidence of disease, and related to severity scores for morbidity effects. The global burden of disease programme of studies involved a combination of quantitative and semi-quantitative methods and the outcomes are quantitative. These methods aim to create a quantitative estimate, described by numbers or probabilities, of disease/pathogen impact.

Critique

These approaches use substantial literature reviewing as a part of their process, and aim to mine associated published evidence for estimates of the different aspects of impact. Other data sources may also be employed where data is missing including the grey literature or web resources. Such analyses can be overly-simplistic and have a detrimental impact on policy decisions, or they can be very complex, data-rich, time-consuming and influenced by the underlying opinion of scientists. In addition, most quantitative methods require some input from experts, giving them a degree of subjectivity, and final outputs may be similar to the results of semi-quantitative studies.

How would it impact the assessment of One Health initiatives?

Data rich analyses such as those utilising quantitative methods work well to answer specific research questions, however as a result of their specificity, it is difficult to analyse the outputs of multiple studies with differing aims; there is a need to develop common data collection standards, so that the results of studies can be more easily compared. One solution is to consider infectious disease epidemiology as an extension of predator-prey interactions following the approach of Anderson and May (Anderson and May, 1978, May and Anderson, 1978), which means that animal-human transmission models can be used to quantify non-linear infectious processes across different host domains (Zinsstag *et al.*, 2005, 2017); such approaches could be extended to ecosystems for example.

Global burden of disease approach

Brief description and outputs

The global burden of disease (GBD) is a systematic effort providing a data-rich framework to quantify the comparative magnitude of health loss due to diseases, injuries, and risk factors by age, sex, and geography over time. Its strengths include: the breath of expertise used to create its analytical mechanisms and outputs, it captures human health outcomes with comparable metrics, and it aims to separate epidemiological assessment from advocacy concerns or entanglement of agendas (Murray *et al.*, 2012). Very high data requirements are needed though the recent framework means that estimates can be calculated relatively quickly (new estimates currently released annually). Data utilised include relevant published and unpublished evidence, field data and internet surveys. GBD estimates for cause-specific and all-cause deaths and disability-adjusted life years (DALYs) are provided annually. Previous iterations of study results have also provided years lived with disability (YLDs).

Critique

The GBD method only estimates the impacts of infectious and non-infectious diseases upon humans; non-human species are not considered, nor are the impacts of disease upon ecosystem health. In addition, there may be biases in the diseases included and in geographical quantification, due to under-reporting and issues of data collection. The GBD approach is a large, complex scientific undertaking and sometimes consensus in outcomes could not be reached by expert groups; in this circumstance final decisions were made by a core team, which potentially can cause biases towards certain diseases. For the comparative risk assessment, an absence of intervention methods e.g. vaccine/preventable, was not considered a risk factor, and other risk factors e.g. total caloric intake or unsafe sexual behaviours could not be included due to an extreme lack of data on exposure.

How would it impact the assessment of One Health initiatives?

The calculations needed to estimate deaths from disease, DALYs, YLDs and effects of risk factors using the GBD manner are complicated, but have been undertaken outside the study itself e.g. (Ao *et al.*, 2015). For researchers (as opposed to policy-makers), the most useful impact is in using GBD estimates as baseline comparators and a starting point to compare against the outcomes and estimates from other work, rather than as an absolute estimate. The GBD approach has been expanded to a multidimensional array of burden and financial dimension, to which environmental or ecological dimensions could be added (Zinsstag *et al.*, 2015c). It is not recommended that the GBD be extended to animals but rather financial valuing of losses to animal production is given in context (Zinsstag *et al.*, 2015a).

Integrated dynamic assessments of disease transmission

Brief description and outputs

One Health approaches in which the transmission of disease is examined often also require economic assessment including comparison of the impact of interventions. This can be achieved by integrating assessments of disease dynamically. For example, three cost scenarios were compared in a study of rabies and administration of post-exposure prophylaxis (PEP) including: (1) PEP alone; (2) dog mass vaccination and PEP; and (3) dog mass vaccination, PEP, and maximal communication between human health and veterinary workers (Zinsstag *et al.*, 2017).

Critique

The data collection phase of large studies needs to be well planned so that results for multiple field teams and time-points are consistent. The costs in relation to both running the study and the effects of interventions must be estimated, often made upon assumptions about factors such as costs being the same in different geographical locations and at different time-points, the number of vaccinations undertaken, and the likelihood of vaccination based on the severity of dog bites. In some cases, there is such an absence of reliable cost data that costs are not included in calculations. Assumptions also need to be made about whether communication will lead to better decision-making. All of these guesses need to be made based on as much prior knowledge as possible, as they can lead to models which under- or over-estimate the impacts of interventions.

How would it impact the assessment of One Health initiatives?

These approaches are important because they create quantitative comparison of One Health versus siloed single-disciple approaches, thereby potentially demonstrating the added value of One Health.

Matrix modelling methods

Brief description and outputs

Many populations are comprised of individuals of different types or at different stages in their life cycle. These heterogeneities can be difficult to capture using conventional modelling techniques, but can be considered as a process of matrix multiplication, which has led to the development of matrix population models (Leslie, 1945, 1948). Matrix modelling approaches can be applied to animal population dynamics and human populations, but also to plant species, which generally enter pronounced developmental and reproductive stages during their lifespan (Crone *et al.*, 2011). As well as offering a useful approach to modelling population growth, matrix population models are also useful for the estimation of the basic reproduction number (R0) of a pathogen, especially in heterogeneous systems, since this has been shown to be equivalent to the dominant eigenvalue of the next generation matrix (Diekmann *et al.*, 1990). This general approach has subsequently been applied to human, animal, and plant diseases (Diekmann, 1991; Dietz, 1993; Roberts and Heesterbeek, 2003; Van den Bosch, 2008). Additionally, the normalised right eigenvector associated with this dominant eigenvalue can be interpreted as the relative number of new infections in the different groups at the steady state, and the normalised left eigenvector as the relative contribution of the different groups to future epidemic growth (Diekmann and Heesterbeek, 2000; Diekmann *et al.*, 2013). By considering one practical interpretation of R0 as the amount of effort required to establish control over the spread of a pathogen, this concept was further adapted in order to effectively estimate R0 for each group in a structured system (termed the 'type reproduction number') (Heesterbeek and Roberts, 2007; Roberts and Heesterbeek, 2003), which can be particularly valuable for the identification of potential pathogen reservoirs. The next generation matrix can also be derived from a system of ordinary differential equations (Diekmann and Heesterbeek, 2000, Diekmann *et al.*, 2010, Van den Driessche and Watmough, 2002), which can also be used to evaluate the stability of a system at a steady state (by inspecting the dominant eigenvalue of the Jacobian matrix, for example (Keeling and Rohani, 2008). These methods can be combined and applied to the evaluation of epidemiological processes within an ecological system (Roberts and Heesterbeek 2003).

Critique

Matrix modelling has been criticised as it may have capacity to adequately cover only restricted spatial and temporal variability (Burkhard *et al.*, 2009), meaning that factors not represented within classifications are ignored.

How would it impact the assessment of One Health initiatives?

Matrix modelling if used for mapping ecosystem services, for example (Jacobs *et al.*, 2015), can use a mixture of qualitative and quantitative methods including expert opinion, to help classify different ecosystems and the health status of organisms within them. As such, this method is potentially of importance in assessing One Health working.

5.3.4.2 Semi-quantitative methods

Brief description and outputs

Semi-quantitative methods lie somewhere in the middle of qualitative and quantitative approaches, using a mix of qualitative terms and/or signs and numbers, for example (Cardoen *et al.*, 2009; McKenzie *et al.*, 2007), with the outcome being semi-quantitative, and often a comparative score for impact (relative to other diseases/pathogens included within the study). Traditional risk assessment approaches to prioritise pathogens or diseases most commonly involve semi-quantitative risk identification, with the criteria used relating to the different steps of hazard identification, the probability of the hazard occurring (release and exposure assessments) and consequences assessment (with the scoring of impacts). Multi-criteria decision tool methods are also semi-quantitative and based on risk. Ideally an evidence-base from the published literature, grey literature or web resources is accessed to provide material for decision-making within the research process, however expert-opinion methods can also be employed.

The outputs of some semi-quantitative studies were reviewed in the Discontools (O'Brien *et al.*, 2016) and ENHanCE projects (Waret-Szkuta *et al.*, 2010). The Discontools project developed a further prioritisation model for 45 diseases, based on 28 criteria split into six modules: disease knowledge, impact on wider society, impact on public health, impact on trade, animal welfare and control tools. Within each disease analysis, criterion were scored depending on the reason for wanting to know the priority, and the modules thereafter weighted to assess the impact of infections (Discontools, 2012).

Critique

Semi-quantitative approaches suffer from the same criticisms as qualitative work but are also criticised due to the large amount of resources they use and because the different diagnostic methods used within work means that the outputs of different studies are rarely comparable.

Commercial quality of meat/ animal products and welfare

Brief description and outputs

Animals suffering stress, malnutrition or chronic illness may have a carcass which reflects their health status. The rate and causes of signs and legions detected in ante- and post-mortem inspection at slaughterhouse could indicate a lack of good production, bad transport practices or low general health status of livestock on farms. A high number of deaths, as well as morbidities such as musculoskeletal injuries and abrasions, are indicative of bad practice. In the EU, livestock carcasses are classified using, for example, the EUROP grid metric (The Commission of the European Communities, 1981). As a part of meat traceability at slaughterhouse, livestock keepers have to provide a Food Chain Information (FCI) certificate which includes a declaration that all medicine withdrawal periods required for animal products to enter the human food chain have been observed (Anon., 2017).

Critique

The EUROP grid metric reflects carcass confirmation and measures market requirements including for breed, age and diet, but has not been validated as a measurement of livestock health.

How would it impact the assessment of One Health Initiatives?
The EUROP grid metric reflects market requirements.

5.3.4.3 Qualitative methods

Brief description and outputs
Qualitative approaches can assess the impact of a large number of diseases with a relatively simple and quick method which is easy to communicate to decision makers but quite subjective. They are useful because a range of bespoke criteria can be developed within the work and then summarised in outputs. The outputs of some qualitative studies were assessed in the Discontools (O'Brien *et al.*, 2016) and ENHanCE projects (Waret-Szkuta *et al.*, 2010). Ideally, an evidence-base is accessed to provide material for decision-making within the research process, however expert-opinion methods can also be employed in which experts discuss the process, evaluate the outcomes and conduct a prioritisation of disease impact according to their expertise, perhaps according to top-priority goals, and from their own perspective.

In qualitative studies to estimate the impact of diseases/pathogens, estimation of parameters and risks is undertaken using ordinal words (relatively high, low…), for example (Valenciano *et al.*, 2002).

Critique
Issues of subjectivity include that it can be difficult to choose the right qualifier (Dufour *et al.*, 2008) however, this can be solved by assessing each disease comparative to the others, and not in an absolute manner. Other criticisms include: the question of how to choose which diseases to examine; who participates in the assessments; the subjectivity of qualitative methods; that the evidence-base used to make judgements may not itself be objective or may be missing or insubstantial (so guesswork must be undertaken); and that the experts used may be unduly influenced by their expertise, and personal or disciplinary differences in gaols. The eventual outcome of an assessment is ideally a unanimous consensus by experts, but it may not always be easy for this to be reached.

How would it impact the assessment of One Health initiatives?
Qualitative methods are often a good solution for decision-making in an infectious disease outbreak circumstance, because experts can be quickly mobilised by the main stakeholders involved in controlling the outbreak. Qualitative solutions can also include measures to quantify the opinion of stakeholders, as well as factual based information provided by experts. As a result, although flawed, they provide an informed and quick decision-making framework.

5.3.4.4 Web-based surveillance systems

As well as traditional approaches, various web-based surveillance systems are available describing infection in humans, animals and plants, for more information see (Madoff and Li, 2014), for example the World Organisation for Animal Health (OIE) World Animal Health Information System database.

World animal health information system

Brief description and outputs

The World Animal Health Information System (WAHIS) provides spatial and temporal information on the presence/absence of infections in animal hosts (domesticated animals including bees, some wild terrestrial species and aquatic species) and zoonotic diseases in humans, using confirmed disease reports.

Critique

WAHIS relies on reporting of infection by OIE Member countries/territories. The reporting details are likely to be affected by issues of data quality, because the availability of surveillance resources will differ, influencing the collection, analysis and dissemination of data. In addition, there are likely to be differences in the likelihood of clinicians/researchers reporting disease, and biases in the willingness to report due to country-level implications e.g. notifiable diseases impacting future trade. The effects of these issues are likely to differ between geographical units such as countries/territories. Biases in reporting of infections are also likely if the infection has significant sub-clinical effects.

How would it impact the assessment of One Health initiatives?

Collation of information on confirmed disease reports is especially useful when the reports are consistently submitted. Although WAHIS includes interfaces to collate information on zoonoses in humans, and wildlife diseases, these are likely to be affected by significant reporting biases. As such, the database outputs are most useful for presence-only rather than presence-absence disease modelling, unless utilised in tandem with other sources of data. At the time of publication, the WAHIS resource is in the process of being redeveloped; at release it will potentially be of greater use to One Health research and decision making.

5.3.4.5 Integrating human and animal surveillance information

Cross-sectoral, integrated interpretation of zoonosis evidence provides an application of surveillance information, describing the prevention, prediction and control of infectious disease (Palmer *et al.*, 2011). Integrating human and animal surveillance can support outbreak investigation, provide 'early warning' detection of disease in animals (or food) before it reaches humans (or vice versa), provide information for source attribution analysis (whereby the source of infections can be identified), allow the undertaking of trend analyses (wherein correlations between disease occurrence in animals and humans are examined over time and space), and aid the monitoring of the effects of surveillance (because surveillance in one sector may influence the results in other sectors); in addition, cost savings are gained when surveillance, including laboratory resources, is combined and becomes more efficient (Zinsstag *et al.*, 2015b).

There are a number of systems which integrate human and animal disease surveillance information, however the details for each varies widely dependent upon surveillance purpose, structure and sources of information. Many systems are described within the grey or unpublished literature. Currently, routinely collected data from established surveillance systems are stored in different databases, depending on their origin (animal health, human health or food safety) and the purpose of data collection. Interfaces between the different sources are rare, and it is therefore difficult to estimate whether data can be effectively linked to enable joint analyses. Such secondary data are consequently challenging to analyse in a cross-sectoral manner (Wendt *et al.*, 2015).

A recent study aimed to identify current surveillance initiatives throughout the world by integrating human and animal zoonotic disease data (Wendt *et al.*, 2015). The integrated systems examined are all unique in purpose but have certain characteristics: most started in the last decade, and approximately half are global; the others covering specific regions including North America and Asia. The systems all have very different surveillance purposes including early or anomaly detection (often using animals as 'sentinels' to detect first cases, or identifying space-time clusters of cases), to monitor the occurrence and geographical extent of a disease/infection, to improve the epidemiological understanding (via cross-sectoral information integration), to predict a disease event for the human population and to identify health risks (in one case, including along the food chain). Three quarters of systems survey many diseases, and one quarter are focussed; two on arboviruses, one on foodborne pathogens, one on parasites and one on Salmonella. Three quarters of systems use predominantly confirmatory diagnostic data, mostly from reports on disease occurrence or laboratory results, with information coming from official reports on notifiable diseases or data collections from hospitals and laboratories. The other five systems rely mostly on pre-diagnostic data from reports on suspected cases collected from electronic news/media sources (syndromic surveillance). Most systems process secondary data originally collected for other purposes.

In some developing world communities e.g. pastoralists in West and Central Africa, improving communication about infectious disease across clinical disciplines has led to integration of surveillance and improvements in health services. Abakar *et al.* (2016) and Jean-Richard *et al.* (2014) discuss using community health and community animal health workers to provide primary health care in remote zones including continuing exchanges concerning quality services and supervision by the health systems as well as patient referral systems. In addition, strong producer organisations/farmer cooperatives can sometimes deliver health services. They argue that mobile communication coverage should be focussed in remote areas, where local communities still have low rates of literacy and require visual supports for education and communication for syndromic surveillance. This is because appropriate syndromic surveillance among pastoralists, using the potential of mobile communication and visual technology (including web-based applications) and utilising their local observations and perceptions of most human and animal diseases, is the best way to extend surveillance in these hard-to-reach populations and include them in national health services.

5.3.4.6 Health impact assessment

Brief description and outputs

Health impact assessment (HIA) is an important contributor to both local and national decision-making processes. Adopting a multi-method approach, it incorporates qualitative and quantitative analyses to determine the various health impacts of policies and projects (Anon., 1999). HIA includes: consideration of evidence about the anticipated relationships between a policy, programme or project and the health of a population; consideration of the opinions, experience and expectations of those who may be affected by the proposed policy, programme or project; provision of more informed understanding by decision makers and the public regarding the effects of the policy, programme or project on health; and proposals for adjustments/options to maximize the positive and minimize the negative health impacts. It is recommended that it be implemented as early as possible in the planning stage of a project.

Critique

HIA is a complex process and the assessment procedure needs to be treated with caution. All policy processes are carried out within a framework of values, goals and objectives that may be more or less explicit in a given society and at a given time. It is essential that such values are taken into account, otherwise HIA runs the danger of being an artificial process, divorced from the reality of the policy environment in which it is being implemented. Whilst providing a framework with which to assess the impacts of policy change upon health, the results of HIA need to be acted upon in order for the process to have value.

How would it impact the assessment of One Health initiatives?

HIA was designed to assess health impacts with a human-centric basis. An extension to this approach has been suggested in One Health impact assessment (OHIA) (Zinsstag *et al.*, 2017), which places particular emphasis on the interlinkages between human, animal and ecosystem health, as well as animal health separately. A further extension of the OHIA is the economic evaluation of health impacts, including the expected cost to human and animal health.

5.4 Health status of plant populations

Plants form the foundation of most terrestrial food webs, meaning that plant disease in both natural and managed ecosystems can have considerable impacts upon ecosystem health, food security, productivity, and livelihoods (Boa *et al.*, 2015). Along with the direct benefits, improvements in plant health can offer considerable benefits to people through food security (Flood, 2010), the potential for plant health to boost service delivery (Bentley *et al.*, 2009; Boa, 2009, 2015; Danielsen, 2013), reduction in the use of potentially dangerous chemicals such as pesticides and herbicides (Yanggen *et al.*, 2004), and minimisation of microbial contamination of plant-based food products (Fletcher *et al.*, 2013). However, disciplinary isolation between medical/veterinary services and plant health, combined with the relative scarcity of plant health extension services (i.e. 'plant doctors') in the field has, to date, limited the integration of plant health into the One Health discipline (Boa *et al.*, 2015). This is further exacerbated by the fact that very few plant pathogens are a direct risk to humans in a similar manner to zoonoses in animal populations (with the notable exception of mycotoxins released by fungi (Peraica *et al.*, 1999).

In order to address these challenges and embed plant health within the One Health framework, an interdisciplinary approach is needed – focussing on human, animal, and plant health and the associated linkages and synergies within the context of natural and managed ecosystems. Rather than focussing on these specific linkages (which are well described elsewhere e.g. (Boa *et al.*, 2015; Danielsen, 2013; Fletcher *et al.*, 2009, 2013), we provide here a brief introduction to the measurement of the health status of plants, in order to better facilitate communication between the specific disciplines.

Brief description and outputs

Ignoring the impacts of phytophagous pests such as locusts, the biotic causes of plant disease are very similar to the causes of animal (including human) disease – with fungi, viruses/viroids, bacteria, protists (oomycetes), and parasites (nematodes) all common causes of disease (Anderson *et al.*, 2004) (although fungi and related organisms have a much greater overall health impact on plants than on mammals (Fisher *et al.*, 2012). The mechanisms of spread of these pathogens are similar to those seen for animal pathogens: with contact-based, aerosol, vector-borne, fomite-based, and vertical transmission recognised, meaning that the epidemiological dynamics are generally similar (although the lesser role of adaptive immunity in plants means that concepts such as herd immunity are not generally seen). Similarly, the methods used for detection of these pathogens are generally very similar to those used for animal disease (Boa *et al.*, 2015; Fletcher *et al.*, 2009), with visual inspection playing a primary role in diagnosis of disease.

Despite the similarities between plant and animal pathogens, the methods used to measure and quantify disease in plants differ from those in animals (Nutter, 1999). Central to these differences are issues of spatial and temporal scale, which for crops is often considered at the level of the plot, over a single growing season. Because of the focus on these spatiotemporal constraints (and due to the sessile nature of plants), the spread of plant pathogens is commonly classified as 'primary' (driven by inoculum from some outside source, resulting in 'monocyclic' epidemics) or 'secondary' (driven by inoculum generated during the epidemic itself, leading to 'polycyclic' epidemics) (Madden *et al.*, 2007).

Another challenge for measuring the level of disease or infection in plants results from the fact that the spread of a pathogen throughout an individual plant generally takes place over much longer temporal scales than that in animals (Rodrigo, 2014; Samuel, 1934). This means that localised disease may be present in the absence of systemic spread, making attempts to classify the disease status of the plant as a whole problematic from an epidemiological perspective (Gilligan, 2008). This within-plant heterogeneity may be addressed to some degree by considering the epidemiological unit of interest as a subunit of the plant – such as a leaf, or a branch, or a main root axis - or through the use of alternative disease intensity metrics such as the 'severity', as described below.

As in the fields of human and veterinary epidemiology, the proportion of infected or symptomatic epidemiological units is a common measure of the intensity of infection or disease. Within the field of botanical epidemiology, this measure is traditionally termed the 'incidence' (there is no recognised equivalent to the incidence risk or incidence rate as defined in human or veterinary epidemiology), although there is currently no clear consensus on terminology (Nutter, 1999; Nutter *et al.*, 1991). In order to capture some of the heterogeneities within a single plant, the 'incidence' may in some cases be better interpreted as a density measurement, with the denominator calculated as an area or a volume rather than a number of plant units within an area of interest (Madden *et al.*, 2007; McRoberts *et al.*, 2003). Other methods of estimating the level of infection at the plot scale and beyond are generally based upon measurement and analysis of reflected electromagnetic radiation (for example, using optical, hyperspectral, or thermal imagery (Calderón *et al.*, 2015; Nilsson, 1995; West *et al.*, 2003).

Another method of capturing some of the within-plant heterogeneity inherent in plant disease is the 'severity' of infection or disease within individuals (Large, 1966; Madden *et al.*, 2007; McRoberts *et al.*, 2003). This may be estimated on a continuous scale, as the proportion of the total area of the plant showing signs of disease, or a discrete scale, as the number of lesions, which can then be presented as a count or density of symptomatic tissue per unit area or volume (Madden *et al.*, 2007; McRoberts *et al.*, 2003). More qualitative measures such as disease scales (in which severity is categorised into distinct groups – commonly either linear or logarithmic – according to the proportion of affected tissue) have historically been popular (Horsfall and Barratt, 1945; Nita *et al*, 2003; Slopek, 1989), but may not be more accurate than estimation on a simple continuous scale (Bock *et al.*, 2010). Despite the potential increased flexibility of severity measurements, a major problem is the variability in performance of different observers. This may be reduced to some degree by training and through the use of standard area diagrams (James, 1971; Nutter and Schultz, 2009; Yadav *et al.*, 2013), or alternative approaches such as reflectance measurements using radiometry or analysis of photographic images (Bock *et al.*, 2010; Nilsson, 1995; Nutter *et al.*, 1993). Another method of quantifying severity is the use of ordinal rating scales, which allow combinations of symptom types to be accounted for. Finally, recent advances in the capabilities of diagnostic tests have also allowed estimates to be made of the amount of pathogen present in tissue, rather than focussing on disease symptoms (Nutter *et al.*, 2006) – for example, by using the threshold cycle (Ct) number from real-time PCR to estimate the level of infection (Stover and McCollum, 2011).

Measurement of plant disease is generally focussed on field inspection – in most cases, based upon visual inspection of plants for symptoms followed by collection of suspect samples for further laboratory-based diagnostic testing. The scale dependency described above means that sampling plans must be carefully formulated in order to ensure that results are generalisable to the plant unit of interest (Binns *et al.* 2000). Simple heuristics have also been developed for planning early detection surveillance which are informed by the biology of the pathogen (Alonso Chavez *et al.*, 2016; Mastin *et al.*, 2017; Parnell *et al*, 2015). However, surveillance for plant pathogens is complicated by the very large numbers of different hosts (including cereal, fruit, or vegetable, crops, ornamental plants, wild flora, and forest trees) and associated pathogens of potential concern (Waage *et al.*, 2007; Wilkinson *et al.*, 2011). Under the International Plant Protection Convention (IPPC), control of plant pathogens is the responsibility of national plant protection organisations (NPPOs) such as Defra in the UK, with regional coordination between NPPOs managed by regional plant protection organisations (RPPOs) such as the European plant protection organisation (EPPO) – which itself works closely with groups such as the Euphresco (European phytosanitary research coordination) network, the European Commission, and the European Food Safety Authority) (MacLeod *et al.*, 2010). Whilst RPPOs may specify pests and pathogens in need of compulsory surveillance and monitoring by NPPOs (in Europe, a list of pests requiring annual survey have been specified (Anonymous, 2000), beyond this, each NPPO has the responsibility for determining which pathogens should be prioritised for surveillance and control. Under this framework, active surveillance for the pest in question is conducted through planned surveys by trained plant health inspectors, following guidance developed by the NPPO and the RPPO. There is also an increasing interest in the use of passive surveillance strategies using data collected by members of the public, especially for forest pathogens (Meentemeyer *et al.*,

2015), or by crop growers, through diagnostic networks (Miller *et al.*, 2009). Moves towards a more neoliberal strategy of sharing disease control responsibilities between governments and industries are also in place, but require further development (Waage *et al.*, 2007).

Critique

Terminology aside, the metrics used to quantify plant disease show particular differences to those in the human and veterinary fields, and therefore care should be taken when interpreting them. Along with the question of the scale of measurement, a particular challenge (and focus of ongoing research) is how best to capture the intensity of disease. Although it has been argued that measures of incidence can be used as measures of severity when interpreted at a higher scale (Seem, 1984), this approach may prove problematic in practice, where underlying heterogeneities in disease severity may complicate interpretation (McRoberts *et al.*, 2003). As a result, the most common approach is to focus on either incidence or severity, according to the pathosystem in question (Madden *et al.*, 2007), although efforts have been made to identify relationships between the two measures (Hughes *et al.*, 1997; Madden *et al.*, 2007; McRoberts *et al.*, 2003; Seem, 1984; Stover and McCollum, 2011; Turechek and Madden, 2001), and between incidence measurements at different scales (Hughes *et al.*, 1997; Madden and Hughes, 1999). Whilst severity has long been considered the more useful metric (Madden *et al.*, 2007), there is an increasing recognition of the potential value of incidence measurements (Madden and Hughes, 1995). For example, direct severity assessment of root disease is often not possible without destructive sampling and therefore attempts to estimate of the average severity within the population as a whole using incidence data may be required (McRoberts *et al.*, 2003).

How would it impact the assessment of One Health initiatives?

As described above, a major impact of plant disease on human and animal health results from the impact of disease on the yield of food crops, and therefore food security. In order to quantify this relationship (James, 1974; Madden *et al.*, 2007), data must be collected from field trials or expert opinion. In the case of non-crop plants (such as forest trees), valuation of the effect of disease is often based upon the concepts of biodiversity and ecosystem services described above (Boyd *et al.*, 2013; Freer-Smith and Webber, 2015).

On a more conceptual level, the field of botanical epidemiology derives largely from ecology (in contrast to the more medicine-driven field of veterinary epidemiology) and thus shares many methodologies with this field. As a result, botanical epidemiology offers a valuable ecological perspective on the study of host-pathogen dynamics which can help to break down community 'silos' (Manlove *et al.*, 2016). In particular, the issues of scale faced when evaluating static and dynamic trends in plant disease (Mikaberidze *et al.*, 2016) lends itself very naturally to consideration of the importance of scale on the spread of human and animal pathogens (Giraudoux *et al.*, 2006) and in ecology more broadly (Wiens, 1989).

5.5 Health status of aquaculture

Brief description and outputs

Aquaculture production has expanded from 13 to 66 million tonnes between 1990 and 2012 (Food and Agriculture Organization of the United Nations, 2014), and this, combined with movements of animals and products on scales ranging from local to global (Rodgers *et al.* 2011) has led to emergence of a range of diseases. It is necessary to rank and compare these diseases in order to prioritise control. Surveillance is vital for targeting controls. The main methods are expert opinion and economic assessment of risk. Expert opinion is widely used and forms the basis of international controls of aquatic diseases (World Organisation for Animal Health, 2016), to support control policies. It is flexible and accepted throughout authority, but it is not objective.

Critique

Economic assessment of pathogen consequences potentially allows objective comparison of disease impacts and is used in supporting disease control (Peeler and Otte, 2016). Unfortunately, methods are generally *ad hoc* for particular cases, and this means outputs of different analyses may not be comparable. A few attempts have been made to systematically compare multiple diseases (Fofana and Baulcomb, 2012; Shinn *et al.*, 2015) and this is a likely area of advance. However economic assessment requires large quantities of quality data, and this may not always be available, making these assessments indirectly dependent on expert opinion. An analysis of scientific literature has been applied to ranking diseases of salmon (Murray *et al.*, 2016), which may prove a way to rank diseases objectively, at least until economic assessments are standardised and data improved.

Surveillance of pathogen presence and/or prevalence is carried out to protect animal health and to allow safe trade. Standards required for internationally notifiable diseases are set by the World Organisation for Animal Health (World Organisation for Animal Health, 2016). Risk-based methods are increasingly used to increase cost-effectiveness (Oidtmann *et al.*, 2013). Technologies to provide quicker, more accurate and cheaper diagnostic methods (Adams and Thompson, 2011), the most effective pooling of samples (Hall *et al.*, 2014) and combining of different data sources (Gustafson *et al.*, 2010) are all important. However, effective surveillance depends on the ability and willingness of farmers to report suspicion or evidence of infection (Brugere *et al.*, 2017); this applies even more strongly in countries with more limited resources for official inspection.

How would it impact the assessment of One Health initiatives?

Metrics to measure the condition of aquaculture systems reflect the health status of the production environment including natural ecosystems in which fish farms are often maintained. They also have a knock-on effect upon the food-chain, and food security.

5.6 Impact of antimicrobial resistance on health status

Brief description and outputs

Antimicrobial usage and resistance work in tandem and are multifactorial challenges (Figure 5.2). For example, it is estimated that 75 to 90% of antibiotics used in humans and food-producing animals are excreted and accumulated in the environment, largely unmetabolized

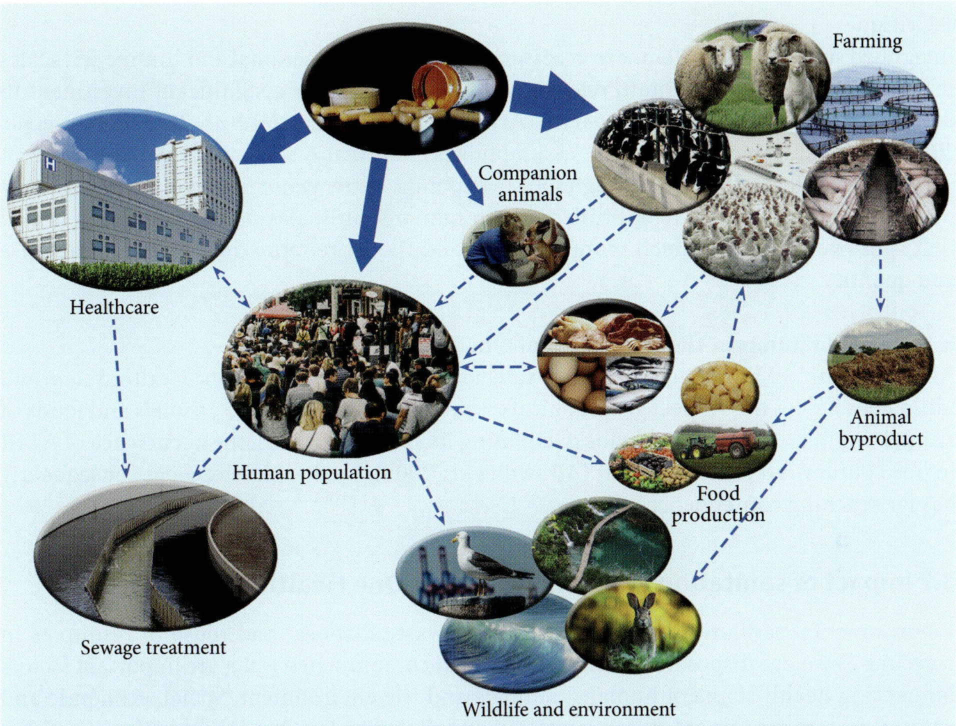

Figure 5.2. The complex ecosystem of antimicrobial resistant bacteria, illustrating different antimicrobial applications (the size of the blue arrows is proportional to antimicrobial use) and potential routes by which resistant bacteria, and resistant genes can spread between animals, humans and in the environment.

(Andersson and Hughes, 2012; Marshall and Levy, 2011) and some molecules such as fluoroquinolones have long half-lives (Gonzalez-Zorn and Escudero, 2012). To quantify the development and transmission of anthropogenic resistance, it is essential to consider complex interactions between: the physical environment (e.g. farms, crops, air, soil, and water), social exchanges (e.g. between animals within a herd, farmers and animals, domestic and wild animals, between clinical settings and the community), the food processing chain (e.g. farming activities, food preparation, transportation and storage), and human use patterns (e.g. meat consumption habits and susceptibility to infection).

No metrics to quantify anthropogenic AMR have yet been developed, however (Gonzalez-Zorn and Escudero, 2012) noted the concept of antimicrobial Resistance Units (RU) and set out a structure of how these could be quantified. In order for any metric to be developed, an important consideration is that of the need to develop integrated surveillance mechanisms which take account of resistance moving through spill-over points between different facets of the environment including host groups (Karesh *et al.*, 2012).

Critique

Integrated surveillance systems are necessary across regional, national and continental scales in order to create and maintain relevant data but will require a significant investment to develop and maintain in the longer-term (Arnold *et al.*, 2016; Carroll *et al.*, 2015; Karesh *et al.*, 2012; Laxminarayan *et al.*, 2013). The assumption is made that AMR spreads between different parts of the ecological system through spill-over points. Better genomic and other assessments might show that AMR develops differently in humans, animals and the environment and that spill-over might be much more complex, and therefore more difficult to characterise and quantify.

How would it impact the assessment of One Health initiatives?

Anthropogenic AMR is ubiquitous within all biotic components of the One Health system and ultimately, it does not respect geographical or biological borders. Food animals and foods of animal origin are traded worldwide, facilitating its spread and thus, the occurrence of AMR in one country is a problem for all (Ahmed *et al.*, 2009); further development consequently has far-reaching consequences.

5.7 Impact of sanitation including waste on One Health

Environmental sanitation including (rubbish) waste disposal and reuse of resources in recycling, sewerage disposal and supply of clean, uncontaminated water are important factors influencing health status of humans, animals and the environment. Social, economic and cultural factors are important for sanitation and therefore for One Health; (Nguyen-viet *et al.*, 2015) provide further discussion and examples.

Particularly in developing countries, the management of human and animal waste is impacted by a lack of appropriate sanitation technologies. In Vietnam, for example, waste from livestock including poultry and ruminants (especially pigs) is reused as feed and fertiliser in agriculture and aquaculture. Whilst this negates the need for chemical fertilisers, it has significant environmental impact in that pathogens within inadequately treated waste go into the environment causing further health risks (Nguyen-viet *et al.*, 2015). Quantitative microbial risk assessment (QMRA) and material flow analysis (MFA) can be used to quantify the impact of sanitation.

Quantitative microbial risk assessment

Brief description and outputs

QMRA estimates infection risk from an exposure and can also estimate disease risk, allowing for assessment of critical control points in food chains (production, transformation and consumption) and sanitation systems (Haas *et al.*, 2014). It has also been used to assess health risks of drinking water (Howard *et al.*, 2006; Van Lieverloo *et al.*, 2007) and wastewater management (Eisenberg *et al.*, 2008; Westrell *et al.*, 2004), including infection and disease risk for populations in contact with wastewater (An *et al.*, 2007; Diallo *et al.*, 2008; Mara *et al.*, 2007; Seidu *et al.*, 2008). QMRA is useful because it allows an *a priori* assessment of the effect of intervention measures along the whole food chain, or combinations of intervention measures, on public health, and because it highlights knowledge gaps. Its value is increased when it is combined with socio-economic analysis (Havelaar *et al.*, 2018).

Critique

It has been suggested that QMRA is extremely data hungry and time consuming, however this is disputed, because the data needs are a reflection of the complexity of the question being investigated and or the degree of certainty required. Some of the issues of data needs for QMRA can be addressed using sensitivity analysis to determine which parameters are important for the mode output so that data collection can be focussed on these parameters (Havelaar *et al.*, 2018). Quantifying the uncertainty of QMRA models is the most significant problem, and difficult or even impossible if the model is complex, partly because of theoretical/model implementation problems associated with computational demands. In addition, if parameter values are fully unknown, uncertainty cannot be estimated around them. The same may hold for estimates based on expert opinion, on data from microbiological results from laboratory experiments or quantification of dose-response based on data from a few strains or tested on a group of health volunteers (Havelaar *et al.*, 2018). Further criticisms include current models not being sufficiently realistic to reflect the complicated technological and biological processes, the risk of wasting resources if questions were not sufficiently focused, the difficulty of adequately reflecting regional differences in food production and consumption, and a lack of harmonisation of models (Havelaar *et al.*, 2018).

How would it impact the assessment of One Health initiatives?

QMRA has lots of potential to be used for quantification of risk for One Health, although the questions being investigated using this technique need to be well thought-out and simplified as much possible in order to fully utilise available data (including collecting new data) and incorporate uncertainty within the models, accounting for other model criticisms. In addition, simplicity in models is more likely to lead to timely outputs which can be developed with the academic resources available (there may be limited expertise in local areas with which to undertake the most complicated QMRA models).

Material flow analysis

Brief description

MFA examines the flows of resources and how they change as they pass through a system. It has been applied as a tool for early recognition to identify environmental and resource management problems, to set priorities, analyse and improve the effectiveness of measures, and to design efficient material management strategies in view of sustainability (Baccini and Brunner, 1991; Brunner and Rechberger, 2004; Hendriks *et al.*, 2000). One of its interesting applications has been in optimising water and nutrient management in environmental sanitation systems in Vietnam and China (Belevi, 2002; Huang *et al.*, 2007; Montangero *et al.*, 2007).

Critique

Despite its potential, assessment of health risks and critical control points is lacking from this tool. This prevents it from providing useful information for the safe use of natural resources and reuse of waste products. For MFA and QMRA, both quantitative and qualitative knowledge are required to comprehensively assess public health risks; specifically information on the human behavioural dimensions, which may not be easy to obtain.

How would it impact the assessment of One Health initiatives?
MFA is a useful aid to understand, quantify and manage the physical process of sanitation management; it needs to be integrated with other tools to obtain a One Health perspective. For example (Nguyen-Viet *et al.*, 2009) developed a conceptual framework for integrated health and environmental assessment, combining health status, physical, socio-economic and cultural environments to improve health and minimise environmental impact. This work provides an excellent basis to visualise and quantify issues of sanitation from a One Health perspective, for further detail please see (Nguyen-viet *et al.*, 2015).

5.8 Link between the environment, economy and society

In order to examine the link between environment economy and society there is a requirement to capture relevant data. There are a range of different data that can be collected to quantify health metrics that, in turn, affect the economy and society. Box 5.1 lists considerations that influence the metrics used to quantify health and are thus, likely to impact policy decisions. The presentation of knowledge on health can be via different forms of information including: surveillance data describing what is changing in the population at a certain time-point, stories or contextual information. Methods to capture the value of knowledge are influenced by certain drivers including the context in which knowledge is acquired and temporal issues arising during the process of knowledge acquisition. In addition, other biases affect the effectiveness of knowledge acquisition and these should be considered, when measuring information (Box 5.1).

Box 5.1. Factors for consideration that influence the metrics used to quantify health and are thus, likely to impact policy decisions.

Biases in the measurement of health status

Whilst quantifying the health of populations, whether aquatic plant, human or animal, other issues which need to be considered include:

- How to conceptualise and measure information.
- The quality and methods used for surveillance.
- The time-frame and spatial scale over which information is measured.
- How data from human, animal and environmental spheres can be combined in One Health approaches.
- How differences in communication impact the collation of information.
- That behavioural changes are likely to affect the measurement of One Health as a part of the process of changing working practices.

1. Contextual circumstances

   Contextual ideas which could influence the capture of knowledge include:

   - The value of knowledge, which is defined by the objective in hand and is context-driven but which may not be ascertained until much later e.g. the USA/USSR space-race led to accelerated developments within automotive engineering.
   - The value of information, which is always relative; nothing has intrinsic value.

>>>

Box 5.1. Continued.

- The need to define the context: the audience, what they need to know, and how they will use information.
- The funders of the action: why they want to fund the acquisition of knowledge and what they are going to do with it.
- The need to think about global value because different stakeholders give different values to subjects.
- Having better information sharing platforms will decrease worldwide duplication of research.
- Compiling data is helpful from information sources, stories, and contextual information.

2. Temporal and spatial issues

   Temporal ideas which could influence the capture of knowledge include:
   - That information sharing has to be relevant within the time-period, for example using historical information for current decision-making may not be a good idea.
   - That the spatial-scale over which information will be collected is influenced by temporal issues, because of the effort involved in collecting greater amounts of data, or data at a higher resolution potentially leading to reduced effective collaboration.
   - Costing information; useful outcomes from information may not be immediately obvious.
   - The quality of data use is often not apparent when it is collected but is part of an iterative process where quality increases through time (through a spectrum of use). Only data that are valued are used; data use is a reflection of the value placed on it.
3. Biases in capturing information

   Biases which could influence the capture of knowledge include:
   - In certain circumstances, objective information can be a threat, for example under-reporting of disease to a surveillance mechanism at a national-level which could influence trading restrictions.
   - The spatial scale information is being considered at, for example country-ownership of data may be likely to create biases in multinational information sharing.
   - Multiple partners within data-capture technologies translate into complexities in the greater dataset and in communicating outcomes.
   - Awareness of risks at certain time-points e.g. improvements in public health are most often made because of epidemics.
   - How data is summarised and communicated, and what detail is lost during this process.
   - Organisational inertia e.g. not improving disease surveillance because basic mechanisms provide the information adequate for clinicians to solve the immediate issue.
4. Considerations on how to measure information

   When measuring information consider:
   - The need for standardisation, randomisation and replication of data-points to create robust research outcomes.
   - No information is still information. Interpretation is important because it means either an absence of information, or that something is not there.
   - The importance of information differs for different stakeholders e.g. Paratuberculosis in cattle, farmers versus veterinary public health professionals.
   - There is a need to measure the value of something but that might change with further knowledge e.g. mitigation of climate effects.
   - The inclusion of an evaluation process to demonstrate the increased acquisition of knowledge.

5.8.1 Social and demographic mechanisms and barriers to surveillance of health

Public health and veterinary public health approaches to characterise the impact of notifiable and not notifiable infectious diseases and target monitoring and interventions include passive surveillance systems, active surveillance and sentinel surveillance (Nsubuga *et al.*, 2006) Once information on the presence or absence of infection has been captured at appropriate temporal and geographical scales, epidemiological estimates quantifying impact can be calculated, such as for deaths from disease, the prevalence and incidence of infection, DALYs, YLDs or the change in risk of infection associated with a risk factor. It is then possible to identify the benefits, including economic benefits, for integrated surveillance of zoonoses (Babo Martins *et al.*, 2016).

The usefulness and quality of a surveillance system is dependent on the quality of data it accesses. The most important factor driving good data is its timeliness (how up-to-date it is), followed by its reliability, completeness and finally by the continuity of coverage. All of these factors are interlinked (Box 5.1).

The reliability and completeness of data is driven by their source. Sources for disease data may range from personal observations of individuals up to government approved test results. The importance given to these different sources depends on the existence, reliability and stability of infrastructures and political systems of the source countries or regions. Thus, informal sources may be especially relevant in developing regions or regions affected by war. Additionally, the informal information is available much quicker than confirmed official data. For the Global Public Health Intelligence Network (GPHIN), for example, more than 60% of reports on initial disease outbreaks may be accounted to informal sources (World Health Organization, 2016b). Governments, on the other hand, may be reluctant to share information on disease outbreaks in a timely manner, as the reporting of certain diseases may result in considerable economic impacts for the reporting country either directly (e.g. by trade restrictions) or indirectly (e.g. by losses in the tourist industry).

The timeliness of disease data depends on disease-associated factors such as the ease of diagnosis, which again is dependent upon the incubation period of disease and therefore, the likelihood of clinical signs being observed, and the specificity of clinical signs. However, the diagnosis of disease strongly depends on the existence and accessibility of a health care system and, further, the availability as well as the sensitivity and specificity of diagnostic measures including laboratory tests.

Integrated surveillance systems face the same factors as solely human or animal centred systems. However, the integration of these systems is facilitated by well-established infrastructure and communication.

5.8.2 Temporal and spatial scales over which information is measured

The systems thinking required to describe the interlinkages within the ecological dimension of One Health, means that its measurement needs to ideally use complex adaptive systems including causal linkages and feedback loops (Rüegg *et al.*, 2017). It is clear that the effects of changing One Health initiatives operate at a range of spatial scales from individual to planetary. The logistical challenges of working at multiple spatial scales include administrative barriers as created by regional, national and continental separation, which may not encapsulate the required spatial dimensions, and thus units may not map on to each other effectively for a One Health analysis.

Systems thinking involves integration of the time dimension, offering a more dynamic and objective understanding of how place affects population health (Kawachi and Berkman, 2003). To date, methods to analyse time geographic data have been largely limited to data visualisation and exploratory analysis.

Consideration of spatial and temporal scales is especially important when consolidating zoonotic information from multiple sources (Asokan and Asokan, 2015). An emerging but not necessarily 'new' field which describes One Health working is that of 'big data' approaches, meaning using a mixture of structured and unstructured data, much of which may have been collated for non-specific purposes, to provide an insight to identify and intervene on the determinants of population health (Mooney *et al.*, 2015). However, such approaches need to be treated with caution, as data is likely to be biased and needs to be properly understood in order to be interpreted correctly (Lazer *et al.*, 2014).

5.8.3 Social and demographic differences in health communication and education

Differences in communication between collaborators as a result of cultural, lexical and idiosyncratic diversity are a particular difficulty when trying to measure health status using a One Health approach. This is illustrated in an excellent manner when we consider zoonoses. A number of the most important zoonoses are reportable to organisations such as the OIE (Ben Jebara *et al.*, 2012), however if examining the overall impact of zoonoses on the global burden of human disease, their impact in humans is superseded by that of ailments such as ischemic heart disease and cerebrovascular disease (Murray *et al.*, 2012), meaning that non-communicable diseases are considered of greater economic importance. The true public health and economic impact of zoonoses is likely underestimated, mainly due to under-reporting of disease events (Grace *et al.*, 2012). In industrialised countries, food-borne zoonoses such as *Campylobacter, Escherichia coli, Listeria* and *Salmonella* are reported most (Anonymous, 2015). However, non-food-borne zoonoses also present an important issue in countries with an absence of animal health monitoring programs (Zhou, 2012).

Continuing and on-going education in health is vital to manage zoonotic disease transmission including disease surveillance, vaccination and other prevention methods (Boete and Morand, 2016; Chomel and Sun, 2011). Higher levels of education increases the health of individuals (via disease avoidance/management or by individuals making healthier life choices e.g. Pennanen *et al.*, 2011), and contributes to greater health-resource allocation (Conti *et al.*, 2010). To achieve a real One Health approach, collaboration among veterinary, human medicine, environmental, wildlife and public health organisations, both public and private, are necessary to establish groups and centres to improve health education, research and training and facilitate behavioural change.

There is an increasing recognition of the ecological drivers of health (Romanelli *et al.*, 2015). We have outlined some of the important components that facilitate the measurement of environmental sustainability through a collection of indices which inform economic sustainability and, in turn, ecological sustainability. There is an acceptance that one of the fundamental drivers of food security and health is the status of the environment in which primary production occurs. This is because the ecological dimension is described by ecological, economic and social elements within an equilateral triangle; overall (One) health is maximised by balancing these elements and all have function in maintaining the equilibrium. Pathogens and disease present a challenge but also opportunity for One Health, since this challenge can stimulate individual and population resilience (Cumming and Cumming, 2015).

5.8.4 Link with ecological aspects of One Health and the economy

The importance of the ecological components of One Health from an economic perspective is acknowledged in a recent operational framework published by the World Bank (Berthe *et al.*, 2018). While, the obvious focus on the human-animal-environment interface relates to the transmission of zoonotic infections, there is also an appreciation of other challenges related to food and water safety and security. Approaches to maximise the benefits for targeted investment in challenges related to prevent, prepare, detect, respond to, and recover from issues like diseases are outlined with the associated One Health benefits.

5.9 Conclusions

One way individuals interact with the environment is by oral uptake of environmental factors i.e. by eating. These substances do not usually exist in pure form, but are contained in more complex structures like plant material or meat from animals (or in mixed solutions such as during the nourishment of plants). Apart from harmful substances, food items may also be contaminated by biological contaminants such as pathogens. The human diet, for example, is to varying degrees depending on sociocultural background, resource availability, and personal preference, based on both plants and animals. As a species, humans can therefore be categorised as omnivores. Consequently, the food sector deals with plant and livestock health, but it does not acknowledge the environment as an integral part of the food web. Therefore, a major challenge to achieving a One Health in the future is to examine the trade-off from producing food (Figure 5.1) and look for synergies where possible. This is related to food quality for both animals and humans but also related to zoonoses emergence (Foley *et al.*, 2005; Jones *et al.*, 2013; Purvis *et al.*, 2012).

Contamination of plants or animals as food sources may occur before harvesting, due to uptake of contaminants during their respective lifecycles, or after harvesting, during processing. Contaminants can include infectious pathogens, but also chemical elements such as heavy metals or chemical compounds (e.g. toxins -algal, bacterial), nanoplastics, pesticides, antibiotics, hormones, etc.). As omnivores, humans are placed at the top of the food chain. Humans may be exposed to contaminants in food of plant or animal origin. These contaminants have either recently entered the food chain through environmental contamination (e.g. soil, fresh- or salt-water or air) or entered at an earlier stage, i.e. at a lower trophic level, and have subsequently been passed on and accumulated via feed plants and prey animals (through the process of biomagnification).

Food security is currently defined as 'a situation that exists when all people, at all times, have physical, social and economic access to sufficient, safe and nutritious food that meets their dietary needs and food preferences for an active and healthy life' (Food and Agriculture Organization of the United Nations, 2002), i.e. a secure supply of food. It indicates that livestock is included as an integral part of global food security. This means that access to healthy livestock is an important factor to be considered when securing food sources for people around the world. Balancing all ecological components (water, air, soil, biodiversity) is critically important for food security, and both food safety and food security should be interlinked for a One Health approach. It is critical that a focus on food quality is not just on outcomes such as food nutrients or food safety (e.g. maximum residue levels) but also on how one-health-ness interacts with the food system at different stages of the value chain (production, processing, transport and consumption of food) to affect food quality.

Given the growing human population, a set of One Health indicators which would capture the link between human health, animal health and ecological health is of importance to inform future global developments (Haines *et al.*, 2018). The criteria which need to be added include factors such as One Health relevance, economic importance, accuracy, amongst others (see Box 5.1). Existing networks do exist, e.g. INDEPTH (Streatfield *et al.*, 2014), however, their scope must broaden and include factors important to ecological aspects of our environment as One Health is going to be particularly important as we enter a period of unprecedented anthropogenic influence on global ecosystems.

Acknowledgements

This chapter is based upon work from COST Action 'Network for Evaluation of One Health' (TD1404), supported by COST (European Cooperation in Science and Technology).

The authors would like to thank Dr Martin Hitziger (University of Zurich) and Prof. Jakob Zinsstag (Swiss Tropical and Public Health Institute) for editorial input to this work. Dr Jarlath Nally (United States Department of Agriculture), Prof. Jim Scudamore (University of Liverpool) and Dr Serge Morand (CIRAD) reviewed and provided useful further suggestions for the work, so the authors would also like to acknowledge their contributions.

References

Abakar, M.F., Schelling, E., Bechir, M., Ngandolo, B.N., Pfister, K., Alfaroukh, I.O., Hassane, H.M. and Zinsstag, J., 2016. Trends in health surveillance and joint service delivery for pastoralists in West and Central Africa. Rev Sci Tech 35: 683-691.

Adams, A. and Thompson, K.D., 2011. Development of diagnostics for aquaculture: challenges and opportunities. Aquacult Res 42: 93-102.

Ahmadi, A., Abbaspour, M., Arjmandi, R. and Abedi, Z., 2015. Air Quality Risk Index (AQRI) and its application for a megacity. Int J Environ Sci Technol 12: 3773-3780.

Ahmed, A.M., Younis, E.E.A., Osman, S.A., Ishida, Y., El-Khodety, S.A. and Shimamoto, T., 2009. Genetic analysis of antimicrobial resistance in Escherichia coli isolated from diarrheic neonatal calves. Vet Microbiol 136: 397-402.

Alonso Chavez, V., Parnell, S. and Van den Bosch, F., 2016. Monitoring invasive pathogens in plant nurseries for early-detection and to minimise the probability of escape. J Theor Biol 407: 290-302.

An, Y.J., Yoon, C.G., Jung, K.W. and Ham, J.H., 2007. Estimating the microbial risk of E. coli in reclaimed wastewater irrigation on paddy field. Environ Monit Assess 129: 53-60.

Anderson, P.K., Cunningham, A.A., Patel, N.G., Morales, F.J., Epstein, P.R. and Daszak, P., 2004. Emerging infectious diseases of plants: pathogen pollution, climate change and agrotechnology drivers. Trends Ecol Evol 19: 535-544.

Anderson, R.M. and May, R.M., 1978. Regulation and stability of host-parasite population interactions. Regulatory processes. J Anim Ecol 47: 219-247.

Andersson, D.I. and Hughes, D., 2012. Evolution of antibiotic resistance at non-lethal drug concentrations. Drug Resist Update 15: 162-172.

Andrews, S.S., Karlen, D.L. and Mitchell, J.P., 2002. A comparison of soil quality indexing methods for vegetable production systems in Northern California. Agric Ecosyst Environ 90: 25-45.

Anonymous, 1999. Health Impact Assessment: main concepts and suggested approach. Gothenburg consensus paper. European Centre for Health Policy, Brussels, Belgium. Available at: https://tinyurl.com/y7mn4to6.

Anonymous, 2000. EC Council Directive 2000/29/EC of 8 May 2000 on protective measures against the introduction into the community of organisms harmful to plants or plant products and against their spread within the community. European Commission, Brussels, Belgium.

Anonymous, 2015. The 2013 joint ECDC/EFSA report on trends and sources of zoonoses, zoonotic agents and food-borne outbreaks published. European Centre for Disease Prevention and Control (ECDC), Stockholm, Sweden.

Anonymous, 2017. Meat traceability, drug residues and withdrawal periods. Available at: https://tinyurl.com/y8crhjww.

Ao, T.T., Feasey, N.A., Gordon, M.A., Keddy, K.H., Angulo, F.J. and Crump, J.A., 2015. Global burden of invasive nontyphoidal salmonella disease, 2010. Emerg Infect Diseases 21: 941-949.

Arnold, K.E., Williams, N.J. and Bennett, M., 2016. 'Disperse abroad in the land': the role of wildlife in the dissemination of antimicrobial resistance. Biol Letters 12(8): 20160137.

Asokan, G.V. and Asokan, V., 2015. Leveraging 'big data' to enhance the effectiveness of 'one health' in an era of health informatics. J Epidemiol Glob Health 5: 311-314.

Babo Martins, S., Rushton, J. and Stark, K.D., 2016. Economic assessment of zoonoses surveillance in a 'One Health' context: a conceptual framework. Zoonos Publ Health 63: 386-395.

Baccini, P. and Brunner, P.H., 1991. Metabolism of the anthroposphere. Springer, Berlin, Germany.

Belevi, H., 2002. Material flow analysis as a strategic planning tool for regional waste water and solid waste management. Available at: https://tinyurl.com/y7lg87pv.

Ben Jebara, K., Caceres, P., Berlingieri, F. and Weber-Vintzel, L., 2012. Ten years' work on the World Organisation for Animal Health (OIE) Worldwide Animal Disease Notification System. Prev Vet Med 107: 149-159.

Bentley, J.W., Boa, E., Danielsen, S., Franco, P., Antezana, O., Villarroel, B., Rodríguez, H., Ferrrufino, J., Franco, J., Pereira, R., Herbas, J., Díaz, O., Lino, V., Villarroel, J., Almendras, F. and Colque, S., 2009. Plant health clinics in Bolivia 2000-2009: operations and preliminary results. Food Secur 1(3): 371-386.

Berthe, F.C.J., Bouley, T., Karesh, W.B., Le Gall, F.G., Machalaba, C.C., Plante, C.A. and Seifman, R.M., 2018. Operational framework for strengthening human, animal and environmental public health systems at their interface. World Bank Group, Washington, DC, USA. Available at: https://tinyurl.com/yco8ak3m.

Binns, M.R., Nyrop, J.P. and Van der Werf, W., 2000. Sampling and monitoring in crop protection: the theoretical basis for developing practical decision guides. CABI, Wallingford, UK.

Boa, E., 2009. How the global plant clinic began. Outlook Pest Manag 20(3): 112-116.

Boa, E., Danielsen, S. and Haesen, S., 2015. Better together: identifying the benefits of a closer integration between plant health, agriculture and One Health. In: Zinsstag, J., Schelling, E., Waltner-Townes, D., Whittaker, M. and Tanner, M. (eds.) One Health: the theory and practice of integrated health approaches. CABI, Wallingford, UK.

Bock, C.H., Poole, G.H., Parker, P.E. and Gottwald, T.R., 2010. Plant disease severity estimated visually, by digital photography and image analysis, and by hyperspectral imaging. Crit Rev Plant Sci 29: 59-107.

Boete, C. and Morand, S., 2016. Bats and academics: how do scientists perceive their object of study? PLoS ONE 11: e0165969.

Boitani, L., Mace, G.M. and Rondinini, C., 2015. Challenging the scientific foundations for an IUCN red list of ecosystems. Conserv Lett 8: 125-131.

Boyd, I.L., Freer-Smith, P.H., Gilligan, C.A. and Godfray, H.C.J., 2013. The consequence of tree pests and diseases for ecosystem services. Science 342(6160): 1235773.

Brevik, E.C. and Sauer, T.J., 2015. The past, present, and future of soils and human health studies. Soil 1: 35-46.

Brugere, C., Onuigbo, D.M. and Morgan, K.L., 2017. People matter in animal disease surveillance: challenges and opportunities for the aquaculture sector. Aquaculture 467: 158-169.

Brunner, P. and Rechberger, H., 2004. Practical handbook of material flow analysis. CRC Press, Boca Raton, FL, USA.

Burkhard, B., Kroll, F., Müller, F. and Windhorst, W., 2009. Landscapes' capacities to provide ecosystem services – a concept for land-cover based assessments. Landscape Online 15: 1-22.

Butchart, S.H.M., Walpole, M., Collen, B., Van Strien, A., Scharlemann, J.P.W., Almond, R.E.A., Baillie, J.E.M., Bomhard, B., Brown, C., Bruno, J., Carpenter, K.E., Carr, G.M., Chanson, J., Chenery, A.M., Csirke, J., Davidson, N.C., Dentener, F., Foster, M., Galli, A., Galloway, J.N., Genovesi, P., Gregory, R.D., Hockings, M., Kapos, V., Lamarque, J.F., Leverington, F., Loh, J., Mcgeoch, M.A., Mcrae, L., Minasyan, A., Morcillo, M.H., Oldfield, T.E.E., Pauly, D., Quader, S., Revenga, C., Sauer, J.R., Skolnik, B., Spear, D., Stanwell-Smith, D., Stuart, S.N., Symes, A., Tierney, M., Tyrrell, T.D., Vie, J.C. and Watson, R., 2010. Global biodiversity: indicators of recent declines. Science 328: 1164-1168.

Calderón, R., Navas-Cortés, J.A. and Zarco-Tejada, P.J., 2015. Early detection and quantification of verticillium wilt in olive using hyperspectral and thermal imagery over large areas. Remot Sens 7: 5584-5610.

Cardoen, S., Van Huffel, X., Berkvens, D., Quoilin, S., Ducoffre, G., Saegerman, C., Speybroeck, N., Imberechts, H., Herman, L., Ducatelle, R. and Dierick, K., 2009. Evidence-based semiquantitative methodology for prioritization of foodborne zoonoses. Foodborn Pathog Dis 6: 1083-1096.

Carroll, D., Wang, J., Fanning, S. and Mcmahon, B.J., 2015. Antimicrobial resistance in wildlife: implications for public health. Zoonos Publ Health 62: 534-542.

Charron, D.F. (ed.), 2012a. Ecohealth research in practice. Springer, New York, NY, USA.

Charron, D.F., 2012b. Ecosystem approaches to health for a global sustainability agenda. Ecohealth 9: 256-266.

Chomel, B.B. and Sun, B., 2011. Zoonoses in the bedroom. Emerg Infect Dis 17: 167-172.

Collen, B., Loh, J., Whitmee, S., Mcrae, L., Amin, R. and Baillie, J.E.M., 2009. Monitoring change in vertebrate abundance: the living planet index. Conserv Biol 23: 317-327.

Commission of the European Communities, 1981. Commission Regulation (EEC) No 2930/81 of 12 October 1981 adopting additional provisions for the application of the Community scale for the classification of carcases of adult bovine animals. Commission of the European Communities, Brussels, Belgium.

Conti, G., Heckman, J. and Urzua, S., 2010. The education-health gradient. Am Econ Rev 100: 234-238.

Crone, E.E., Menges, E.S., Ellis, M.M., Bell, T., Bierzychudek, P., Ehrlen, J., Kaye, T.N., Knight, T.M., Lesica, P., Morris, W.F., Oostermeijer, G., Quintana-Ascencio, P.F., Stanley, A., Ticktin, T., Valverde, T. and Williams, J.L., 2011. How do plant ecologists use matrix population models? Ecol Lett 14: 1-8.

Cumming, D.H.M. and Cumming, G.S., 2015. One Health: an ecological and conservation perspective. In: Zinsstag, J., Schelling, E., Waltner-Toews, D., Whittaker, M. and Tanner, M. (eds.) One Health: the theory and practice of integrated health approaches. CABI, Wallingford, UK.

Danielsen, S., 2013. Including plant health in the 'One Health' concept – in theory and in Uganda. In: Olson, A., Ørnbjerg, N. and Winkel, K. (eds.) In a success story in Danish development aid (1964-2012). University of Copenhagen, Copenhagen, Denmark.

Diallo, M.B., Anceno, A.J., Tawatsupa, B., Houpt, E.R., Wangsuphachart, V. and Shipin, O.V., 2008. Infection risk assessment of diarrhea-related pathogens in a tropical canal network. Sci Total Environ 407: 223-232.

Diekmann, O. and Heesterbeek, J.A.P., 2000. Mathematical epidemiology of infectious diseases: model building, analysis and interpretation. John Wiley and Sons, Chichester, UK.

Diekmann, O., Dietz, K. and Heesterbeek, J.A., 1991. The basic reproduction ratio for sexually transmitted diseases: I. theoretical considerations. Math. Biosci. 107(2): 325-339.

Diekmann, O., Heesterbeek, H. and Britton, T., 2013. Mathematical tools for understanding infectious disease dynamics. Princeton University Press. Princeton, NJ, USA.

Diekmann, O., Heesterbeek, J.A. and Metz, J.A., 1990. On the definition and the computation of the basic reproduction ratio R0 in models for infectious diseases in heterogeneous populations. J Math Biol 28: 365-382.

Diekmann, O., Heesterbeek, J.A. and Roberts, M.G., 2010. The construction of next-generation matrices for compartmental epidemic models. J R Soc Interface 7: 873-885.

Dietz, K., Heesterbeek, J.A. and Tudor, D.W., 1993. The basic reproduction ratio for sexually transmitted diseases. Part 2. effects of variable HIV infectivity. Math. Biosci. 117 (1-2): 35-47.

Discontools, 2012. Disease database. Available at: https://www.discontools.eu/database.html.

Dudka, S. and Miller, W.P., 1999. Accumulation of potentially toxic elements in plants and their transfer to human food chain. J Environ Sci Health B 34: 681-708.

Dufour, B., Moutou, F., Hattenberger, A.M. and Rodhain, F., 2008. Global changes: impact, management, risk approach and health measures – the case of Europe. Rev Sci Tech 27: 529-540.

Eisenberg, J.N., Moore, K., Soller, J.A., Eisenberg, D. and Colford, J.M., 2008. Microbial risk assessment framework for exposure to amended sludge projects. Environ Health Perspect 116: 727-733.

European Parliament and Council of the European Union, 2000. Directive 2000/60/EC of the European Parliament and of the Council of 23 October 2000 establishing a framework for Community action in the field of water policy. Available at: https://tinyurl.com/ya5vj6sg.

Fisher, M.C., Henk, D.A., Briggs, C.J., Brownstein, J.S., Madoff, L.C., Mccraw, S.L. and Gurr, S.J., 2012. Emerging fungal threats to animal, plant and ecosystem health. Nature 484: 186-194.

Fletcher, J., Franz, D. and Leclerc, J.E., 2009. Healthy plants: necessary for a balanced 'One Health' concept. Vet Italiana 45: 79-95.

Fletcher, J., Leach, J.E., Eversole, K. and Tauxe, R., 2013. Human pathogens on plants: designing a multidisciplinary strategy for research. Phytopathol 103: 306-315.

Flood, J., 2010. The importance of plant health to food security. Food Secur 2: 215-231.

Fofana, A. and Baulcomb, C., 2012. Counting the cost of farmed salmonid diseases. J Appl Aquacult 24: 118-136.

Foley, J.A., Defries, R., Asner, G.P., Barford, C., Bonan, G., Carpenter, S.R., Chapin, F.S., Coe, M.T., Daily, G.C., Gibbs, H.K., Helkowski, J.H., Holloway, T., Howard, E.A., Kucharik, C.J., Monfreda, C., Patz, J.A., Prentice, I.C., Ramankutty, N. and Snyder, P.K., 2005. Global consequences of land use. Science 309: 570-574.

Food and Agriculture Organization of the United Nations (FAO), 2002. The state of food insecurity in the world 2001. FAO, Rome, Italy. Available at: https://tinyurl.com/je3jrne.

Food and Agriculture Organization of the United Nations (FAO), 2014. The state of world fisheries and aquaculture. Opportunities and challenges. FAO, Rome, Italy. Available at: https://tinyurl.com/p5w5yhl.

Fosse, J., Seegers, H. and Magras, C., 2008. Foodborne zoonoses due to meat: a quantitative approach for a comparative risk assessment applied to pig slaughtering in Europe. Vet Res 39: 1.

Freer-Smith, P.H. and Webber, J.F., 2015. Tree pests and diseases: the threat to biodiversity and the delivery of ecosystem services. Biodivers Conserv 26: 3167-3181.

Gilbert, W. and Rushton, J., 2016. Incentive perception in livestock disease control. J Agric Econ 69: 243-261.

Gilligan, C.A., 2008. Sustainable agriculture and plant diseases: an epidemiological perspective. Philos Trans R Soc Lond B Biol Sci 363: 741-759.

Giraudoux, P., Pleydell, D., Raoul, F., Quere, J.P., Wang, Q., Yang, Y., Vuitton, D.A., Qiu, J., Yang, W. and Craig, P.S., 2006. Transmission ecology of *Echinococcus multilocularis*: what are the ranges of parasite stability among various host communities in China? Parasitol Int 55: S237-246.

Gonzalez-Zorn, B. and Escudero, J.A., 2012. Ecology of antimicrobial resistance: humans, animals, food and environment. Int Microbiol 15: 101-109.

Grace, D., Gilbert, J., Randolph, T. and Kang'ethe, E., 2012. The multiple burdens of zoonotic disease and an ecohealth approach to their assessment. Trop Anim Health Pro 44: S67-S73.

Groombridge, B., 1992. Global biodiversity: status of the world's living resources. Chapman & Hall, London, UK.

Gustafson, L., Klotins, K., Tomlinson, S., Karreman, G., Cameron, A., Wagner, B., Remmenga, M., Bruneau, N. and Scott, A., 2010. Combining surveillance and expert evidence of viral hemorrhagic septicemia freedom: a decision science approach. Prev Vet Med 94: 140-153.

Haas, C.N., Rose, J.B. and Gerba, C.P., 2014. Quantitative microbial risk assessment. Available at: https://tinyurl.com/y7wo7wzr.

Haines, A., Hanson, C. and Ranganathan, J., 2018. Planetary health watch: integrated monitoring in the anthropocene epoch. Lancet Planet Health 2: e141-e143.

Hajat, A., Kaufman, J.S., Rose, K.M., Siddiqi, A. and Thomas, J.C., 2010. Do the wealthy have a health advantage? Cardiovascular disease risk factors and wealth. Soc Sci Med 71: 1935-1942.

Hall, L.M., Munro, L.A., Wallace, I.S., Mcintosh, R., Macneish, K. and Murray, A.G., 2014. An approach to evaluating the reliability of diagnostic tests on pooled groups of infected individuals. Prev Vet Med 116: 305-312.

Halpern, B.S., Longo, C., Hardy, D., Mcleod, K.L., Samhouri, J.F., Katona, S.K., Kleisner, K., Lester, S.E., O'leary, J., Ranelletti, M., Rosenberg, A.A., Scarborough, C., Selig, E.R., Best, B.D., Brumbaugh, D.R., Chapin, F.S., Crowder, L.B., Daly, K.L., Doney, S.C., Elfes, C., Fogarty, M.J., Gaines, S.D., Jacobsen, K.I., Karrer, L.B., Leslie, H.M., Neeley, E., Pauly, D., Polasky, S., Ris, B., St Martin, K., Stone, G.S., Sumaila, U.R. and Zeller, D., 2012. An index to assess the health and benefits of the global ocean. Nature 488: 615-620.

Havelaar, A.H., Evers, E.G. and Nauta, M.J., 2018. Challenges of quantitative microbial risk assessment at EU level. Trends Food Sci Tech 19: S26-S33.

Heesterbeek, J.A. and Roberts, M.G., 2007. The type-reproduction number T in models for infectious disease control. Math Biosci 206: 3-10.

Hendriks, C., Obernosterer, R., Müller, D., Kytzia, S., Baccini, P. and Brunner, P.H., 2000. Material flow analysis: a tool to support environmental policy decision making. case-studies on the city of Vienna and the Swiss lowlands. Local Environ 5(3): 311-328.

Horsfall, J.G. and Barratt, R.W., 1945. An improved grading system for measuring plant diseases. Phytopathology 35(8): 655.

Howard, G., Pedley, S. and Tibatemwa, S., 2006. Quantitative microbial risk assessment to estimate health risks attributable to water supply: can the technique be applied in developing countries with limited data? J Water Health 4: 49-65.

Huang, D.B., Bader, H.P., Scheidegger, R., Schertenleib, R. and Gujer, W., 2007. Confronting limitations: new solutions required for urban water management in Kunming City. J Environ Manage 84: 49-61.

Hudson, P., Rizzoli, A., Grenfell, B., Heesterbeek, H. and Dobson, A., 2002. The ecology of wildlife diseases. Oxford University Press, Oxford, UK.

Hughes, G., Mcroberts, N., Madden, L.V. and Gottwald, T.R., 1997. Relationships between disease incidence at two levels in a spatial hierarchy. Phytopathol 87: 542-550.

International Union for Conservation of Nature and Natural Resources, 2017. The IUCN red list of threatened species. Version 2017-1. Available at: http://www.iucnredlist.org.

Jacobs, S., Burkhard, B., Van Daele, T., Staes, J. and Schneiders, A., 2015. 'The Matrix Reloaded': a review of expert knowledge use for mapping ecosystem services. Ecol Model 295: 21-30.

James, W.C., 1971. An illustrated series of assessment keys for plant diseases, their preparation and usage. Canadian Plant Disease Survey 51.

James, W.C., 1974. Assessment of plant diseases and losses. Annu Rev Phytopathol 121: 27-48.

Jean-Richard, V., Crump, L., Moto Daugla, D., Hattendorf, J., Schelling, E. and Zinsstag, J., 2014. The use of mobile phones for demographic surveillance of mobile pastoralists and their animals in Chad: proof of principle. Glob Health Action 7: 23209.

Jones, B.A., Grace, D., Kock, R., Alonso, S., Rushton, J., Said, M.Y., Mckeever, D., Mutua, F., Young, J., Mcdermott, J. and Pfeiffer, D.U., 2013. Zoonosis emergence linked to agricultural intensification and environmental change. Proc Natl Acad Sci USA 110: 8399-8404.

Karesh, W.B., Dobson, A., Lloyd-Smith, J.O., Lubroth, J., Dixon, M.A., Bennett, M., Aldrich, S., Harrington, T., Formenty, P., Loh, E.H., Machalaba, C.C., Thomas, M.J. and Heymann, D.L., 2012. Zoonoses 1 ecology of zoonoses: natural and unnatural histories. Lancet 380: 1936-1945.

Kawachi, I. and Berkman, L.F., 2003. Neighborhoods and health date. Available at: https://tinyurl.com/yd4wubxp.

Keeling, M.J. and Rohani, P., 2008. Modeling infectious diseases in humans and animals. Princeton University Press, Princeton, NJ, USA.

Keesing, F., Belden, L.K., Daszak, P., Dobson, A., Harvell, C.D., Holt, R.D., Hudson, P., Jolles, A., Jones, K.E., Mitchell, C.E., Myers, S.S., Bogich, T. and Ostfeld, R.S., 2010. Impacts of biodiversity on the emergence and transmission of infectious diseases. Nature 468: 647-652.

Keith, A.M., Schmidt, O. and Mcmahon, B.J., 2016. Soil stewardship as a nexus between ecosystem services and One Health. Ecosyst Serv 17: 40-42.

Keith, D.A., Rodriguez, J.P., Rodriguez-Clark, K.M., Nicholson, E., Aapala, K., Alonso, A., Asmussen, M., Bachman, S., Basset, A., Barrow, E.G., Benson, J.S., Bishop, M.J., Bonifacio, R., Brooks, T.M., Burgman, M.A., Comer, P., Comin, F.A., Essl, F., Faber-Langendoen, D., Fairweather, P.G., Holdaway, R.J., Jennings,

M., Kingsford, R.T., Lester, R.E., Mac Nally, R., Mccarthy, M.A., Moat, J., Oliveira-Miranda, M.A., Pisanu, P., Poulin, B., Regan, T.J., Riecken, U., Spalding, M.D. and Zambrano-Martinez, S., 2013. Scientific foundations for an IUCN red list of ecosystems. PLoS ONE 8: e62111.

Kock, R., Queenan, K., Garnier, J., Rosenbaum Nielsen, L., Buttigieg, S., De Meneghi, D., Holmberg, M., Zinsstag, J., Rüegg, S. and Häsler, B., 2018. Health solutions: theoretical foundations of the shift from sectoral to integrated systems. Chapter 2. In: Rüegg, S.R., Häsler, B. and Zinsstag, J. (eds.) Integrated approaches to health: a handbook for the evaluation of One Health. Wageningen Academic Publishers, Wageningen, the Netherlands, pp. 22-37.

Krebs, C.J., 1972. Ecology. Harper & Row, New York, NY, USA.

Lambin, E.F., Tran, A., Vanwambeke, S.O., Linard, C. and Soti, V., 2010. Pathogenic landscapes: interactions between land, people, disease vectors, and their animal hosts. Int J Health Geogr 9: 54.

Large, E.C., 1966. Measuring plant disease. Annu Rev Phytopathol 4(1): 9-26.

Laxminarayan, R., Duse, A., Wattal, C., Zaidi, A.K.M., Wertheim, H.F.L., Sumpradit, N., Vlieghe, E., Hara, G.L., Gould, I.M., Goossens, H., Greko, C., So, A.D., Bigdeli, M., Tomson, G., Woodhouse, W., Ombaka, E., Peralta, A.Q., Qamar, F.N., Mir, F., Kariuki, S., Bhutta, Z.A., Coates, A., Bergstrom, R., Wright, G.D., Brown, E.D. and Cars, O., 2013. Antibiotic resistance – the need for global solutions. Lancet Infect Dis 13: 1057-1098.

Lazer, D., Kennedy, R., King, G. and Vespignani, A., 2014. Big data. The parable of Google flu: traps in big data analysis. Science 343: 1203-1205.

Leslie, P.H., 1945. On the use of matrices in certain population mathematics. Biometrika 33: 183-212.

Leslie, P.H., 1948. Some further notes on the use of matrices in population mathematics. Biometrika 35: 213-245.

Lodge, D.M., Turner, C.R., Jerde, C.L., Barnes, M.A., Chadderton, L., Egan, S.P., Feder, J.L., Mahon, A.R. and Pfrender, M.E., 2012. Conservation in a cup of water: estimating biodiversity and population abundance from environmental DNA. Mol Ecol 21: 2555-2558.

Macleod, A., Pautasso, M., Jeger, M.J. and Haines-Young, R., 2010. Evolution of the international regulation of plant pests and challenges for future plant health. Food Secur 2: 49-70.

Madden, L.V. and Hughes, G., 1995. Plant-disease incidence – distribution, heterogeneity, and temporal analysis. Annu Rev Phytopathol 33: 529-564.

Madden, L.V. and Hughes, G., 1999. An effective sample size for predicting plant disease incidence in a spatial hierarchy. Phytopathol 89: 770-781.

Madden, L.V., Hughes, G. and Bosch, F.V.D., 2007. The study of plant disease epidemics. American Phytopathological Society, St. Paul, MN, USA.

Madoff, L.C. and Li, A., 2014. Web-based surveillance systems for human, animal, and plant diseases. Microbiol Spectr 2: Oh-0015-2012.

Manlove, K.R., Walker, J.G., Craft, M.E., Huyvaert, K.P., Joseph, M.B., Miller, R.S., Nol, P., Patyk, K.A., O'Brien, D., Walsh, D.P. and Cross, P.C., 2016. 'One Health' or three? Publication Silos among the One Health disciplines. PLoS Biol 14: e1002448.

Mara, D.D., Sleigh, P.A., Blumenthal, U.J. and Carr, R.M., 2007. Health risks in wastewater irrigation: comparing estimates from quantitative microbial risk analyses and epidemiological studies. J Water Health 5: 39-50.

Marshall, B.M. and Levy, S.B., 2011. Food animals and antimicrobials: impacts on human health. Clin. Microbiol Rev 24: 718-733.

Mastin, A.J., Van den Bosch, F., Gottwald, T.R., Alonso Chavez, V. and Parnell, S.R., 2017. A method of determining where to target surveillance efforts in heterogeneous epidemiological systems. PLoS Comput Biol 13(8): e1005712.

May, R.M. and Anderson, R.M., 1978. Regulation and stability of host-parasite population interactions. Destabilizing processes. J Anim Eco 47: 249-267.

Mcbratney, A., Field, D.J. and Koch, A., 2014. The dimensions of soil security. Geoderma 213: 203-213.

Mckenzie, J., Simpson, H. and Langstaff, I., 2007. Development of methodology to prioritise wildlife pathogens for surveillance. Prev Vet Med 81: 194-210.

Mcroberts, N., Hughes, G. and Madden, L.V., 2003. The theoretical basis and practical application of relationships between different disease intensity measurements in plants. Ann Appl Biol 142: 191-211.

Meentemeyer, R.K., Dorning, M.A., Vogler, J.B., Schmidt, D. and Garbelotto, M., 2015. Citizen science helps predict risk of emerging infectious disease. Front Ecol Environ 13: 189-194.

Mikaberidze, A., Mundt, C.C. and Bonhoeffer, S., 2016. Invasiveness of plant pathogens depends on the spatial scale of host distribution. Ecol Appl 26: 1238-1248.

Miller, S.A., Beed, F.D. and Harmon, C.L., 2009. Plant disease diagnostic capabilities and networks. Annu Rev Phytopathol 47: 15-38.

Montangero, A., Le, C., Nguyen, V.A., Vu, D.T., Pham, T.N. and Belevi, H., 2007. Optimising water and phosphorus management in the urban environmental sanitation system of Hanoi, Vietnam. Sci Total Environ 384: 55-66.

Mooney, S.J., Westreich, D.J. and El-Sayed, A.M., 2015. Commentary: epidemiology in the era of big data. Epidemiology 26: 390-394.

Morelli, J., 2011. Environmental sustainability: a definition for environmental professionals. J Environ Sustain 1: 2.

Munoz-Erickson, T.A., Aguilar-Gonzalez, B. and Sisk, T.D., 2007. Linking ecosystem health indicators and collaborative management: a systematic framework to evaluate ecological and social outcomes. Ecol Soc 12: 6.

Murray, A.G., Wardeh, M. and Mcintyre, K.M., 2016. Using the H-index to assess disease priorities for salmon aquaculture. Prev Vet Med 126: 199-207.

Murray, C.J.L., Vos, T., Lozano, R., Naghavi, M., Flaxman, A.D., Michaud, C., Ezzati, M., Shibuya, K., Salomon, J.A., Abdalla, S., Aboyans, V., Abraham, J., Ackerman, I., Aggarwal, R., Ahn, S.Y., Ali, M.K., Alvarado, M., Anderson, H.R., Anderson, L.M., Andrews, K.G., Atkinson, C., Baddour, L.M., Bahalim, A.N., Barker-Collo, S., Barrero, L.H., Bartels, D.H., Basanez, M.G., Baxter, A., Bell, M.L., Benjamin, E.J., Bennett, D., Bernabe, E., Bhalla, K., Bhandari, B., Bikbov, B., Bin Abdulhak, A., Birbeck, G., Black, J.A., Blencowe, H., Blore, J.D., Blyth, F., Bolliger, I., Bonaventure, A., Boufous, S.A., Bourne, R., Boussinesq, M., Braithwaite, T., Brayne, C., Bridgett, L., Brooker, S., Brooks, P., Brugha, T.S., Bryan-Hancock, C., Bucello, C., Buchbinder, R., Buckle, G., Budke, C.M., Burch, M., Burney, P., Burstein, R., Calabria, B., Campbell, B., Canter, C.E., Carabin, H., Carapetis, J., Carmona, L., Cella, C., Charlson, F., Chen, H.L., Cheng, A.T.A., Chou, D., Chugh, S.S., Coffeng, L.E., Colan, S.D., Colquhoun, S., Colson, K.E., Condon, J., Connor, M.D., Cooper, L.T., Corriere, M., Cortinovis, M., De Vaccaro, K.C., Couser, W., Cowie, B.C., Criqui, M.H., Cross, M., Dabhadkar, K.C., Dahiya, M., Dahodwala, N., Damsere-Derry, J., Danaei, G., Davis, A., De Leo, D., Degenhardt, L., Dellavalle, R., Delossantos, A., Denenberg, J., Derrett, S., Des Jarlais, D.C., Dharmaratne, S.D., Dherani, M., Diaz-Torne, C., Dolk, H., Dorsey, E.R., Driscoll, T., Duber, H., Ebel, B., Edmond, K., Elbaz, A., Ali, S.E., Erskine, H., Erwin, P.J., Espindola, P., Ewoigbokhan, S.E., Farzadfar, F., Feigin, V., Felson, D.T., Ferrari, A., Ferri, C.P., Fèvre, E.M., Finucane, M.M., Flaxman, S., Flood, L., Foreman, K., Forouzanfar, M.H., Fowkes, F.G., Fransen, M., Freeman, M.K., Gabbe, B.J., Gabriel, S.E., Gakidou, E., Ganatra, H.A., Garcia, B., Gaspari, F., Gillum, R.F., Gmel, G., Gonzalez-Medina, D., Gosselin, R., Grainger, R., Grant, B., Groeger, J., Guillemin, F., Gunnell, D., Gupta, R., Haagsma, J., Hagan, H., Halasa, Y.A., Hall, W., Haring, D., Haro, J.M., Harrison, J.E., Havmoeller, R., Hay, R.J., Higashi, H., Hill, C., Hoen, B., Hoffman, H., Hotez, P.J., Hoy, D., Huang, J.J., Ibeanusi, S.E.,

Jacobsen, K.H., James, S.L., Jarvis, D., Jasrasaria, R., Jayaraman, S., Johns, N., Jonas, J.B., Karthikeyan, G., Kassebaum, N., Kawakami, N., Keren, A., Khoo, J.P., King, C.H., Knowlton, L.M., Kobusingye, O., Koranteng, A., Krishnamurthi, R., Laden, F., Lalloo, R., Laslett, L.L., Lathlean, T., Leasher, J.L., Lee, Y.Y., Leigh, J., Levinson, D., Lim, S.S., Limb, E., Lin, J,K., Lipnick, M., Lipshultz, S.E., Liu, W., Loane, M., Ohno, S.L., Lyons, R., Mabweijano, J., MacIntyre, M.F., Malekzadeh, R., Mallinger, L., Manivannan, S., Marcenes, W., March, L., Margolis, D.J., Marks, G.B., Marks, R., Matsumori, A., Matzopoulos, R., Mayosi, B.M., McAnulty, J.H., McDermott, M.M., McGill, N., McGrath, J., Medina-Mora, M.E., Meltzer, M., Mensah, G.A., Merriman, T.R., Meyer, A.C., Miglioli, V., Miller, M., Miller, T.R., Mitchell, P.B., Mock, C., Mocumbi, A.O., Moffitt, T.E., Mokdad, A.A., Monasta, L., Montico, M., Moradi-Lakeh, M., Moran, A., Morawska, L., Mori, R., Murdoch, M.E., Mwaniki, M.K., Naidoo, K., Nair, M.N., Naldi, L., Narayan, K.M., Nelson, P.K., Nelson, R.G., Nevitt, M.C., Newton, C.R., Nolte, S., Norman, P., Norman, R., O'Donnell, M., O'Hanlon, S., Olives, C., Omer, S.B., Ortblad, K., Osborne, R., Ozgediz, D., Page, A., Pahari, B., Pandian, J.D., Rivero, A.P., Patten, S.B., Pearce, N., Padilla, R.P., Perez-Ruiz, F., Perico, N., Pesudovs, K., Phillips, D., Phillips, M.R., Pierce, K., Pion, S., Polanczyk, G.V., Polinder, S., Pope 3rd, C.A., Popova, S., Porrini, E., Pourmalek, F., Prince, M., Pullan, R.L., Ramaiah, K.D., Ranganathan, D., Razavi, H., Regan, M., Rehm, J.T., Rein, D.B., Remuzzi, G., Richardson, K., Rivara, F.P., Roberts, T., Robinson, C., De Leòn, F.R., Ronfani, L., Room, R., Rosenfeld, L.C., Rushton, L., Sacco, R.L., Saha, S., Sampson, U., Sanchez-Riera, L., Sanman, E., Schwebel, D.C., Scott, J.G., Segui-Gomez, M, Shahraz, S., Shepard, D.S., Shin, H., Shivakoti, R., Singh, D., Singh, G.M., Singh, J.A., Singleton, J., Sleet, D.A., Sliwa, K., Smith, E., Smith, J.L., Stapelberg, N.J., Steer, A., Steiner, T., Stolk, W.A., Stovner, L.J., Sudfeld, C., Syed, S., Tamburlini, G., Tavakkoli, M., Taylor, H.R., Taylor, J.A., Taylor, W.J., Thomas, B., Thomson, W.M., Thurston, G.D., Tleyjeh, I.M., Tonelli, M., Towbin, J.A., Truelsen, T., Tsilimbaris, M.K., Ubeda, C., Undurraga, E.A., Van der Werf, M.J., Van Os, J., Vavilala, M.S., Venketasubramanian, N., Wang, M., Wang, W., Watt, K., Weatherall, D.J., Weinstock, M.A., Weintraub, R., Weisskopf, M.G., Weissman, M.M., White, R.A., Whiteford, H., Wiebe, N., Wiersma, S.T., Wilkinson, J.D., Williams, H.C., Williams, S.R., Witt, E., Wolfe, F., Woolf, A.D., Wulf, S., Yeh, P.H., Zaidi, A.K., Zheng, Z.J., Zonies, D., Lopez, A.D., AlMazroa, M.A. and Memish, Z.A., 2012. Disability-adjusted life years (DALYs) for 291 diseases and injuries in 21 regions, 1990-2010: a systematic analysis for the global burden of disease study 2010. Lancet 380: 2197-2223.

Nguyen-Viet, H., Pham-Duc, P., Nguyen, V., Tanner, M., Odermatt, P., Vu-Van, T., Hoang, M.V., Zurbrügg, C., Schelling, E. and Zinsstag, J., 2015. A One Health perspective for integrated human and animal sanitation and nutrient recycling. In: Zinsstag, J., Schelling, E., Waltner-Toews, D., Whittaker, M. and Tanner, M. (eds.) One Health: the theory and practice of integrated health approaches. CABI, Wallingford, UK.

Nguyen-Viet, H., Zinsstag, J., Schertenleib, R., Zurbrugg, C., Obrist, B., Montangero, A., Surkinkul, N., Kone, D., Morel, A., Cisse, G., Koottatep, T., Bonfoh, B. and Tanner, M., 2009. Improving environmental sanitation, health, and well-being: a conceptual framework for integral interventions. Ecohealth 6: 180-191.

Nilsson, H.E., 1995. Remote-sensing and image analysis in plant pathology. Annu Rev Phytopathol 33: 489-527.

Nita, M., Ellis, M.A. and Madden, L.V., 2003. Reliability and accuracy of visual estimation of phomopsis leaf blight of strawberry. Phytopathology 93(8): 995-1005.

Nsubuga, P., White, M.E., Thacker, S.B., Anderson, M.A., Blount, S.B., Broome, C.V., Chiller, T.M., Espitia, V., Imtiaz, R., Sosin, D., Stroup, D.F., Tauxe, R.V., Vijayaraghavan, M. and Trostle, M., 2006. Public health surveillance: a tool for targeting and monitoring interventions. In: Jamison, D.T., Breman, J.G., Measham, A.R., Alleyne, G., Claeson, M., Evans, D.B., Jha, P., Mills, A. and Musgrove, P. (eds.) Disease control priorities in developing countries. World Bank, Washington, DC, USA.

Nutter, F.W., 1999. Understanding the interrelationships between botanical, human, and veterinary epidemiology: the Ys and Rs of it all. Ecosyst Health 5: 131-140.

Nutter, F.W., Gleason, M., Jenco, J.H., Christians, N.H., 1993. Assessing the accuracy, intra-rater repeatability, and inter-rater reliability of disease assessment systems. Phytopathology 83: 806-812.

Nutter, F.W.J. and Schultz, P.M., 2009. Improving the accuracy and precision of disease assessments: selection of methods and use of computer-aided training programs. Can J Plant Pathol 17: 174-184.

Nutter, F.W.J., Esker, P.D. and Coelho Netto, R.A., 2006. Disease assessment concepts and the advancements made in improving the accuracy and precision of plant disease data. Plant disease epidemiology: facing challenges of the 21st century. Springer Netherlands, Dordrecht, the Netherlands.

Nutter, F.W.J., Teng, P.S. and Shokes, F.M., 1991. Disease assessment terms and concepts. Iowa State University, Ames, IA, USA.

O'Brien, D., Scudamore, J., Charlier, J. and Delavergne (2016) DISCONTOOLS: a database to identify research gaps on vaccines, pharmaceuticals and diagnostics for the control of infectious diseases of animals. BMC Veterinary Research 13: 1.

O'Brien, E.C., Taft, R., Geary, K., Ciotti, M. and Suk, J.E., 2016. Best practices in ranking communicable disease threats: a literature review, 2015. Euro Surveill 21.

Oidtmann, B., Peeler, E., Lyngstad, T., Brun, E., Bang Jensen, B. and Stark, K.D., 2013. Risk-based methods for fish and terrestrial animal disease surveillance. Prev Vet Med 112: 13-26.

Oliver, M.A. and Gregory, P.J., 2015. Soil, food security and human health: a review. Eur J Soil Sci 66: 257-276.

Ontario Ministry of the Environment and Climate Change, 2010. Air quality health index. Available at: https://tinyurl.com/y8sus5vh.

Ostrom, E., 2007. A diagnostic approach for going beyond panaceas. Proc Natl Acad Sci USA 104: 15181-15187.

Palmer, S.R., Lord, S., Paul, T. and David, W.G.B., 2011. Oxford textbook of zoonoses: biology, clinical practice, and public health control. Oxford University Press, Oxford, UK.

Parnell, S., Gottwald, T.R., Cunniffe, N.J., Alonso Chavez, V., Van den Bosch, F., 2015. Early detection surveillance for an emerging plant pathogen: a rule of thumb to predict prevalence at first discovery. Proc Biol Sci 282: 1814.

Peeler, E.J. and Otte, M.J., 2016. Epidemiology and economics support decisions about freedom from aquatic animal disease. Transbound Emerg Dis 63: 266-277.

Pennanen, M., Haukkala, A., De Vries, H. and Vartiainen, E. 2011. Longitudinal study of relations between school achievement and smoking behavior among secondary school students in Finland: results of the ESFA study. Subst Use Misuse 46: 569-579.

Peraica, M., Radić, B., Lucić, A. and Pavlović, M., 1999. Toxic effects of mycotoxins in humans. Bull World Health Organ 77: 754-766.

Pertsemlidis, A. and Fondon 3rd, J.W., 2001. Having a BLAST with bioinformatics (and avoiding BLASTphemy). Genome Biol 2: 1-10.

Purvis, G., Downey, L., Beever, D., Doherty, M.L., Monahan, F.J., Sheridan, H. and Mcmahon, B.J., 2012. Development of a sustainably-competitive agriculture. In: Lichtfouse, E. (ed.) Agroecology and strategies for climate change. Springer Netherlands, Dordrecht, the Netherlands.

Purvis, G., Louwagie, G., Northey, G., Mortimer, S., Park, J., Mauchline, A., Finn, J., Primdahl, J., Vejre, H., Vesterager, J.P., Knickel, K., Kasperczyk, N., Balazs, K., Vlahos, G., Christopoulos, S. and Peltola, J., 2009. Conceptual development of a harmonised method for tracking change and evaluating policy in the agri-environment: the agri-environmental footprint index. Environ Sci Policy 12: 321-337.

Roberts, M.G. and Heesterbeek, J.A., 2003. A new method for estimating the effort required to control an infectious disease. Proc Biol Sci 270: 1359-1364.

Rodgers, C.J., Mohan, C.V. and Peeler, E.J. 2011. The spread of pathogens through trade in aquatic animals and their products. Rev. Sci. Tech. Off. Int. Epiz. 30(1): 241-256.

Rodrigo, G., 2014. Onset of virus systemic infection in plants is determined by speed of cell-to-cell movement and number of primary infection foci. J R Soc Interface 11(98): 20140555.

Romanelli, C., Cooper, D., Campbell-Lendrum, D., Maiero, M., Karesh, W.B., Hunter, D. and Golden, C.D., 2015. Connecting global priorities: biodiversity and human health: a state of knowledge review. World Health Organisation, Geneva, Switzerland.

Rüegg, S.R., Mcmahon, B.J., Hasler, B., Esposito, R., Nielsen, L.R., Ifejika Speranza, C., Ehlinger, T., Peyre, M., Aragrande, M., Zinsstag, J., Davies, P., Mihalca, A.D., Buttigieg, S.C., Rushton, J., Carmo, L.P., De Meneghi, D., Canali, M., Filippitzi, M.E., Goutard, F.L., Ilieski, V., Milicevic, D., O'shea, H., Radeski, M., Kock, R., Staines, A. and Lindberg, A., 2017. A blueprint to evaluate One Health. Front Publ Health 5: 20.

Salkeld, D.J., Padgett, K.A. and Jones, J.H., 2013. A meta-analysis suggesting that the relationship between biodiversity and risk of zoonotic pathogen transmission is idiosyncratic. Ecol Lett 16: 679-686.

Samuel, G., 1934. The movement of tobacco mosaic virus within the plant. Ann Appl Biol 21(1): 90-111.

Schwabe, C.W., 1984. Veterinary medicine and human health. Williams & Wilkins, Baltimore, MD, USA.

Seem, R.C., 1984. Disease incidence and severity relationships. Annu Rev Phytopathol 22: 133-150.

Seidu, R., Heistad, A., Amoah, P., Drechsel, P., Jenssen, P.D. and Stenstrom, T.A., 2008. Quantification of the health risk associated with wastewater reuse in Accra, Ghana: a contribution toward local guidelines. J Water Health 6: 461-471.

Shinn, A.P., Pratoomyot, J., Bron, J.E., Paladini, G., Brooker, E.E. and Brooker, A.J., 2015. Economic costs of protistan and metazoan parasites to global mariculture. Parasitology 142: 196-270.

Slopek, S.W., 1989. An improved method of estimating percent leaf area diseased using a 1 to 5 disease assessment scale. Can J Plant Pathol 11(4): 381-387.

Smith, K.R., Corvalan, C.F. and Kjellstrom, T., 1999. How much global ill health is attributable to environmental factors? Epidemiology 10: 573-584.

Stover, E. and Mccollum, G., 2011. Incidence and severity of Huanglongbing and Candidatus Liberibacter Asiaticus titer among field-infected citrus cultivars. HortScience 46: 1344-1348.

Streatfield, P.K., Khan, W.A., Bhuiya, A., Alam, N., Sié, A., Soura, A.B., Bonfoh, B., Ngoran, E.K., Weldearegawi, B., Jasseh, M., Oduro, A., Gyapong, M., Kant, S., Juvekar, S., Wilopo, S., Williams, T.N., Odhiambo, F.O., Beguy, D., Ezeh, A., Kyobutungi, C., Crampin, A., Delaunay, V., Tollman, S.M., Herbst, K., Chuc, N.T.K., Sankoh, O.A., Tanner, M. and Byass, P., 2014. Cause-specific mortality in Africa and Asia: evidence from INDEPTH health and demographic surveillance system sites. Glob Health Action 7: 25362.

Suter, G.W., 1993. A critique of ecosystem health concepts and indexes. Environ Toxicol Chem 12: 1533-1539.

Tabor, G.M., 2002. Defining conservation medicine. In: Alonso Aguirre, A., Ostfeld, R.S., Tabor, G.M., House, C. and Pearl, M.C. (eds.) Conservation medicine: ecological health in practice. Oxford University Press, New York, NY, USA.

Taylor-Robinson, D., Diggle, P., Smyth, R. and Whitehead, M., 2015. Health trajectories in people with cystic fibrosis in the UK: exploring the effect of social deprivation. In: Burton-Jeangros, C., Cullati, S., Sacker, A. and Blane, D. (eds.) A life course perspective on health trajectories and transitions. Springer, Cham, Switzerland.

TEEB, 2015. TEEB for agriculture and food: an interim report. United Nations Environment Programme, Geneva, Switzerland. Available at: https://tinyurl.com/y99z65co.

Turechek, W.W. and Madden, L.V., 2001. Effect of scale on plant disease incidence and heterogeneity in a spatial hierarchy. Ecol Model 144: 77-95.

Valenciano, M., Dufour, B., Garin-Bastuji, B., La Vieille, S., André-Fontaine, G., Manfredi, D., Ancelle, T. and Deshayes, F., 2002. Definition des priorités dans le domaine des zoonoses non alimentaires 2000-2001 Rapport de l'InVS, AFSSA, ENVN. La direction générale de la santé, le centre hospitalier universitaire Cochin et la cellule interregionale d'epidemiologie EST. Available at: https://tinyurl.com/yd8555gv.

Van den Bosch, F., McRoberts, N., Van den Berg, F. and Madden. L.V., 2008. The basic reproduction number of plant pathogens: matrix approaches to complex dynamics. Phytopathology 98(2): 239-249.

Van den Driessche, P. and Watmough, J., 2002. Reproduction numbers and sub-threshold endemic equilibria for compartmental models of disease transmission. Math Biosci 180: 29-48.

Van Lieverloo, J.H., Blokker, E.J. and Medema, G., 2007. Quantitative microbial risk assessment of distributed drinking water using faecal indicator incidence and concentrations. J Water Health 5, Suppl. 1: 131-149.

Waage, J.K., Mumford, J.D., Leach, A.W. and Quinlan, M., 2007. Responsibility and cost sharing options for quarantine plant health. Centre for Environmental Policy, London, UK.

Waret-Szkuta, A., Morand, S., Mcintyre, K.M. and Baylis, M., 2010. ENHanCE Position Paper #1 – quantitative and qualitative approaches to the prioritisation of diseases. CIRAD/University of Liverpool. Available at: https://tinyurl.com/yd7fv25a.

Wendt, A., Kreienbrock, L. and Campe, A., 2015. Zoonotic disease surveillance – inventory of systems integrating human and animal disease information. Zoonos Publ Health 62: 61-74.

West, J.S., Bravo, C., Oberti, R., Lemaire, D., Moshou, D. and Mccartney, H.A., 2003. The potential of optical canopy measurement for targeted control of field crop diseases. Annu Rev Phytopathol 41: 593-614.

Westrell, T., Schonning, C., Stenstrom, T.A. and Ashbolt, N.J., 2004. QMRA (quantitative microbial risk assessment) and HACCP (hazard analysis and critical control points) for management of pathogens in wastewater and sewage sludge treatment and reuse. Water Sci Technol 50: 23-30.

Wiens, J.A., 1989. Spatial scaling in ecology. Funct Ecol 3: 385-397.

Wilkinson, K., Grant, W.P., Green, L.E., Hunter, S., Jeger, M.J.,Lowe, P., Medley, G.F., Mills, P., Phillipson, J., Poppy, G.M. and Waage, J., 2011. Infectious diseases of animals and plants: an interdisciplinary approach. Philos Trans R Soc Lond B Biol Sci 366(1573): 1933-1942.

Woolhouse, M.E.J. and Gowtage-Sequeria, S., 2005. Host range and emerging and reemerging pathogens. Emerg Infect Dis 11: 1842-1847.

World Health Organisation, 2016a. Environment and health. Available at: http://www.euro.who.int.

World Health Organisation, 2016b. Epidemic intelligence – systematic event detection. Available at: https://tinyurl.com/yb93f6v8.

World Organisation for Animal Health, 2016. Aquatic animal health code. Available at: https://tinyurl.com/ycwonju7.

Wright, J.F., Moss, D., Armitage, P.D. and Furse, M.T., 1984. A preliminary classification of running-water sites in Great Britain based on macro-invertebrate species and the prediction of community type using environmental data. Freshwater Biol 14: 221-256.

Yadav, N.V.S., De Vos, S.M., Bock, C.H. and Wood, B.W., 2013. Development and validation of standard area diagrams to aid assessment of pecan scab symptoms on fruit. Plant Pathol 62(2): 325-335.

Yanggen, D., Cole, D.C., Crissman, C. and Sherwood, S., 2004. Pesticide use in commercial potato production: reflections on research and intervention efforts towards greater ecosystems health in northern Ecuador. EcoHealth 1: SU72-SU83.

Zhou, X.N., 2012. Prioritizing research for 'One Health – One world'. Infect Dis Poverty 1: 1.

Zinsstag, J., Choudhury, A., Roth, F. and Shaw, A., 2015a. One Health economics. In: Zinsstag, J., Schelling, E., Waltner-Toews, D., Whittaker, M. and Tanner, M. (eds.) One Health: the theory and practice of integrated health approaches. CABI, Wallingford, UK.

Zinsstag, J., Lechenne, M., Laager, M., Mindekem, R., Naissengar, S., Oussiguere, A., Bidjeh, K., Rives, G., Tessier, J., Madjaninan, S., Ouagal, M., Moto, D.D., Alfaroukh, I.O., Muthiani, Y., Traore, A., Hattendorf, J., Lepelletier, A., Kergoat, L., Bourhy, H., Dacheux, L., Stadler, T. and Chitnis, N., 2017. Vaccination of dogs in an African city interrupts rabies transmission and reduces human exposure. Sci Transl Med 9(421): eaaf6984.

Zinsstag, J., Mahamat, M.B. and Schelling, E., 2015b. Measuring added vaue from integrated methods. In: Zinsstag, J., Schelling, E., Waltner-Toews, D., Whittaker, M. and Tanner, M. (eds.) One Health: the theory and practice of integrated health approaches. CABI, Wallingford, UK.

Zinsstag, J., Roth, F., Orkhon, D., Chimed-Ochir, G., Nansalmaa, M., Kolar, J. and Vounatsou, P., 2005. A model of animal-human brucellosis transmission in Mongolia. Prev Vet Med 69: 77-95.

Zinsstag, J., Schelling, E., Waltner-Toews, D. and Tanner, M., 2011. From 'one medicine' to 'one health' and systemic approaches to health and well-being. Prev Vet Med 101: 148-156.

Zinsstag, J., Schelling, E., Waltner-Toews, D., Whittaker, M. and Tanner, M., 2015c. One Health: the theory and practice of integrated health approaches. CABI, Wallingford, UK.

Chapter 6

The economic evaluation of One Health

Photo: Alf Ribeiro/Shutterstock.com

Massimo Canali[1*], Maurizio Aragrande[1], Soledad Cuevas[2], Laura Cornelsen[2], Mieghan Bruce[3], Cristina Rojo-Gimeno[4] and Barbara Häsler[5]

[1]Department of Agricultural and Food Sciences, University of Bologna, Viale Giuseppe Fanin 44, 40127 Bologna, Italy; [2]Department of Global Health and Development, London School of Hygiene and Tropical Medicine, 15-17 Tavistock Place, London, WC1H9SH, United Kingdom; [3]College of Veterinary Medicine, Murdoch University, 90 South Street, Perth, 6150, Australia; [4] Flanders Research Institute for Agriculture, Fisheries and Food, Social Sciences Unit, Burgemeester Van Gansberghelaan 115, box 2, 9820 Merelbeke, Belgium; [5]Department of Pathobiology and Population Sciences, Royal Veterinary College, Hawkshead Lane, North Mymms, Hatfield, Hertfordshire, AL9 7TA, United Kingdom; massimo.canali2@unibo.it

Abstract

This chapter provides an overview of the main methods and techniques available for the economic evaluation of One Health initiatives to introduce scientists and professionals from backgrounds other than economics to key considerations and implications of such assessments. The first part of the chapter describes the main analytical tools currently used in economic evaluations and discusses their potential and limitations when applied in a One Health context. A critical assessment is provided in particular to issues dealing with complexity of interrelations between human and animal health, and effective management of environmental resources. The second part of the chapter introduces and describes a range of pragmatic approaches to economic evaluation which have been inspired from the need to deal with and account for such complexity. It also investigates how systems approaches and methods used in One Health can enhance the capacity of economic evaluations to support informed decision making. With this chapter we are making a contribution to develop One Health economics as a scientific trans-disciplinary topic and stimulate further economic evaluations of One Health activities from a broader range of disciplines.

Keywords: One Health, economic evaluation, economic evaluation methods, economics and complexity, systems transdisciplinary approaches

6.1 Introduction: brief rationale behind economic evaluations of One Health initiatives

The initiatives to promote human, animal and environmental health have direct and indirect economic implications through the consumption of scarce resources and the generation of outcomes. Economic evaluations intend to understand these implications and support decisions on the implementation of interventions. The general motivation for economic evaluations is that resources are limited and can be devoted to different competing uses: a systematic analysis and comparison of the costs and the outcomes associated with an intervention can inform the decisions on the most efficient allocation.

In this chapter, we describe the main methods that can be used for the economic evaluation of One Health initiatives. With this text we aim to introduce researchers and professionals involved in One Health from backgrounds other than economics to the basics of economic analysis. The text should help to improve their awareness about the multifaced implications of an economic evaluation process, by showing and discussing the main theoretical and operational tools available, their potentials and their limitations. The development of economics as a discipline has not been straightforward. The theoretical constructs of this branch of knowledge have been affected by dogmatism, and the tendency to methodological reductionism, although inspiring pragmatic approaches to economic analysis, has narrowed its capacity of seizing complex phenomena. The recognition of such limits, confronted by the One Health vision, has highlighted the need in economic evaluations to account for the complexity of interrelations between a sound environmental resource management and human and animal health. Thus, we explore how the fields of systems approaches and methods can enhance the potential of economic evaluations to support informed decision-making through the same analytical tools that One Health scientists and practitioners commonly use. We think that this specific feature should characterise One Health economics and its development as a scientific trans-disciplinary topic (e.g. Zinsstag *et al.*, 2015a). We also hope that researchers and professionals from other disciplines working in One Health initiatives will be motivated to conduct practical economic evaluations for their One Health activities.

After this introduction, Section 6.2.1 presents the main concepts and theoretical aspects of economic evaluations and Section 6.2.2 is dedicated to the cost-benefit analysis (CBA), a method that translates all the effects of an initiative into monetary values by exemplifying most of the problems connected to economic assessments: e.g. the identification of the initiative's effects (Section 6.2.2.2), the attribution of values and their monetization over time (Section 6.2.2.3-6.2.2.5), and the choice of appropriate indicators for decision making (Section 6.2.2.6). The following sections show other methods that find wide application in health-related studies (Section 6.2.3), i.e. the cost-effectiveness, the cost-utility, and the cost-consequence analysis, and the issue of uncertainty in economic evaluations (Section 6.2.4). Next, Section 6.2.5 summarizes the limitations and challenges of economic evaluation techniques in the context of One Health, by introducing the subject of Section 6.3: the trade-off between One Health complexity and the reductionist approaches of economists. After showing how the economic thought evolved to deal with complex phenomena (Section 6.3.2), the final sections present a variety of methods and models, mainly of systemic type, that can contribute to account for the diversified and intangible values, and added values, created by One Health initiatives.

One Health is a wide concept and the related initiatives can vary in terms of complexity, context, characteristics, and objectives, and there is no 'one size fits all' for economic evaluations (Häsler *et al.*, 2012): each assessment must be context-specific. Nonetheless, One Health itself has been defined as 'any added values in terms of health of humans and animals, financial savings or environmental services achievable by the cooperation of human and veterinary medicine' with respect to the approaches of the two medicines working separately (Zinsstag *et al.*, 2015b). This definition of One Health has a strong economic connotation that takes the shape of technical achievements producing monetizable benefits, for example: reduced loss of lives and reduced suffering for humans and animals, reduced time of disease detection, full assessment of disease burden, increased awareness of cross-sectoral costs of diseases and cross-sectoral benefits from disease control, shared costs of health interventions, capacity to identify interventions of higher leverage, easier access to health care, improved food security and ecosystem services (Zinsstag *et al.*, 2015c).

A classification of economic evaluation for different types of One Health initiatives, whether possible, is beyond the scope of this work, but in Box 6.1 we provide a brief outline of few examples that illustrate the complexity of the effects and how far economic evaluations can reach through an adequate set of concepts and tools. We will refer to them throughout this chapter to illustrate specific aspects of the economic evaluation of One Health initiatives.

Box 6.1. Examples of One Health initiatives.

A. Human and animal vaccination delivery to remote nomadic families, Chad (Schelling *et al.*, 2007):
The implementation of a joint vaccination program for humans and cattle led to considerable cost savings while also increasing uptake in humans, particularly among nomadic populations, women, and children. The cost per immunised children or women is calculated, in comparison to a non-integrated intervention.

B. Integrated surveillance for the prevention and management of *Escherichia coli* cases (Elbasha *et al.*, 2000):
This study evaluates a surveillance mechanism for the detection and management of *E. coli* cases. The authors take a cost-benefit approach, attempting to value all costs and outcomes in monetary terms. They then calculate the net benefits depending on the number of cases prevented, compared to a scenario with no surveillance mechanism, where emergent cases are individually traced back.

C. Health, agricultural, and economic effects of adoption of healthy diet recommendations (Lock *et al.*, 2010):
The assessments of policies aimed at improving health and reducing greenhouse gas emissions through reduced livestock consumption estimate the potential impact of diminished meat consumption on human health and gross domestic product, both in Brazil and the UK, considering the consumption patterns and trade links involved. The impacts are further disaggregated by sector by acknowledging the relevance of the geographic and sectoral distribution of the impacts.

D. Comparison of human post exposure prophylaxis (PEP) versus a mass dog rabies vaccination and culling policy to control human rabies (Zinsstag *et al.*, 2009):
This study assessed the economic impact of using PEP versus a mass dog rabies vaccination in the capital of Chad. The results show that a mass dog rabies vaccination is more cost-effective over a six year period than PEP alone.

6.2 Tools for economic evaluation

6.2.1 What do we understand by economic evaluation?

6.2.1.1 A definition of economic evaluations

According to a definition very popular in the field of health economics, the economic evaluation of an intervention is 'the comparative analysis of alternative courses of action in terms of both their costs and consequences' (Drummond *et al.*, 2005). An economic evaluation is then subject to two main conditions:

- a comparison between two or more alternatives, and no action could be one of the alternatives examined;
- the identification, measurement and valuation of the costs and consequences (or outcomes) of each alternative examined.

Drummond *et al.* (2005) also indicate a typology of evaluations for health interventions based on the partial or full compliance with such conditions, as shown below:

- Partial evaluations:
  - Only one course of action is examined:
    - Only consequences are examined: outcome description.
    - Only costs are examined: cost description.
    - Costs and consequences are examined: cost-outcome description.
  - Two or more alternatives are examined (no action could be one of the alternatives):
    - Only consequences are examined: efficacy or effectiveness evaluation.
    - Only costs are examined: cost analysis (CA) or cost minimisation analysis (CMA).
- Full economic evaluations:
  - Two or more alternatives are examined (no action could be one of the alternatives):
    - Costs and consequences are examined:
      - Cost-effectiveness analysis (CEA).
      - Cost-utility analysis (CUA).
      - Cost-benefit analysis (CBA).

Such a setting is related to the scarcity of resources and the need to optimise choices for their use: hence, the main objective of economic evaluations is to inform decision making.

6.2.1.2 Optimisation of resource use and social choices: welfarist and extra-welfarist approaches

Optimisation of resource use is connected to the notion of efficiency, which is rooted in the marginalist-neoclassical economic theory, the mainstream current of thought in economics, and represents its first goal: e.g. 'The problem of Economics may [...] be stated thus: Given, a certain population, with various needs and powers of production, in possession of certain lands and other sources of material: required, the mode of employing their labour which will maximise the utility of the produce' (Jevons, 1871); 'Economics is the science which studies human behaviour as a relationship between ends and scarce means which have alternative uses' (Robbins, 1932).

Utility is a concept of economics that indicates the satisfaction gained by individuals from the consumption of goods and services. The maximisation of utility, through an optimal use of scarce resources, depends on individual preferences and individual choices, but interventions in the health sector, as in many other sectors, are mostly determined by decisions involving whole communities of citizens at different levels: local, national, global, etc. Social choices are a topic of welfare economics. For this branch of the marginalist thought, social choices should be driven by the concept of Paretian dominance or Paretian efficiency, in which one solution is preferable to another if it provides a better off for someone without implying any worse off for someone else (Pareto, 1894). An extension of the Paretian principle, the so-called Kaldor-Hicks compensation, which is more usable in the practice of economic evaluations, is based on the hypothesis that if a solution implies the outcome of gainers and losers and the gainers may compensate the losers and also maintain some benefit, then this solution is Pareto efficient (Hicks 1939, 1943; Kaldor, 1939).

The Paretian approach and the Kaldor-Hicks compensation have been severely criticized, as well as other major elements of the welfare economics' theory (e.g. Arrow, 1951; De Scitovszky, 1941; Samuelson, 1942; Sen, 1970, 1999). Extra-welfarist approaches focus on societal objectives as separated by maximisation of subjective utility of individuals and enable the optimisation of resource use within the framework of socially relevant outcomes: e.g. to impose additional taxes for the improvement of the public health system beyond the utility of those single individuals that could prefer to keep these resources for other personal uses. The judgement on the relevance of the outcomes might be indicated by the affected individuals as in the welfarist approach, but also by a representative sample of the general public, by an expert or by an authoritative decision-maker, and may integrate ethical considerations (Brouwer *et al.*, 2008). In particular, the decision-maker approach emphasises the function of decision makers in defining societal objectives and relevant outcomes, by acting as representatives of the individual members of society (Coast, 2004; Sugden and Williams, 1978). Within this approach, the definition of societal goals can also be achieved through transdisciplinary initiatives engaging with stakeholders to find consensus for problem solving (Schelling and Zinsstag, 2015).

6.2.1.3 Economic evaluations in health interventions and the One Health approach

The last four methods listed in Section 6.2.1.1 (CA or CMA, CEA, CUA, and CBA) are the most common approaches in the economic evaluation of health interventions. All four methods assess the cost of interventions in monetary values but differ in the ways outcomes are measured. In cost analysis or cost minimisation analysis (CMA) comparison of cost is made between two or more alternative interventions implying different costs but producing equivalent outcomes. Cost-effectiveness analysis (CEA) compares two or more alternatives that produce the same type of effect, but to a different extent: the comparison has to be made in terms of monetary cost per unit of effect measurable in natural metrics, for example: number of cases treated appropriately, lives saved, life years gained, days free of pain or symptoms, cases successfully diagnosed, etc. (Robinson, 1993). In cost-utility analysis (CUA) the effects are measured by indicators of the utility provided to patients in terms of health-related life quality along the time: e.g. quality-adjusted life years (QALY) or disability-adjusted life years (DALY). With respect to CEA, the CUA approach is broader, since it allows comparing treatments that produce different effects in terms of both mortality and morbidity on patients, as well as the costs and the outcomes of health programmes targeting completely different objectives.

Both CEA and CUA are based on extra-welfarist/decision-maker approaches. CBA is rooted in welfare economics and aims to evaluate, in monetary terms, all the possible effects, i.e. both the direct and indirect benefits and costs, of two or more alternative interventions, to identify the one which provides the highest net benefit to society. The use of a single parameter, the monetary value, for the assessment of any kind of outcome makes CBA potentially adaptable to economic evaluations in any sector of human activities and to comparisons between and across sectors. Furthermore, the CBA of an intervention in one specific sector, e.g. health, can also consider all the indirect costs and benefits affecting other sectors involved. This feature can be particularly suited for the evaluation of One Health initiatives, which in general have wide cross-sectoral impacts involving areas such as human, animal and ecosystem health, agricultural and livestock production, food industries and services, etc. (e.g. Häsler *et al.*, 2014; Roth *et al.*, 2003).

The use of money as single parameter in CBA evaluations is however greatly discussed (Ackerman and Heinzerling, 2005; Anderson, 1988; Hansson, 2007; Kelman, 1981; Sagoff, 1988; Sunstein, 2005). On the one side, this implies to attribute monetary value to goods and services that are not tradable in the market (also defined as intangible goods and services), while the market price is the main reference to establish the actual economic value of goods. On the other side, the monetary appraisal of many intangible goods, like absence of pain and suffering, human health and lives, biodiversity conservation, protection of endangered species, integrity of ecosystems, animal welfare, etc., raise significant ethical concerns. The latter issue seems to be the main reason why CBA has not found extensive application in the human health sector (Coast, 2004; Drummond *et al.*, 2005). CEA and CUA in part avoid the problem, but they limit the possibility to take into consideration the cross-sectoral benefits of an intervention. Recent contributions have however tried to widen the potential of CEA and CUA techniques in animal health evaluations with promising results (Shaw *et al.*, 2017; Torgerson *et al.*, 2018).

In other sectors, including industry, infrastructures, public services and utilities, and the environment, CBA has been increasingly recommended and adopted by government and international institutions to assess projects and programmes of public interest, despite concerns for monetisation of intangible goods and the criticisms against its welfarist background. This contradictory outcome has been explained with the lack of effective competing approaches and evaluation methods within mainstream economics and with the fact that the criticisms were not so relevant for the ordinary practice of economic evaluations and for the needs of the real-world decision makers (Pearce *et al.*, 2006).

While CBA seems the most appropriate tool for the economic evaluation of the wide cross-sectoral outcomes targeted by the One Health approach, the monetisation of intangible goods and the notions of efficiency that underpins economic evaluations may be questionable or hard to interpret from a One Health perspective. For example, Garnett *et al.* (2015) offer an analysis on the relativeness and contradictions which may result from the application of the concept of efficiency in livestock production.

6.2.2 The cost-benefit analysis

6.2.2.1 The procedure of the cost-benefit analysis and its basic elements

Any economic evaluation requires a coherent procedure consisting in a logical progression of subsequent steps, which identify the objectives of the analysis, the examined alternatives, the effects or impacts in terms of resource use and outcomes obtained over time, the methods for the economic evaluation of the effects, and the indicators addressing the final choice.

Figure 6.1 summarises a CBA procedure. The first step of the process defines the essential hypothesis needed for the evaluation: the goals of the initiatives, the operational alternatives taken into consideration and the point of view of the analysis are necessary to formulate the economic evaluation questions, which detail the specific objectives of the assessment as well

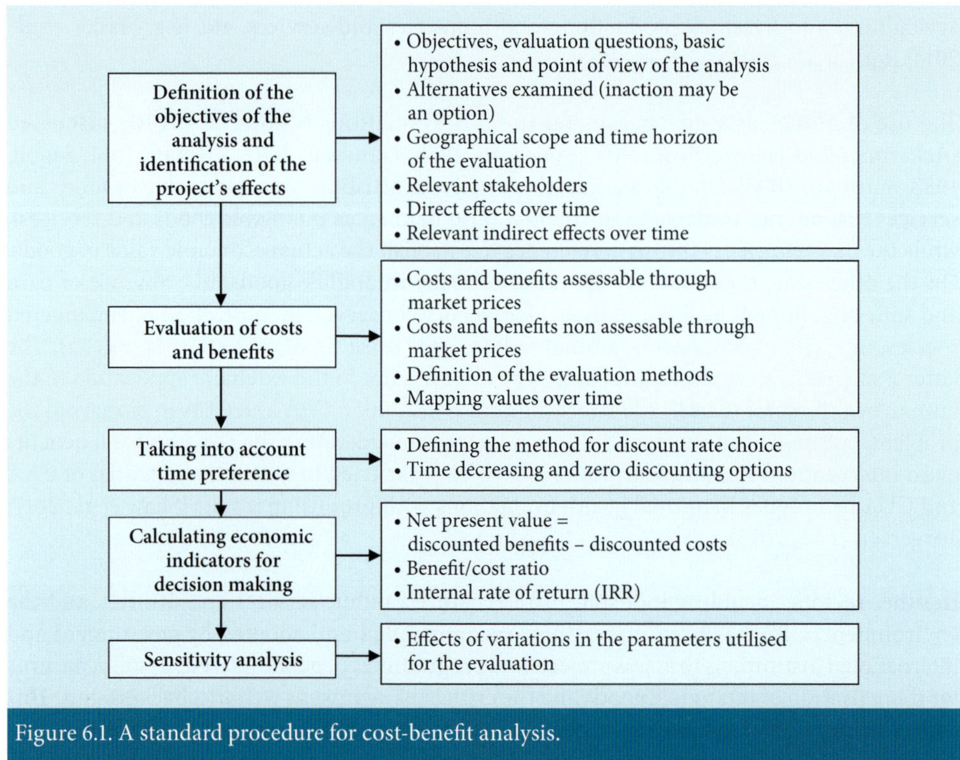

Figure 6.1. A standard procedure for cost-benefit analysis.

as its scale and boundaries. An *ex-ante* evaluation aims to establish the best choice from a set of different alternatives for an initiative under consideration, while an *ex-post* evaluation investigates the results of an already implemented initiative to verify the hypothesis that justified the choice. In the case of an interim analysis, the *ex-post* evaluation of the part of the initiative that has been already implemented may help to correct the original hypotheses in the light of new data and objectives. CBA can be used in all these contexts: *ex-ante, ex-post*, and interim analyses.

The definition of the point of view, or perspective, of the evaluation is crucial, since a cost for one stakeholder may represent a benefit for another and this affects the classification of costs and benefits (Box 6.2A). In CBA, the evaluation perspective is, by definition, societal-welfarist (Section 6.2.1.2). As already seen (Section 6.2.1.1), the comparison between two or more alternatives for an intervention, or counterfactual (Box 6.2B) is a pre-requisite of economic evaluations, considering that also doing nothing may be one of the alternatives to be considered. In this case, the effects of inaction should be compared with the effect of a planned intervention.

Box 6.2. 'Perspectives' and 'counterfactual' in economic analyses.

A. Perspectives, multi-sector and multi-stakeholder analysis

The perspective of an economic evaluation determines whose costs and whose outcomes are included in the assessment. A societal perspective implies that all costs and outcomes should be included regardless of who sustains them. Another option is to adopt the perspective of the provider or funder of the intervention. In the context of health interventions, for example, evaluations often take the perspective of the health care system. Health care costs paid privately by patients, however, are usually considered, although a stricter funder perspective might not include them. A third approach acknowledges that a decision maker will have to deal with several concerns, apart from efficiency. It attempts to provide information that can reflect this variety of concerns which can be interpreted, prioritized, and given different weights reflecting context. This approach, known as decision maker perspective (Coast, 2004) is not rooted in welfare economics, although it can incorporate elements and methods of welfarist analysis (Drummond *et al.*, 2005). In the case of One Health initiatives, usually there will be several providers and/or funders involved, with different and often competing priorities. In addition, substantial costs or economic impacts of the intervention might fall on stakeholders in different economic sectors. In these cases, the allocation of costs and benefits across providers, funders, sectors, or subpopulations can be an important aspect of the economic evaluation. The 'separable costs-remaining benefits' method (Gittinger, 1985), for example, is a technique to allocate joint cost often used in the assessment of development projects that found application for the evaluation of One Health initiatives (Roth *et al.*, 2003). Such challenges are also encountered in the public health context (Claxton and Culyer, 2006). Although research in the area of multi-sector and multi-stakeholder economic evaluation remains scarce, this is a topic that attracts increasing attention (Remme *et al.*, 2014; Weatherly *et al.*, 2009). There is no one-size-fits-all recommendation on this issue. However, in the table below we provide some examples of how different authors have dealt with the issue of multiple perspectives.

Study	Topic	Perspectives adopted	Method proposed for allocation of costs and benefits
Roth *et al.*, 2003	Animal and human brucellosis	Private agricultural and public health sectors of animal and human brucellosis.	Application of the 'separable costs-remaining benefits' method: costs allocated in proportion to the savings for each sector derived from reduced incidence.
Schelling *et al.*, 2007	Joint human and animal vaccination	Health or agricultural ministries	In proportion to the number of personnel sent and rounds carried out respectively for human and animal vaccination. Outcomes of interest are hypothesized.
Remme *et al.*, 2014	Nutrition interventions	Public organisms in charge or addressing HIV and nutrition	Based on sector-specific assumed outcomes of interest, valued based on willingness to pay thresholds for interventions
Tiwari *et al.*, 1999	Irrigation intervention	Government officials, local communities	Use of participatory methodologies and multi-criteria decision analysis to determine outcomes of interest and costs

>>>

Box 6.2. Continued.

B. Counterfactual

The full economic evaluation of an initiative implies the comparison of its costs and outcomes with an alternative course of action. In the context of One Health, this alternative could be a non-integrated approach to the initiative or a situation with no intervention. In the example of Box 6.1A, the authors compare the integrated vaccination programme with the non-integrated approach because they have data on vaccination associated to both. In the example of Box 6.1D the cumulative costs of investment in public health through mass dog vaccination and PEP (the integrated approach) is compared with the non-integrated approach (only the PEP treatment). Both cases show the cost-effectiveness of the integrated approaches. Given the call for integration of disciplines and sectors in One Health, economic evaluations of One Health initiatives are generally likely to include a comparison with a non- or less integrated approach and thereby provide information on the economic efficiency of an integrative approach to health.

6.2.2.2 Identification of effects

The geographical scope and the time horizon of the evaluation are connected to the identification of the relevant effects or impacts of the initiative. The increasing awareness of the complexity of interactions between animal, human, and environmental health tends to expand the need to account for the spatial and time extent of One Health economic evaluations. In CBA, the definition of the geographical scope of the evaluation has also consequences on the point of view of the analysis, since an intervention may impact differently on the territories inhabited by different communities, for example when transboundary effects of pollution or disease spreading are concerned (Bond *et al.*, 2016; Otte *et al.*, 2004), while the time factor affects values and the evaluation of health and environmental effects through the operation of discounting.

For direct effects, we consider the consumption of resources directly related to the implementation of the intervention and the gains of the direct beneficiaries. Figure 6.2 describes the sectors potentially affected by an intervention in livestock farms to prevent the outbreak and spreading of zoonoses. In this case, the direct effects are represented by the costs for the intervention and the gains obtained by farms due to a reduction in livestock losses, improvements in animal welfare and productivity, savings on animal health expenditure and diminished health risks for farm workers. Indirect relevant effects are the benefits (and the possible costs) associated to population health, environmental health and biodiversity, food supply chain and related outcomes on national economy and the State budget, and the implications for the cultural values that the society may attribute to elements like: the environment, biodiversity, human and animal welfare, food traditions, etc. For each of the alternatives taken into consideration, it is necessary to identify the occurrence of the effects of the initiative and quantify their intensity.

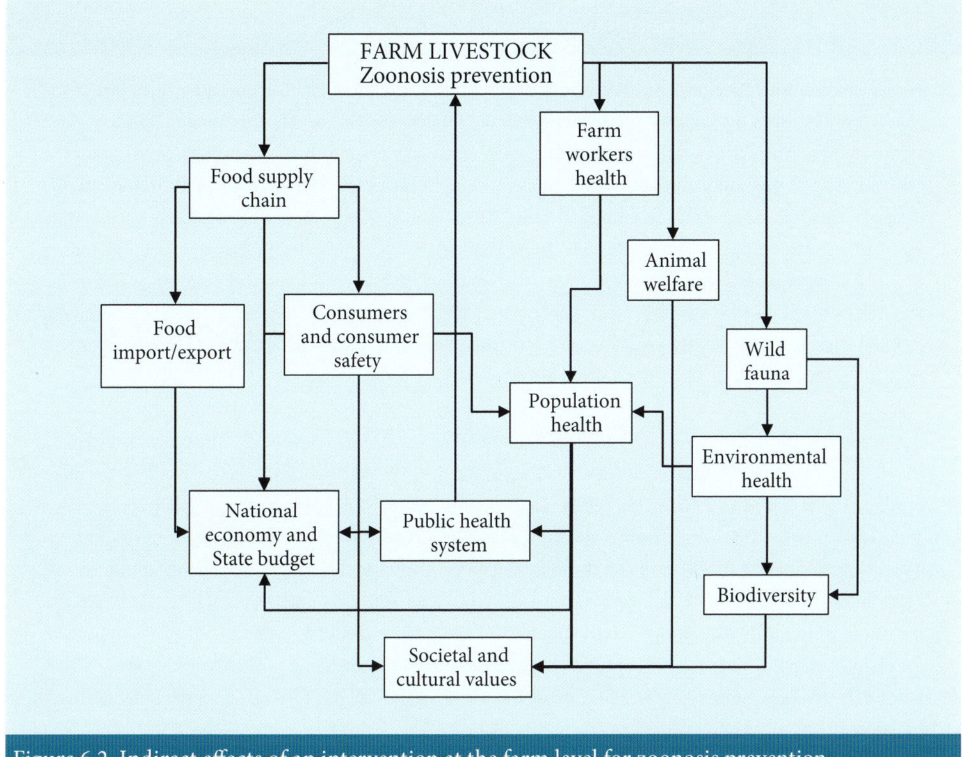

Figure 6.2. Indirect effects of an intervention at the farm level for zoonosis prevention.

The identification of effects is a crucial step of the evaluation of an intervention and often technical competences are more deeply involved in this process than economic expertise since the economic assessment is a subsequent step. Box 6.3 provides a brief picture of some problems to be faced in the identification and quantification of effects for economic evaluations.

Box 6.3. Identification and quantification of effects in economic evaluations.

A crucial issue when performing an economic evaluation is the identification and quantification of the effects resulting from the initiative analysed. In the comparative evaluation of health treatments, the highest standard in measuring the efficacy of the therapies is obtained by randomised controlled trials, but for the analysis of socio-economic impacts, as it may be needed in the evaluation of One Health initiatives, the design of this type of study is often impracticable. The methods that make use of non-experimental data are an alternative (Blundell and Costa Dias, 2005; Weatherly *et al.*, 2009). The propensity scores matching (PSM), for example, is a statistical technique which makes it possible to obtain unbiased estimates of the magnitude of the disease impacts (Heckman *et al.*, 1997; Ichimura and Taber, 2000). In PSM a treatment group is matched after propensity scores are estimated based on observable baseline characteristics to a control group with as similar as possible propensity score.

Applications of this technique in animal health economics can be found in Birol *et al.* (2010), who investigated the effect of market shocks caused by highly pathogenic avian influenza outbreaks on the livelihood of small producers of poultry in 4 African countries. To this end a PSM in which the treatment group (households suffering from a market shock) were matched with a control group (households which have not been affected by the market shock) based on the following matching variables: household demographics, assets, regional characteristics such as location, poverty status, number of income sources. The size effect was measured in terms of the average treatment effect of the treated group. An example of PSM application for socio-economic issues related to animal health in developed countries is in Rojo-Gimeno *et al.* (2016), who analysed the economic impact of reducing the use of antimicrobials through improved management practices in pig farms of Flanders.

6.2.2.3 Individual preferences, values, and willingness to pay

In CBA, the preferences of individuals are considered the only judgement of value. If one individual prefers the situation A to the situation B, it can be assumed that:

- his or her utility or welfare is higher in the situation A than in B;
- he or she values A more than B.

Any decrease in human welfare related to an initiative (e.g. the passage from the situation A to the situation B) implies a decrease of value or a cost, while any increase in human welfare represents a benefit. The preferences of one individual are measurable in terms of his willingness to pay (WTP) to obtain a benefit or to avoid a cost[1].

A further assumption relates to the fact that individual preferences can be aggregated: hence, the sum of all the individual benefits of an initiative corresponds to its social benefit and the sum of all the individual costs to the social cost. As seen in Section 6.2.1.2, an initiative is a

[1] For a reduction in welfare it is more appropriate to speak of willingness to accept (WTA) a given compensation for the cost suffered. Here we avoid the distinction between WTP and WTA since in most cases such a distinction is not relevant for the evaluation (Pearce *et al.*, 2006).

Pareto-efficient solution if, with respect to other possible solutions and to the status quo, it provides an increase of net social benefit and the gainers may compensate the losers.

The economic assessment of an initiative consists in a monetary evaluation of the WTP of all the concerned individuals to obtain the welfare increase and to avoid the welfare losses related to the effects of such initiative (Figure 6.3).

For the effects that correspond to the creation or consumption of goods and services traded in the market, also indicated as tangible effects in literature, the respective market prices are taken as indicators to appraise the WTP. For intangible effects, i.e. those related to the creation of non-tradable goods and services (e.g. a given health condition, a certain level of environmental quality, the beauty of a landscape, a cultural value, etc.), the WTP of individuals must be assessed through the analysis of the prices of other marketed goods and services that incorporate some aspects of the intangible goods examined or, alternatively, through surveys conducted on the population involved (Section 6.2.2.4). The use of the WTP concept in the economic evaluation of health interventions implies the attribution of

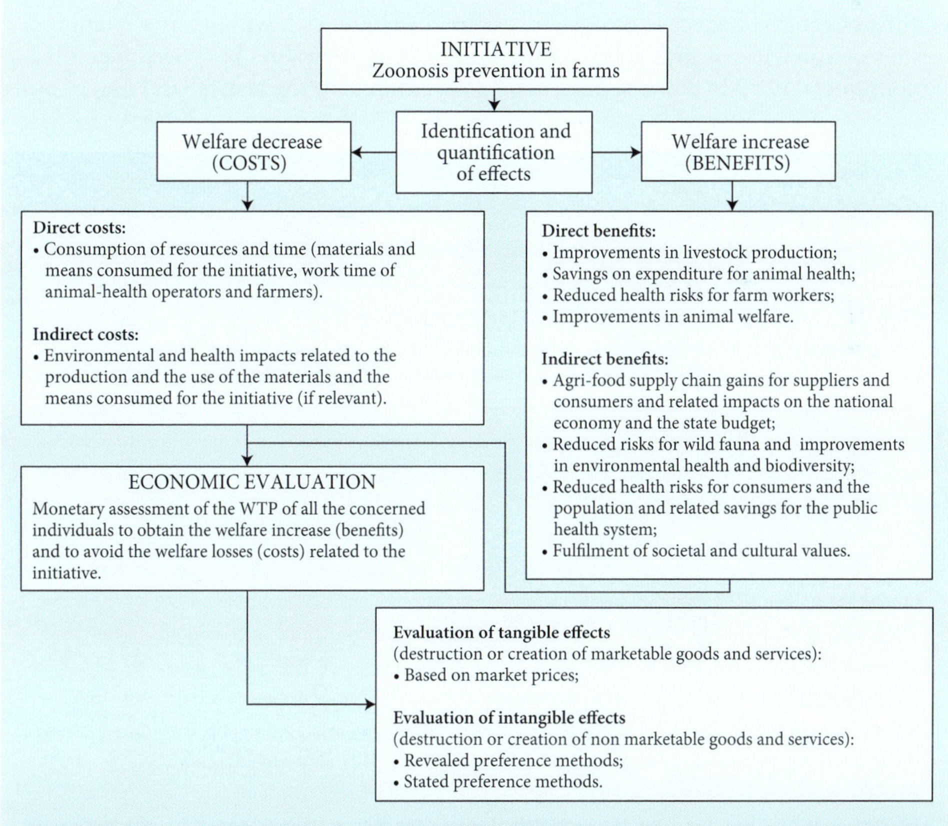

Figure 6.3. Identification of effects and evaluation of willingness to pay for a public health initiative in farms for zoonosis prevention.

monetary values to health states and lives of people, which rises important ethical concerns (Box 6.4).

The WTP is used to estimate the total economic value of environmental goods and services and this notion is also relevant in the One Health context, for example in evaluations related to environmental health, biodiversity, and animal welfare evaluations (Hansson and Lagerkvist, 2015, 2016; Laurila-Pant *et al.*, 2015; McInerney, 2004; Schreiner and Hess, 2017). The literature identifies the total economic value of a given environmental asset as formed by two types of values: use values and non-use values.

Use values are values related to the WTP of consumers for actual or possible uses of an environmental good or service. This includes option values, i.e. the possibility to avoid the actual consumption of an environmental good, to maintain the option of using it in the future.

Non-use values are related to the consumers' WTP to maintain the simple presence of an environmental good, independently on any possible use of it, actual or in the future. Existence values represent a common type of non-use values. Examples of existence values are those related to the consumers' WTP for maintenance of biodiversity, landscape beauties, wildlife, endangered species, ecosystems, animal welfare, etc., without any relation of a possible personal use of such goods. Other types of non-use values have been identified in the consumers' WTP to preserve environmental resources to the benefit (and use) of other

Box 6.4. Willingness to pay, prices, incomes, and monetization of health.

The use of the willingness to pay (WTP) concept to assign a monetary value to outcomes is consistent with the theoretical framework of welfare economics, which considers that every individual is the best judge of his own preferences and well-being. The amount of money that an individual is willing to pay to obtain something is then used to approximate how he or she values it. This would, in theory, be reflected by market prices in a perfectly competitive market. In practice, however, the outcomes of a One Health initiative are also non-marketed goods, so we do not have a market price to approximate their value. For example, we do not have market prices for quality of air or for health improvements of a collectively funded vaccination programme. In this case, we use other methods to estimate the hypothetical willingness to pay of individuals for these outcomes.

WTP of single individuals, however, depends on their income. In economic evaluations, theoretically, this implies acceptance of the status quo and of income inequalities among individuals (Jan, 1998; Johannesson, 1995), with the consequence of skewing allocation of resources towards the interventions that are most valued by the wealthiest part of the population. Other inconveniences of WTP monetization are related to the moral objections to the notion itself that life and health could be valued in monetary terms. Also for these reasons, cost-benefit analysis (CBA) evaluations are less commonly used in the context of human health than cost-effectiveness analysis and cost-utility analysis (Coast, 2004; Drummond *et al.*, 2005). Although similar concerns could be raised about placing a monetary value on the environment or on animal welfare (Ackerman and Heinzerling, 2005; Hansson, 2007; Heinzerling *et al.*, 2005), CBAs are far more widespread and accepted in this context (Pearce *et al.*, 2006).

people (altruistic values) or of future generations (bequest values) (McClelland *et al.*, 1992; Vázquez Rodríguez and León, 2004). There are authors who assume that the environment and other types of goods may have value by themselves, independently on consumers' WTP. In these cases, they speak of intrinsic values (e.g. Attfield, 1998; O'Neil, 1993).

6.2.2.4 Methods to estimate willingness to pay

According to mainstream economics, the value of a good corresponds to its equilibrium price in a perfectly competitive market, but such a condition is defined by a number of unrealistic assumptions (the supply or the demand of one single supplier or consumer is irrelevant with respect to the whole market demand and supply; the choices of one market operator do not influence market price and the choices of any other operator; all the operators are perfectly informed and act rationally and selfishly, with the only aim to maximize personal utility; products proposed by one supplier can be perfectly substituted by products proposed by any other supplier; etc.) that cannot be found in the real-world markets. Nonetheless, lacking perfect markets and better theoretical alternatives, in the current practice of economic evaluations, market prices are considered the main indicators to approximate the value of goods and services.

Further evaluation problems arise for the goods and the services that create utility to individuals without being the object of market transactions, the so-called 'externalities' (Section 6.3.2). To attribute monetary values to non-marketed goods, economists have developed specific methodologies, which are grouped under two main categories: the revealed preference methods and the stated preference methods.

Revealed preference methods

In the absence of market prices, the WTP can be approximated indirectly by analysing the consumer behaviour on surrogate markets. The basic principle is to make use of information from marketed goods and services to infer information on related non-marketed goods and services. These methods are commonly employed for the assessment of use values (Pearce *et al.*, 2006). The hedonic price method, for example, estimates the value of the environmental quality of residential areas by comparing the prices of buildings, under the hypothesis that consumers are willing to pay more for the properties located in the best areas (Brid Gleeson, 2007; Currie *et al.*, 2015; Portney, 1981). The travel cost method establishes the value of locations used for recreational purposes (parks, beaches, lakes, forests, wildlife, etc.) on the basis of the travel costs sustained by visitors to visit them, but it has also been used in the health sector to evaluate, for example, the benefits for impacted population of mobile health care units (Clarke, 1998) and of free distribution of vaccines (Jeuland *et al.*, 2010). Defensive expenditure and averting behaviour methods evaluate the environmental quality of an asset on the basis of the cost to be sustained to avoid the damage suffered from a diminution of such quality (Bresnahan and Dickie, 1995; Cullino, 1996; Nirmala, 2014): e.g. the value of an unpolluted lake is at least equal to the expenditure to be sustained to clean the lake from some type of pollution. Similarly, cost of illness and lost output methods evaluate the loss of environmental quality on the basis of the medical expenditure derived from related illness in humans (Freeman III *et al.*, 2014), or reduction in the output of livestock and crop production (Pearce *et al.*, 2006).

Stated preference methods

Stated preference methods identify the consumers' WTP for non-market goods and services by setting hypothetical markets for such goods. These methods can estimate both use and non-use values. Thanks to its wide adaptability, contingent valuation is the most used stated preference method: it consists of investigating the consumers' WTP for changes in the availability of non-market goods and services through a questionnaire-based survey (Carson, 2000; Halasa *et al.*, 2012; Klose, 1999; Venkatachalam, 2004). The conditions of the hypothetical market under analysis are described in the questionnaire submitted to the interviewees and should be realistically obtainable. The consumers' preferences should be expressed in monetary terms. Choice modelling is a group of stated preference methods based on surveys that allow to value multidimensional changes in environmental goods and services (Hanley *et al.*, 2002; Pearce *et al.*, 2006). They are particularly useful when the measure under analysis implies changes in different elements of an environmental asset and each variation needs a distinct valuation. Choice modelling can estimate marginal or unit changes in the elements affected by variations. Types of choice modelling are: choice experiments, in which respondents are asked to choose between two or more alternatives, including a status quo situation, under the assumption that the total utility provided by the environmental asset results from the utility of its constitutive elements subject to variations; contingent ranking, in which respondents indicate priorities among different alternatives characterised by a number of attributes at different levels; contingent rating, in which respondents rate the proposed alternatives on a numeric or semantic scale without making direct comparisons among them; paired comparison exercises, in which respondents indicate a priority between two alternatives by rating the level of their preference on a numeric or semantic scale (Hanley *et al.*, 2002).

The economic evaluations based on the assessment of revealed and stated preferences have their own pros and cons. A study of Dürr *et al.* (2008) compares results obtained from both types of techniques to estimate the WTP of the citizens of N'Djaména (Chad) for the antirabies vaccination of their dogs and the probability to have one dog vaccinated in function of the vaccination price charged to the owner. The aim is to set the price allowing the maintenance of dog vaccination rates at the WHO-recommended rate of 70%, which represents a public good whose cost of production has to be shared between the dog owners and the public health service.

6.2.2.5 Time preference and discounting

The evaluation of the effects allows mapping (for each alternative examined) of the values consumed and generated by the intervention over time, but a direct comparison among values occurring in different times is not possible. Individuals prefer to receive benefits immediately rather than delayed: for example, a gain of 1000 euros now is better than a gain of 1000 euros tomorrow or in one year. On the other hand, a delayed cost is preferred to a cost to be supported immediately. The comparison among values occurring in different times can be made only after the conversion of the identified costs and benefits to present values through the operation of discounting. Discounted costs and benefits make it possible to calculate the net present value (NPV) of each examined alternative, which is the first indicator provided by CBA to decision makers (Section 6.2.2.6).

The discounting of a future value is obtained as follows:

$$\frac{Future\ value}{(1+r)^t} = Present\ value$$

Where, *r* is the discount rate applied and is the time span between the present and the creation of value in the future: it can be observed that it corresponds to the reverse process of adding interest at the discount rate. As shown in Figure 6.4, such factors substantially influence the present value attributed to the costs and the benefits of an intervention. Thus, the choice of the discount rate is critical since it can significantly affect the amount of the NPV and the final decision on the feasibility of an intervention.

There are different approaches that may be adopted in CBA for the choice of the discount rate, also called 'social discount rate' (Kula, 2006; Zhuang *et al.*, 2007), for example:

- The social rate of time preference (SRTP) assumes to compensate the diversion of resources from consumption caused by an investment in the public sector. The SRTP is calculated by estimating a coefficient indicating the social time preference summed up to the expected growth rate of domestic consumption during the life cycle of the intervention.

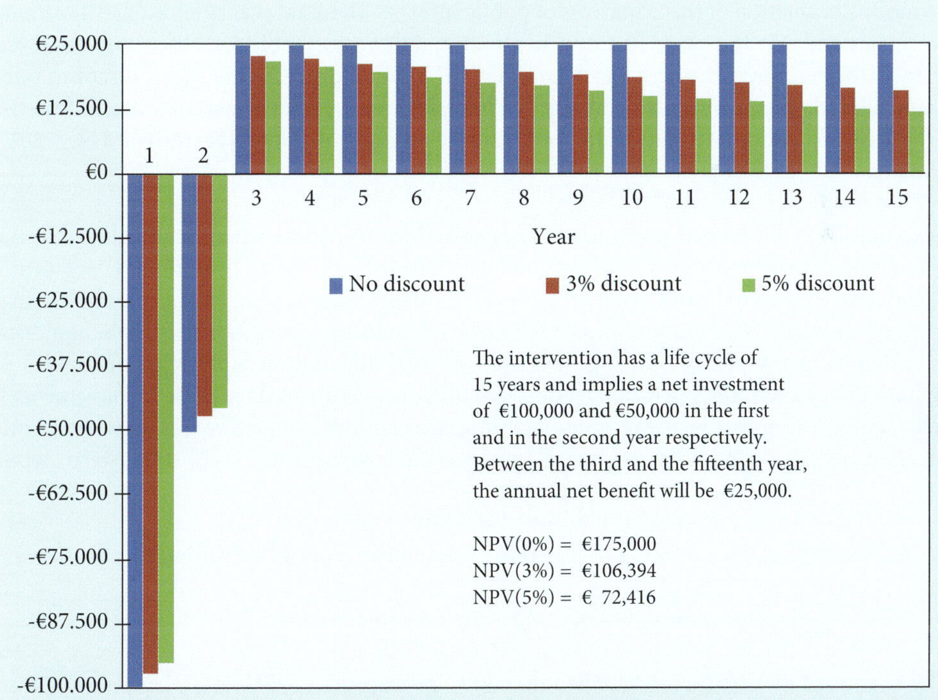

Figure 6.4. Present value of future costs and benefits of an intervention resulting from the application of three different discount rates (0, 3 and 5%).

- The social opportunity cost approach (SOC) assumes that an investment in the public sector implies reduction of resources available for private investments. Therefore, the social discount rate to be adopted should correspond, at least, to the return rate of private-sector projects that have an economic risk of the same order of the public-sector intervention under examination.
- The weighted average approach focuses on the possibility that investments for interventions of public interest may come from various sources including public, private, and foreign capitals. Thus, a weighted average between SRTP, SOC and the rate of the foreign borrowing is proposed, by considering the proportion of funding provided by the different sources.
- The shadow price of capital (SPC) approach assumes that resources for public investments are in part dislocated from direct consumption and in part from private investments. The latter are expected to generate an increased value of consumption in the future, which determines the SPC. This approach implies that one euro diverted from private investments reduces the discounted present value of a public intervention more than one euro diverted from direct consumption. Likewise, the gains from the intervention benefits flowing into private investments are valued more than the intervention benefits flowing into direct consumption (Boardman and Greenberg, 1998; Lyon, 1990).

Many government agencies indicate or suggest the social discount rate to be utilised in the economic evaluation of the initiatives of public interest. These rates vary significantly among countries and, for the same country, in the different years (Table 6.1). It can be observed that, in the recent years, there has been a general trend to decrease the social discount rates applicable for the economic evaluation of public interest projects and this has followed a progressive lowering of inflation and interest rates at the global level (Ferrero and Neri 2017; Holston *et al.*, 2017).

Discounting is a financial operation that raises ethical concerns when is introduced in the evaluation of interventions with effects on human health and the environment. Discounting social costs related to health or environmental damages occurring in the future implies that most of the burden due to their impact would be left to future generations. This phenomenon is called 'tyranny of discounting' (Figure 6.5) and does not comply with principles of sustainability and intergenerational equity. In order to avoid or reduce these inconveniences, many authors propose the use of time-declining rates or undiscounted values in the economic assessment of health and environmental impacts (Kula, 2006; Pearce *et al.*, 2006; Stern, 2008).

6.2.2.6 Indicators for decision making and sensitivity analysis

A positive NPV is the condition of feasibility for an intervention, according to Equation 1:

$$NPV = \sum Present\ value\ of\ benefits - \sum Present\ value\ of\ costs > 0 \qquad (1)$$

The examined alternatives should be ranked by the respective NPV.

Another indicator for decision making is the benefit/cost ratio (BCR), which is subject to Equation 2:

Table 6.1. Social discount rates indicated by government agencies of selected countries (Zhuang *et al.*, 2007).

Country	Social discount rates	Approach for estimation[1]
Australia	1991: 8%; current: SOC rate annually reviewed	SOC
Canada	10%	SOC
China	8% for short and medium-term projects; lower than 8% rate for long-term projects	Weighted average
France	Real discount rate set since 1960; set at 8% in 1985 and 4% in 2005	1985: To keep a balance between public and private sector investment. 2005: SRTP approach
Germany	1999: 4% 2004: 3%	Based on federal refinancing rate, which over the late 1990s was 6% nominal; average gross domestic product deflator (2%) was subtracted giving 4% real
India	12%	SOC
Italy	5%	SRTP
New Zealand – Treasury	10% as a standard rate whenever there is no other agreed sector discount rate	SOC
Norway	1978: 7% 1998: 3.5%	Government borrowing rate in real terms
Pakistan	12%	SOC
Philippines	15%	SOC
Spain	6% for transport; 4% for water	SRTP
United Kingdom	1967: 8% 1969: 10% 1978: 5% 1989: 6% 2003: 3.5% Different rates lower than 3.5% for long-term projects over 30 years	SOC approach until early 1980s; thereafter SRTP
		>>>

Table 6.1. Continued.

Country	Social discount rates	Approach for estimation[1]
US – Office of Management and Budget	Before 1992: 10% after 1992: 7%	Mainly SOC approach with the rate being derived from pre-tax return to private sector investment. Other approaches (SPC, treasury borrowing rates) are also mentioned
US – Congressional Budget Office and General Accounting Office	Rate of marketable treasury debt with maturity comparable to project span	SRTP
US – Environmental Protection Agency	Intra-generational discounting: 2-a3% subject to sensitivity analysis in the range of 2-3% and at 7%, as well as presentation of undiscounted cost and benefit streams. Intergenerational discounting: presentation of undiscounted cost and benefit streams subject to sensitivity analysis in the range of 0.5-3% and at 7%	SRTP

[1] SOC = social opportunity cost approach; SPC = shadow price of capital; SRTP = social rate of time preference.

$$\text{B/C ratio} = \frac{\sum \textit{Present value of benefits}}{\sum \textit{Present value of costs}} > 1 \qquad (2)$$

If two or more alternatives have the same NPV, the one with the highest BCR is the best choice since it implies the least investment.

The internal rate of return (IRR) corresponds to the discount rate that reduces to zero the NPV of an intervention. Then the IRR is the discount rate that resolves Equation 3:

$$\sum \textit{Present value of benefits} - \sum \textit{Present value of costs} = 0 \qquad (3)$$

The examined alternatives should be ranked according to the higher IRR. With respect to the NPV and the BCR, the IRR has the advantage to avoid the choice of the discount rate, but other shortcomings strongly limit its utilisation, especially for ranking mutually exclusive interventions (HM Treasury, 2003; Kelleher and MacCormack, 2004; Pearce *et al.*, 2006; The World Bank, 2007): for example, the distribution of costs and benefits over time may result in multiple IRR or in illogical or misleading results (Kelleher and MacCormack, 2004; The World Bank, 2007). Figure 6.6 shows how the highest IRR may not result in the highest NPV. In general, the most accepted approach to the use of decision making indicators refers

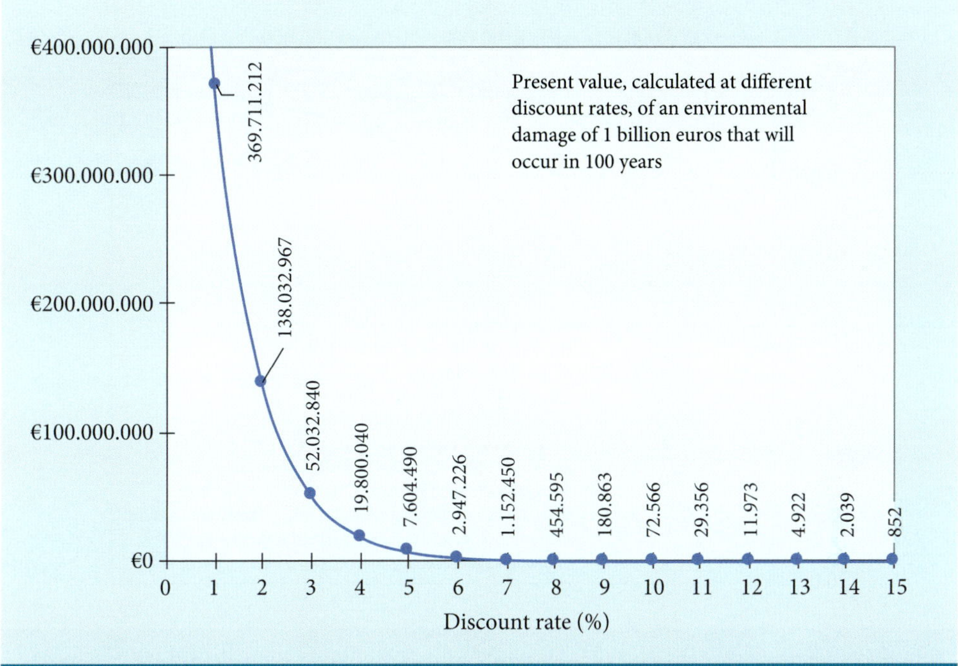

Figure 6.5. The tyranny of discounting.

to a positive NPV for the feasibility of an initiative and to the higher NPV for ranking the alternatives (Pearce *et al.*, 2006).

After the calculation of the NPV, the sensitivity analysis represents a test of reliability on the profitability of the intervention. It consists in changing the parameters previously used in the CBA (prices, currency exchange rates, materials used, production techniques, discount rate, etc.). The aim is to examine how these changes have an impact on the estimated values and therefore on the final indicators of the economic evaluation (Section 6.2.4).

6.2.3 Other economic evaluation methods in the health sector: cost-effectiveness and cost-utility analyses

6.2.3.1 Cost-effectiveness analysis

CEA evaluates the cost of the outcomes of an initiative expressed in natural units, along a one-dimensional scale, e.g. number of individuals treated, cases prevented or detected, deaths avoided, years of life gained, etc. Like in CBA, in CEA evaluations costs and outcomes should be compared to counterfactuals, which may correspond to alternative initiatives and/or a status-quo situation (no initiative). CEA is most used in health economics, but it finds also relevant application in studies that investigate environmental impacts and environmental policies (Görlach *et al.*, 2006) and in animal health evaluations (Babo Martins and Rushton, 2014).

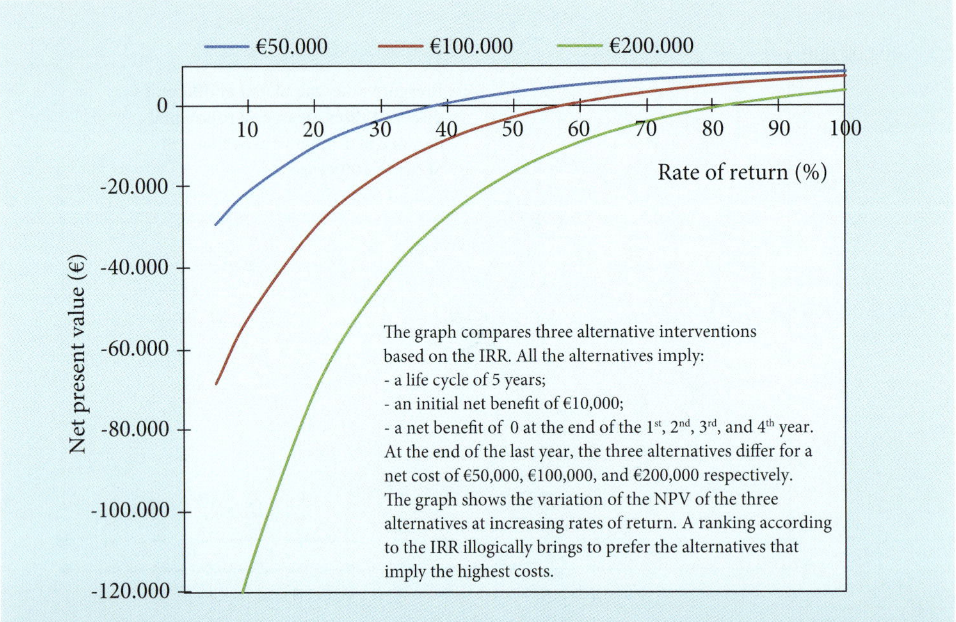

Figure 6.6. An example of possible illogical consequences of ranking alternative initiatives based on the internal rate of return (IRR).

CEA results are commonly presented in the form of incremental cost effectiveness ratio (ICER), which is the incremental cost per additional unit of outcome of the analysed initiative and indicates the efficiency with respect to the counterfactual. For example, if the outcome of an initiative is measured in terms of the number of immunized individuals, the corresponding ICER indicates the extra cost per additional individual immunized vis-à-vis the counterfactual scenario.

In CEAs of initiatives impacting on environment a variety of effectiveness indicators can be found: e.g. Berbel *et al.* (2011) made use of Mm^3/year of underground water extracted as an outcome variable, and compared the cost per avoided Mm^3/year of extraction, across a set of water management initiatives; Harrington *et al.* (1999) calculated the cost-effectiveness of Arizona's vehicle inspection and maintenance regulations based on the cost of vehicle control per reduced ton of pollutants (HC, CO, and NO_X); Wynn (2002) evaluated cost-effectiveness of biodiversity management in Scottish heather, wetlands, and herb-rich grasslands through indicators of habitat suitability of species and other indicators of biodiversity based on the portion of the total area covered by the different species. A wide review of CEA studies with the respective indicators of effectiveness utilized is presented in Babo Martins and Rushton (2014).

Table 6.2 shows the CEA of a hypothetical integrated *Escherichia coli* surveillance programme. The example is inspired by one of the studies mentioned in Box 6.1 (Elbasha *et al.*, 2000).

Table 6.2. Cost-effectiveness analysis of a hypothetical integrated *Escherichia coli* surveillance programme, figures are hypothetical.[1]

Items	I_1	I_0	Incremental (I_1–I_0)
Direct costs	€ 2,500,000	€ 2,000,000	€ 500,000
Indirect costs	€ 5,900,000	€ 6,000,000	-€ 100,000
Total costs (TC)	€ 8,400,000	€ 8,000,000	€ 400,000
Health effects (HE) = cases prevented	400	300	100
Incremental cost effectiveness ratio (ICER) = $(TC_1 - TC_0) / (HE_1 - HE_0)$			€ 4,000

[1] I_1 = integrated surveillance programme; I_0 = tracking reported cases, no integrated surveillance (counterfactual).

All the figures of the table are hypothetical. The introduction of an integrated surveillance system increases direct costs, which include personnel, set-up, and coordination among others. However, it reduces indirect costs, including health care costs associated to treating infected patients. The ICER represents the incremental cost per additional case of *E. coli* infection prevented.

The results of ICER calculation can be represented in a cartesian diagram called 'cost-effectiveness plan' (Figure 6.7), where the examined alternatives are identified as points of the plan and are confronted with the counterfactual, which is positioned at the origin of the axes. The horizontal axis indicates the increase (or decrease) in efficiency of the examined alternatives with respect to the counterfactual, whereas the vertical axis refers to the incremental cost of the alternatives per additional unit of the efficiency indicator.

The initiatives positioned in the fourth quadrant are more efficient and cheaper than the counterfactual, therefore they are Pareto superior: i.e. there is an absolute dominance of the alternative initiative with respect to the counterfactual. On the contrary, the initiatives positioned in the second quadrant are less efficient and costlier than the counterfactual, which is dominant in this case.

The alternatives positioned in the first quadrant (as in the case described by Table 6.2 and reported in Figure 6.7) are more efficient but also costlier. In this case, the initiative can be compared with one or more economic thresholds (e.g. budget constraints, opportunity cost for other projects, societal WTP for the initiative, etc.), which act as benchmarks to orient the decision-making process. In Figure 6.7, the hypothetical integrated surveillance programme (I_1) results more efficient and costlier than the counterfactual tracking of reported cases without integrated surveillance (I_0): but its position below the economic threshold indicates that it could be adopted. Finally, the third quadrant collects the initiatives that result less

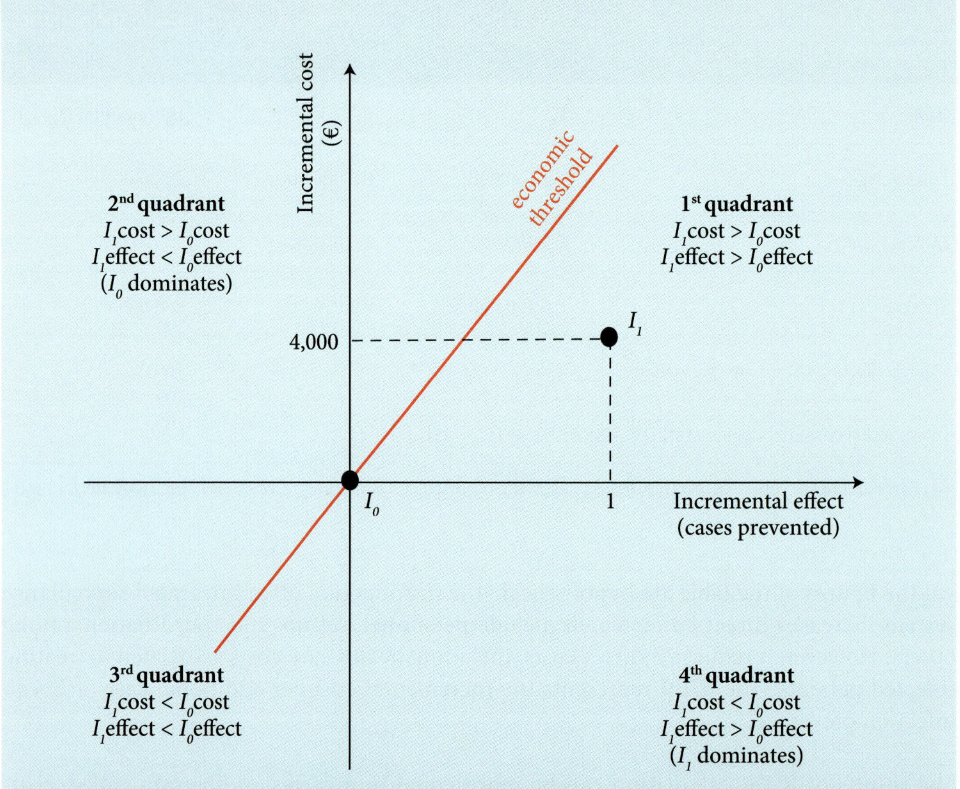

Figure 6.7. Cost-effectiveness plan of a hypothetical integrated *Escherichia coli* surveillance programme, figures are hypothetical.
I_1 = integrated surveillance programme; I_0 = tracking reported cases, no integrated surveillance (counterfactual).

efficient and less costly than the counterfactual. Also in this case, the setting of economic threshold can help decision making.

One of the advantages of CEA is that it is a relatively simple method. Outcome variables in natural units can be measured more accurately and involve fewer assumptions than the more complex approaches used in CBA (Section 6.2.2) or in CUA (Section 6.2.3.2). For this reason, results can be appreciated by a wider audience beyond health economists and managers. Terms such as 'additional cost per case prevented' or 'additional cost per person treated' can be easily understood by people not familiar with economic evaluations. However, this simplicity can be misleading and thus hinder the usefulness of CEA in informing decision-making.

An important limitation of CEA is the one-dimensional nature of the effectiveness indicator, which prevents a direct comparison of resource allocation among alternatives that cannot be referred to the same physical parameter of efficiency: e.g. to evaluate the opportunity of financing a given health programme with respect to another programme that operates in a completely different field of health cares. Furthermore, it may occur that the indicator of efficiency chosen is imperfectly correlated with the real objective of the initiative (Diamond and Kaul, 2009; Garber and Phelps, 1997; Raftery, 1999; Weintraub and Cohen, 2009). For example, an intervention might be the most efficient at increasing the number of persons treated, but if the treatment is less targeted, it might not be efficient at reducing the incidence of the disease. It can also be difficult to compare across settings or populations: for example, we can find that a vaccination campaign costs £20 per case prevented in one region and £40 per case prevented in another more remote region. However, the population affected in the second area might have lower access to the health care services in general, so that the mortality and morbidity associated to each case might be higher. In that case, it could be preferable to fund the second intervention, which in a simplistic CEA would seem less efficient.

These limitations apply to CEA in general. In the context of One Health there are often several outcomes of interest, including environmental and health interventions. Unlike CBA, CEA cannot be used to integrate different types of outcomes into the analysis, since it does not provide a common metric. Multi-criteria approaches, however, can be used in combination with economic analysis tools to integrate multiple outcomes and take into account the preferences and trade-offs across these. For example, CEA can be used to inform multi-criteria decision analysis (Section 6.3.7).

6.2.3.2 Cost-utility analysis

Several health-related utility measures have been developed by economists to overcome the limitations of CEA. This approach describes the outcomes of a health initiative through generic indicators that map the multi-dimensional concept of health onto a one-dimensional cardinal index, which integrates both life expectancy and quality of life (Figure 6.8). Such indicators, which permit a higher comparability of outcome variables, have brought CEA to a relevant evolution by allowing, for example, the comparison of resource allocation between initiatives impacting on distinct areas of the health sector on the basis the cost of additional utility provided to patients or the whole society in terms of quality of life and lifetime gained.

The evaluations that make use of these generic metrics can be considered a specific type of CEA and are commonly indicated in the scientific literature as cost-utility analyses (CUAs) (Birch and Gafni, 1992; Drummond *et al.*, 2015; Garber and Phelps, 1997; Leung, 2016; Tengs, 2004): where the term 'utility' is derived from the von Neumann-Morgenstern's utility theorem (von Neumann and Morgenstern, 1953) and refers to the preference of single individuals or society for a given health state (Drummond *et al.*, 2015). The most commonly used CUA indicators, taken as measures of efficiency for the examined initiatives, include QALYs, DALYs and Healthy Year Equivalents (HYEs), while the designation of Health Adjusted Life Years represents a general umbrella definition for such indices (Gold *et al.*, 2002). Evaluation results are presented in the form of the incremental cost per additional DALY, QALY, or HYE (as shown in Figure 6.8), also defined by some authors as incremental cost-utility ratio (Jakubiak-Lasocka and Jakubczyk, 2014).

The various indicators differ for methodology used to set the metrics and the underlying assumptions. Therefore, the most suitable index depends on the context and aims of the evaluation. QALYs, for example, are based on large population surveys. DALYs, on the other hand, are based on expert judgement on the morbidity and loss in quality of life associated to different diseases. DALYs are therefore not appropriate for measuring relatively small changes in the health status of individuals but are easier and cheaper to obtain. These and other differences explain why global studies of burden of disease and evaluations in low and middle-income countries normally make use of DALYs. On the other hand, QALYs are more frequently used to assess health-care interventions in high income countries, although data availability and resources allocated to evaluation of health in low-and-middle income countries are improving (Bleichrodt, 1995; Devleesschauwer *et al.*, 2014a; Gold *et al.*, 2002; Murray and Acharya, 1997; Weinstein *et al.*, 2009).

The construction of these metrics involves many assumptions that may raise ethical and technical concerns, and theoretical and practical limitations can make the results harder to be interpreted than in CEA, and limit comparability across population groups and regions (Barker and Green, 1996; Devleesschauwer *et al.*, 2014a; Weyler and Gandjour, 2011). Furthermore, the use of health-related quality of life indicators does not avoid CUA being affected by some key shortcomings of the CEA approach. One relevant limitation, for instance, has been identified in the use of the ICER (also called incremental cost-utility ratio – in CUA) and of economic thresholds as parameters for decision making, since they do not assure the achievement of a

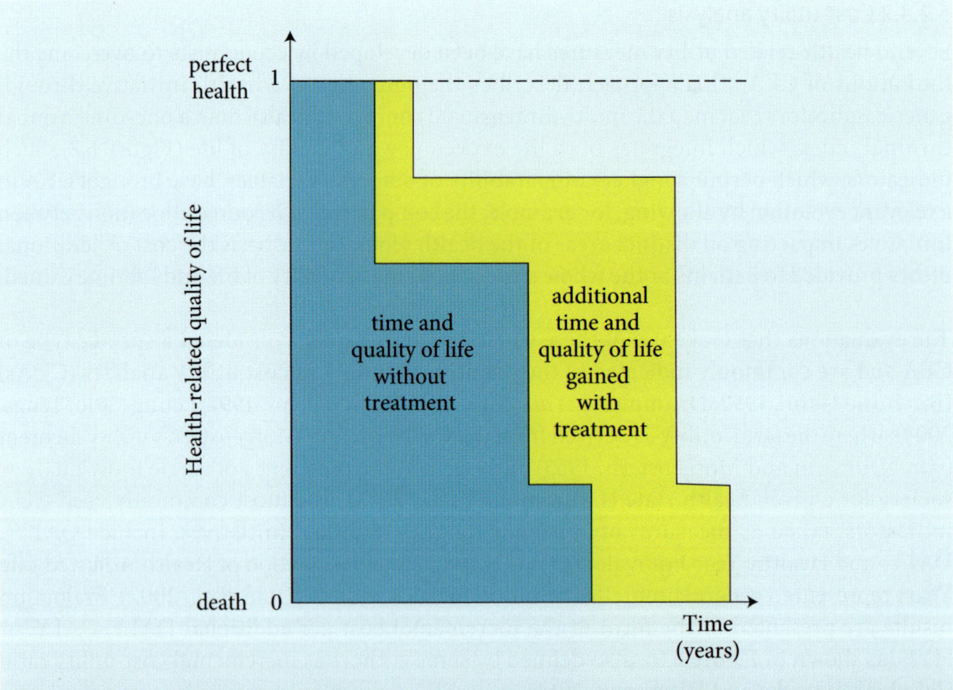

Figure 6.8. Perspectives of time and quality in the lives of patients receiving and not receiving treatment: the yellow area represents the incremental life quality and expectancy (QALYs) provided by the treatment.

societal optimum (Garber and Phelps 1997; Diamond and Kaul 2009; Weintraub and Cohen 2009). In the example of Figure 6.7, the solution I_1 results more effective, but also costlier than I_0 and is chosen because its incremental cost does not exceed the economic threshold. However, in the practice of health economic evaluations, the justifications for setting this kind of thresholds are in general weak and do not incorporate all the societal preferences needed to guarantee that the decision-making process issue an optimal solution (Nuijten and Dubois, 2011). For a discussions on the different CUA indices, see also: Bleichrodt (1995), Morrow and Bryant (1995), Anand and Hanson (1997), Murray and Acharya (1997), Dolan (2000), Gold *et al.* (2002), Mathers *et al.* (2003), and Devleesschauwer *et al.* (2014b).

DALYs, QALYs, and HYEs are generic indicators of health-state preference adaptable to assess interventions in almost all areas of health-care, which has made of CUA the most utilized economic evaluation tool in this sector (Drummond *et al.*, 2015). Beyond the described health adjusted life years-type measures, also disease-specific indicators have been elaborated to evaluate the particular state preference settings in many pathologies (Bowling, 1995; Fayers *et al.*, 2002; Guyatt *et al.*, 1986; Kirkley and Griffin, 2003), but their application is limited to the comparison between interventions targeting the disease concerned.

CUA indicators of generic health-state preference are already widely used to capture the effects on human health of initiatives addressed to improve animal and/or environmental health, as in Roth *et al.* (2003), Hutton (2008), Zinsstag *et al.* (2009) and Babo Martins and Rushton (2014). Therefore, they may result very useful also for One Health evaluations, where they could also be integrated in broader frameworks with other decision-making tools, e.g. multicriteria analysis, and weighed against other relevant outcomes (Hitziger *et al.*, 2018).

On this perspective, animal loss equivalents have been also proposed (Shaw *et al.*, 2017). They are calculated as a ratio between the sum of the monetary values of animal losses (mortality, fertility, production, weight, etc.), plus the animal health expenditure incurred by livestock owners and public veterinary services, on the one side, divided by the gross domestic income per capita, on the other side. This procedure intends to give account of the livestock contribution to economy and of the timework needed by an average worker to create the value that could replace the animal loss. According to the authors, in a zoonosis assessments animal loss equivalents could be added to the DALYs derived from human health impacts to issue a combined metric indicated as zoonotic DALY.

6.2.3.3 Cost-consequence analysis

Cost-consequence analysis (CCA) consists of a systematic presentation of the outcomes considered relevant for the decision-making process of a given initiative. These studies do not indicate the most efficient solutions, but just show the significant information and leave independent choice to decision makers (Coast 2004; Jacklin *et al.*, 2003; Kaufman and Watkins, 1996; Kaufman *et al.*, 1997). Although less popular than CEA, CUA, and CBA in disciplinary assessments, the flexibility of this approach can offer some advantages in the complex and multidisciplinary framework of the evaluation of One Health initiatives. Costs and outcomes can be calculated with respect to a counterfactual or can be merely a description of the current situation, resulting in partial evaluations (Section 6.2.1.1). CCA does not necessarily imply an efficiency assessment, and can incorporate different types of information, including qualitative.

This method is not rooted in the welfarist theory, which is seen as a strong limitation by some authors (Birch and Gafni, 2004; Claxton, 2005; Wilkinson, 1999), who point out that there is a risk of producing ad hoc analyses lacking theoretical background. Other authors argue that this methodology has the advantage of relative simplicity and transparency and recognizes the decision maker's role in weighting and prioritizing different outcomes (Coast, 2004). Table 6.3 provides a hypothetical example of a CCA of an *E. coli* integrated surveillance system. This integrates a range of costs and outcomes deemed relevant, highlighting who incurs specific costs. The decision maker then has an explicit role in establishing preferences across these outcomes and weighing them against each other, relying on either formal or informal decision-making processes (including multicriteria decision analysis; MCDA). Another example is in Mindekem *et al.* (2017), who compare cumulative costs of conventional and One Health approaches in initiatives against rabies in N'Djamena, Chad.

6.2.4 Uncertainty, sensitivity analysis and reporting results

Economic evaluations always imply many assumptions regarding correctness and relevance of both the data and methodologies utilised, which is cause of uncertainty about the results. The main types of uncertainty relate to the data requirements of the study, the extrapolation to outcomes beyond the primary data sources (e.g. it is assumed that CO_2 emission reduction has effects on global warming), generalisability of outcomes from a specific context to other contexts, and soundness of methodologies used (e.g. is it correct to discount future values related to human health or environmental states?) (Briggs, 1995, 2001). A common practice

Table 6.3. Cost-consequence analysis of a hypothetical integrated surveillance programme for *Escherichia coli*. All figures are hypothetical. For a published example of an *E. coli* integrated surveillance programme evaluation see Elbasha *et al.* (2000).

Outcomes	**Current practice**	**Integrated surveillance**
Human morbidity	200 cases/year	100 cases/year
Costs to small farmers	20 €/head of cattle; total € 500,000	40 €/head of cattle; total € 1,000,000
Other costs of surveillance	€ 1,000,000	€ 2,500,000
Costs to health-care sector	€ 6,000,000	€ 5,900,000
Involvement of farmers in surveillance mechanism	Farmer report low trust and engagement with surveillance mechanisms	Increased trust on health regulators and institutions. Increased engagement with surveillance

in economic evaluation is to analyse the impact of these uncertainties on the outcomes of the initiative by using sensitivity analysis. There are two main approaches to sensitivity analysis: deterministic and probabilistic.

In a deterministic sensitivity analysis, the examined parameters are changed by a certain value (e.g. assume intervention is effective in 75% of cases instead of 80%). These can be done for one parameter at a time (univariate) or for multiple parameters combined (multivariate). One-way deterministic analysis is easy to perform and to understand but has the limitation that it cannot provide information on the level of uncertainty and on which parameters the uncertainty depends for the most. Multivariate deterministic analysis brings only partial solutions to these problems since just two-ways analyses can be easily conducted and represented (although they are difficult to be interpreted when the two parameters examined are interdependent). With more than two variables this exercise becomes cumbersome to be executed and increasingly difficult to be illustrated and interpreted (Drummond *et al.*, 2015; Taylor, 2009; Walker and Fox-Rushby, 2001).

Scenario analysis is a type of multivariate analysis where multiple parameters are simultaneously set at values that are considered relevant for the investigation: e.g. in the best/worst case analysis the examined parameters are set at the levels respectively considered the most advantageous and the most disadvantageous with the aim of observing the sensitivity of the outcome under extreme circumstances. However, in general, the probability that the best or the worst combination of variables take place is very scarce, therefore even if the outcome results sensitive to the extreme combinations, this does not provide a useful information about the level of uncertainty of the evaluation (Drummond *et al.*, 2015; Walker and Fox-Rushby, 2001).

Probabilistic sensitivity analysis may overcome some of these shortcomings. In this type of exercise, a computer software attributes a probabilistic distribution to the examined parameters and estimates the model outcome a great number of times (e.g. between 1000 and 10,000) by picking sample values at random from the distribution of each parameter. The results of such iterations are graphically visualised to assess the impact of the parameter variability on the outcome (Baio and Dawid, 2008; Briggs, 2001; Drummond *et al.*, 2015; Edlin *et al.*, 2015; Taylor, 2009).

For example, if two alternative solutions are compared in a CEA or CUA, the findings can be presented as points of a scatter plot in a cost-effectiveness plane (Section 6.2.3.1), which allows to observe the possible variability of the outcome with respect to the positions of the plane's cartesian quadrants and the economic threshold. Figure 6.9 shows an example where the probability that the examined alternative results cost-effective (I_0 is the counterfactual) is represented by the points of the scatter located below the economic threshold (the yellow point represents the ICER). With more than two alternatives to be examined, a comparison with a scatter plot is not practicable. In these cases, evaluators make use of cost-effectiveness acceptability curves, which show the probability that each alternative is cost-effective by varying the economic threshold (Baio and Dawid, 2008; Drummond *et al.*, 2015; Edlin *et al.*, 2015).

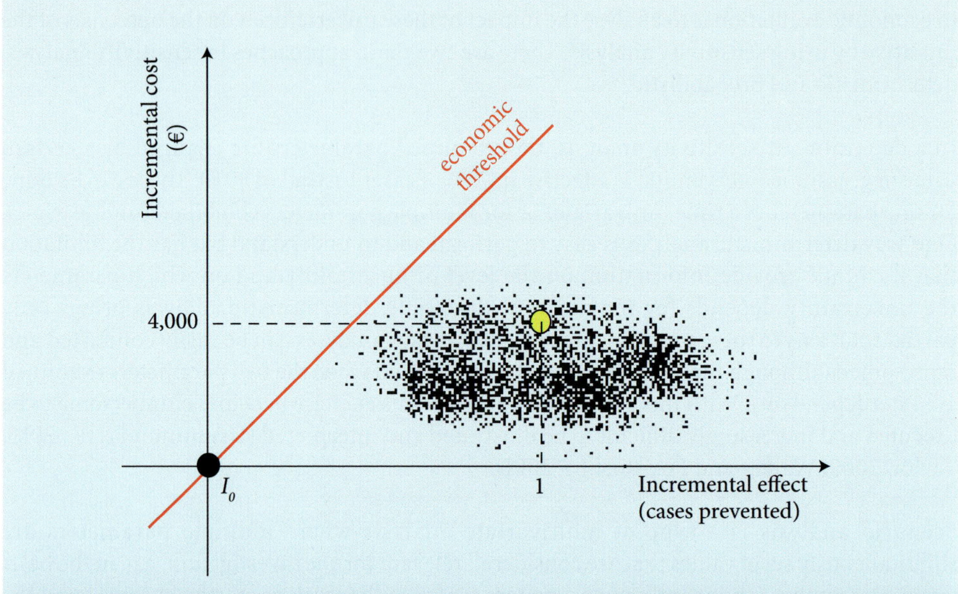

Figure 6.9. Results of a probabilistic sensitivity analysis of a hypothetical integrated surveillance programme for *Escherichia coli*. All figures are hypothetical. For a published example of an *E. coli* integrated surveillance programme evaluation see Elbasha *et al.* (2000).

6.2.5 Limitations and challenges for the use of economic evaluation techniques in the context of One Health

CEA, CUA and CBA offer a structured yet relatively flexible framework for economic evaluations. Additionally, there is a large body of theoretical and applied research on within these methodologies, in the fields of human, animal, and environmental health economics, which provides a solid base for their use to evaluate One Health initiatives. Nevertheless, many of the core assumptions of mainstream economic evaluation methods can be problematic in the context of One Health.

The framework for economic evaluation is mathematically formalised, designed to deal with quantitative information, and ill-equipped to incorporate qualitative information. Relevant outcomes of One Health evaluations include environmental attributes or elements of process, such as community engagement and trust, or knowledge, which are difficult to quantify. In addition, according to the *ceteris paribus* assumption, central to economic evaluations, all factors other than those directly considered as the variables for efficiency assessment (i.e. effects, costs, and benefits) are taken as fixed. In particular, the institutional framework is assumed to be unchanged by the initiative or during the course of it, as are the societal and individual preferences and cultural norms.

This type of assumption can be a reasonable simplification of reality in many cases. One Health initiatives, however, often consist precisely of institutional changes, encouraging cooperation, for example, between health systems and institutions concerned about biodiversity, animal health or land use, or with actors along the food supply chain. In Box 6.1B, for example, the setup of an integrated surveillance system for *E. coli* involves the collaboration between meat producers and health care providers. Increased collaboration between local institutions, communities and national government actors is also often a fundamental aspect of the initiative, leading to important changes in preferences, knowledge, and attitudes. In the case of vaccination of nomadic communities (Box 6.1A), researchers acknowledge the relevance of increased trust of nomad communities on the health care system as an outcome of the initiative, and comment on it, but do not include these process measures explicitly in the evaluation framework, which is restricted to quantitative information. Narrod *et al.* (2012) propose addressing the issue in a more formalised way, carrying out a KAP (knowledge, attitudes, and perceptions) assessment after the economic evaluation, to determine the factors that can affect the uptake of desirable, efficient One Health initiatives. This is one alternative, but we might also consider the possibility that these factors should determine what we consider to be a desirable initiative and should be incorporated into the evaluation itself. A cost-consequence framework could be appropriate for this type of analysis.

The question of how to interpret efficiency in One Health is a cross cutting issue which emerges from considering the previous limitations concerning multi-dimensional outcomes, the relevance of processes, the centrality of changes in population needs and institutional framework, the possibility of inter-generational and irreversible impacts, and the existence of various stakeholders representing potentially conflicting interests. Although efficiency is certainly an important criterion for evaluation, it should also be considered that this term has been trapped in the sustainability debate and is currently used with different and often contrasting meanings by scientists depending on their disciplinary and ideological backgrounds and aims (Garnett *et al.*, 2015). Hence, the use of efficiency as a general guiding principle in One Health evaluations should be carefully appraised. This does not mean to reduce the relevance of economic evaluation for this type of initiatives. Rather, it calls for a broader interpretation of the concepts and practices of economic evaluations and for conscious methodological choices to be undertaken case by case. Rather than considering different tools mutually exclusive, evaluations can gain explicative potential by using different methods complementarily, while being aware of the differing assumptions that might be implicit.

6.3 Dealing with complexity

6.3.1 Introduction

The aim of this section is to provide an overview on the main analytical tools that may allow economists to deal with complexity to investigate the relationships and interactions working among the economic, the environmental and social dimensions of One Health initiatives in the perspective of their evaluation.

Qualitative and multidimensional evaluation methods can provide a powerful tool for dealing with complexity, complementing quantitative information to provide a holistic assessment. Several authors (Jan, 1998; Ostrom, 2008) have argued for an increased use of mixed quantitative and qualitative methods for the evaluation of health and environmental interventions from an institutionalist perspective. In economics, the institutionalist school of thought places a strong emphasis on complexity, context, and non-linear causation (Menard and Shirley, 2005; Rutherford, 2001). Nevertheless, qualitative approaches remain infrequent in the economic evaluation literature and even when mixed (quantitative and qualitative) methods for evaluation are used, these are often not explicitly identified as such. The study indicated in Box 6.1A (Schelling *et al.*, 2007) could be considered an example of mixed evaluation, combining quantitative information of certain aspects of the evaluation (cost per women or child immunized in the nomadic communities), while relying on qualitative information for other aspects of the assessment, structured in a narrative form: e.g. improved reliance of local populations towards livestock and human vaccination programmes and public health services; contribution of population surveys to identify health services organisational failures causing disease outbreaks; enhanced collaboration between public health services and private veterinary services, etc.

Such considerations imply that economic evaluations understand the effects of One Health initiatives in their whole complexity. Some characteristics of One Health complexity that are relevant for economic evaluation have been already outlined (the non-linearity of the relationships occurring in health systems and the existence of feedbacks and loops, the large array of outcomes across sectors and their critical time-value profile, and the emerging properties of the context where the One Health initiatives are implemented), as well as the aspects that economics, and especially the mainstream marginalist approach, may disregard, mainly because of conceptual and methodological limits. In few words, the traditional economic approach, often resulting in formally impeccable models of the real world, may fail to grasp exactly what we need to know about the effects of One Health initiatives. On the other side, economics has however undergone a conceptual and methodological evolution, also allowing for a better consideration of complex situations.

Before entering the core of this section, we retrace the ways economists looks at the complexity of the real world and how economics has evolved to overcome, in part, its own limits. This is an unusual exercise: complexity as such has been rarely a main concern of the economic thinking, which prefers simplifications and reduces the reality to basic patterns to focus on specific analytical issues. In the following pages we provide a brief insight about some aspects of such evolution that may help to understand the limits and the possibilities of the economics to capture the complexity of One Health.

6.3.2 Economists and the complexity of the real world

Few milestones of economic reasoning and methodologies may well represent the way economists, and especially the neoclassic school, look at the understanding of economic phenomena: according to some postulates (e.g. scarcity of resources, and hedonism, rationality, and perfect information of economic agents) the analysis on consumers' and producers' behaviour is based on the so-called methodological individualism (Box 6.5). The relevant economic interaction between economic agents (i.e. producers and consumers) is the market, where the value of goods, identified with their price, is established by trading.

Box 6.5. Economic analysis and methodological individualism.

Economic analysis is mainly based on methodological individualism, an approach that explains social phenomena as the result of individual actions (Arrow, 1994). In economics, the methodology consists in setting elementary units such as the consumer or the producer and considering each of them as isolated from any other individual. This method is at the core of the analysis of producers' and consumers' choices and market dynamics. One of most known representation of this approach is the 'homo economicus', a virtual individual who reflects and conforms to all the fundamental hypothesis determining the behaviour of economic agents and is assumed to act perfectly rationally in order maximize an individual function of hedonistic benefit.

The equilibria of the economic agents are mathematically identified through the marginal analysis, which allows an optimal allocation of scarce resources to maximize an individual index (i.e. profit maximization for a producer or utility for consumer) according to specific functions representing the state of the nature (e.g. a production function or a cost function for the producer; a utility function for the consumer. Optimization is identified by the confrontation of incremental ratios: for a producer, for example, the equality between incremental (say marginal) cost and incremental (marginal) revenues.

The individual choices of economic agents about the use of resources are supposed to aim, hedonistically, at the maximization of individual benefits.

A consumer maximizes his benefit when he or she obtains with the available resources the maximum of utility, or satisfaction, from the consumption of goods and services. Thus, his/her purchases are driven by the ratio between the satisfaction attainable from the consumption of goods and services and the respective prices, under the income constraints. A producer, on the other side, aims to maximize profit as difference between the revenues from selling products and the cost of production. His/her choices are driven by the ratio between the productivity of inputs, their prices, and the selling price of products.

In a perfect market, where all consumers and producers are perfectly informed, and act rationally as described, and without influencing each other, the competitive behaviour of agents brings the price of the traded good to an equilibrium. At the equilibrium price all the supply of the good is sold, all the demand is satisfied, and all the agents obtain the maximum of benefit achievable. This corresponds to a Paretian optimum, i.e. the best possible solution that grants the optimal use of available resources, through individual choices of all the economic agents acting freely and selfishly. This marginalist interpretation of the Adam Smith's Invisible Hand concept is mainly due to Alfred Marshall (1890) and the so-called neoclassical school of economics, whose methods and approaches to economics have become largely prevailing in the academia worldwide.

Before Marshall, the marginalist economist Léon Walras (1874) had tried a mathematical formalisation of a general equilibrium theory. Since all sectors of the economy are interrelated and all goods and services can be somewhat subrogated by other goods and services, a price adjustment taking place for any reason (e.g. a bad harvest, a plant or animal disease outbreak, etc.) in the market of a given good also affects the market prices of all the other traded goods. This means that the condition of equilibrium may be defined only simultaneously in all the markets of all the traded goods and makes a practical application of the Walras' theory unlikely. To overcome this problem, Marshall introduced the *ceteris paribus* assumption allowing the analysis of one market by considering only its relevant variables under the hypothesis that all the other possible variables remain unchanged.

Few considerations about the way the marginalist-neoclassical approach looks at complexity emerge from the picture above:

- The mutual relationships among individuals are reduced to the market mechanisms that also regulate, through simple adjustments in the demand and supply functions, the transmission of economic effects across the different markets and industries.
- The concept of market equilibrium and the optimisation goals limit the analysis to a deterministic approach, which eliminates the possibility to consider evolutionary adaptation.
- The *ceteris paribus* assumption, largely adopted at any level of the theoretical and empirical analysis, results in the elimination of a series of variables determining the complexity, namely: the context (thrown out from the field of economic analysis by definition), the institutional framework and its evolution and the existence of feedbacks among actors.
- The social dimension of the analysis is reduced to a linear aggregation of individual behaviours (microeconomic equilibria) generating a macroeconomic equilibrium that involves the whole economy.

Such simplifications stem from the definition of the scientific field of economics and the analytical methods developed by marginalism. Reasoning in term of complexity, this approach could be considered a step back from the former political economy of the classical economists (Screpanti and Zamagni, 2005), but further conceptual developments occurred during the 20th century that improved the degree of complexity of the economic analysis.

A few pillars of this evolution should be mentioned because of their relevance in relation to the objective of this section, namely:

- The theories of welfare economics admitted the possibility that the market may not succeed in the optimal allocation of resources and opened to the idea that a greater social and economic welfare can be reached by correcting market failures through adequate incentives provided by the state (Pigou, 1920).
- The concept of externality, a particular case of market failure (Arrow, 1970; Pigou, 1920) showed that other mechanisms beside market can transmit the value of tangible and intangible goods across sectors, economic system and the whole society (e.g. environmental models, epidemiological models, social behaviour models are exemplification of alternative ways of transmitting values -usually not coinciding with market prices- which fit with the need to analyse complexity of effects occurring in the One Health context).

- The institutionalism and related schools of thought (Hodgson, 2000; Menard and Shirley, 2005; Rutherford, 2001) focused their attention on the economic nature and role of the institutions (i.e. any kind of rule governing the behaviour of the economic agents in a mutual evolving relationship), enlarging the traditional analysis to understand the effects of the *context* on the behaviour of the economic agents.
- The application of the game theory to economic analysis has revealed the dynamic nature of the economic equilibrium as the result of mutual strategic behaviours of economic agents (Nash, 1951), while the role of information in economic relationships and the removal of the rationality assumption funded the evolutionary approach in economics (Krugman, 1996).
- As in many other disciplines, several essays were made to re-build the economic analysis according to the fundamentals of the system thinking. This means to investigate the intrinsic complexity of economic phenomena through a system approach, which has been applied in different areas, from the level of the business units (Thompson and Valentinov, 2017), to the meso- and macroeconomic levels (Dopfer *et al.*, 2004), as well as in the teaching of economics (Colander, 2000; Moscardini *et al.*, 1999; Wheat, 2007). Thanks to this trend, rooted economic models have been reshaped according to system analysis, and new models created according to a system vision (Foster, 2005; Radzicki, 1990, 2009).

Many other developments would deserve attention to understand the conceptual complexity of the current economic analysis. In general, the strict assumptions of the marginalist-neoclassical approach have been revised and criticised along the last century, adding shades, or weakening the traditional foundations of economists' reasoning.

Though simplified and incomplete, the picture outlined above shows that economics evolved to deal with complexity. Reductionism of original economic models have different reasons and undeniable empirical convenience and the neoclassical approach remains at the core of the current economic evaluation tools, with its set of concepts, methods, and criteria (e.g. individual utility, social utility, efficiency; the conceptual background of optimisation criteria; the partial equilibrium approach applied in many evaluation; the CBA). But models are just tools to understand the reality, not the reality itself: 'In short, I believe that economics would be a more productive field if we learned something important from evolutionists: that models are metaphors, and that we should use them, not the other way around' (Krugman, 1996).

As problems evolve, economic concepts and theories should also evolve to face new situations. One Health initiatives create indeed new situations and problems that do require a methodological innovation to adequate the existing tools (or create new ones) to perform evaluations in complex frameworks. In the next sections we will see how economic evaluation can take advantage of new concepts and tools to comply with the needs of One Health evaluations. We will comply with this task going through a limited number of models that we deem relevant for the purpose, namely: the socio-ecological system framework, the agrarian system and the food supply chain analyses, the bio-economic models, the dynamic transmission models, and the multi-criteria analysis. A common trait shared by almost all these models is the reference to systems approach and systems thinking.

6.3.3 Social-ecological system framework

The premise of systems theory is the recognition that the structure of any system, the many interconnected relationships among its components, is as important in determining its behaviour as the individual components themselves (Pinstrup-Andersen and Watson, 2011). A systems approach to One Health places emphasis on the different types of structures that shape human, animal, plant and environmental health, including the geophysical and biological systems, the organisational systems in which people work and the political systems that govern public policies (Leischow and Milstein, 2006). Social-ecological systems (SESs) incorporate individuals and their environment by relating outcomes, such as health and wellbeing, to systemic interactions, which are influenced by a person's own behaviour, as well as the institutions and resources available within a given social, economic and political setting (Ostrom, 2007).

SESs are dynamic systems that are continuously changing (Schlüter *et al.*, 2014). They co-evolve through interactions between people, institutions, and resources constrained and shaped by a given social-ecological context (Holling and Gunderson, 2002). A SES can be defined as a comparatively bounded structure consisting of interacting, interrelated, or interdependent elements that form a whole and generally consists of a community that is situated within an environment, such as health systems and food systems. The term 'social-ecological' explicitly incorporates the social, institutional and cultural contexts of people-environment relations (Stokols, 1996). This perspective emphasises the multiple dimensions (physical, biological, and social) and multiple levels (individuals, groups, organisations) that interact within a complex system, which are inherent properties of One Health.

One of the major constraints of systems thinking in the context of One Health is the inherent interdisciplinary nature of One Health problems. Different disciplines use entirely different frameworks, theories, and models to analyse various parts of the complex multilevel whole. Ostrom (2009) posited that: 'Without a common framework to organize findings, isolated knowledge does not cumulate.' To address this constraint, this author and colleagues developed the now well-established multi-tier SES framework, which aims to allow knowledge accumulation (McGinnis and Ostrom 2014).

The framework, presented in Figure 6.10, conceptualises SESs into four highest-tier variables: (1) resource systems (e.g. a designated protected park with a distinct human-animal-wildlife interface); (2) resource units (e.g. trees, wildlife, amount and flow of water); (3) governance systems (e.g. organisations that manage the park, the specific rules related to how the park is used, and how these rules are made); and (4) actors (e.g. individuals that use the park for subsistence, recreation, or commercial purposes). All highest tiers affect and are affected by action situations, which denote, on the one side, the transformation of inputs, by actions of multiple actors, into outcomes. On the other side, feedback occurs from action situations to the tier categories. Each of these core subsystems contain multiple second-level variables, which are then further composed of lower-tier variables. The SES framework provides a common set of variables for organising research, so that isolated knowledge acquired from studies of diverse resource systems in different countries by bio-physical and social scientists can cumulate. Intuitively, by sharing data and information, synergy in the organisation of research should be more effective and the use of resources allocated to research more efficient.

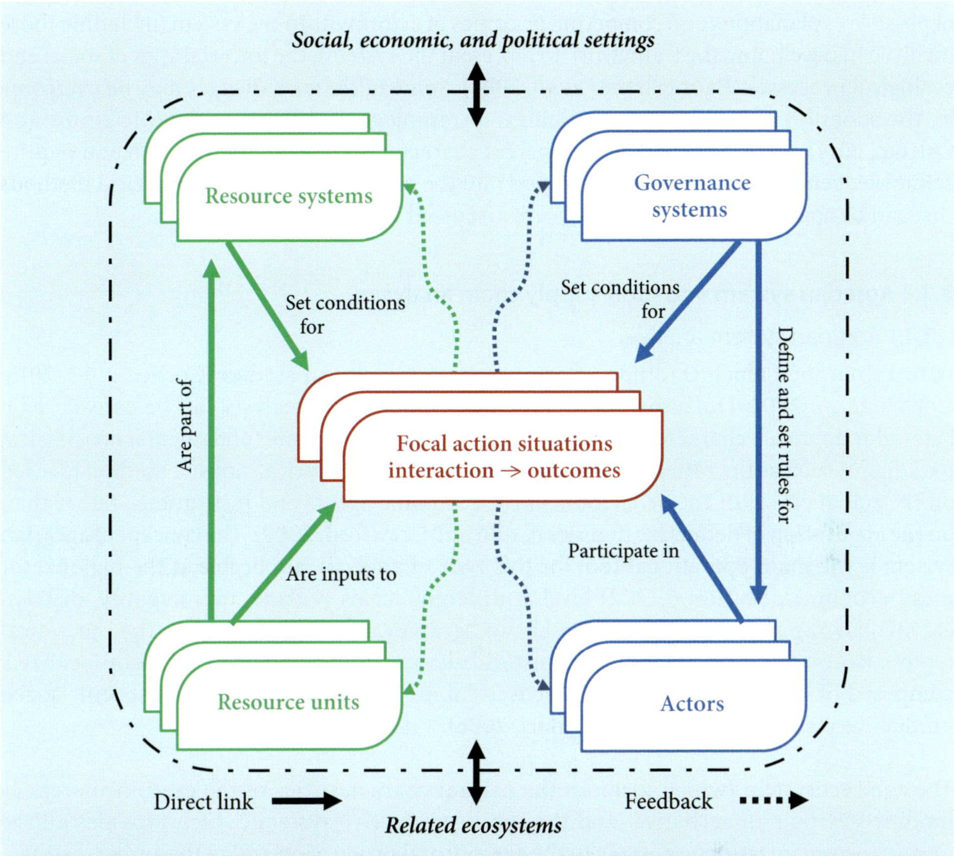

Figure 6.10. A depiction of the social-ecological system (SES) framework with multiple first-tier components (reproduced from McGinnis and Ostrom, 2014).

Zinsstag *et al.* (2011) have introduced human and animal health as quantitative and qualitative interaction and outcome of SESs in what they call 'health in social-ecological systems':

> Analogous to 'systems biology' which focuses mostly on the interplay of proteins and molecules at a sub-cellular level, a systemic approach to health in social-ecological systems (HSES) is an inter- and trans- disciplinary study of complex interactions in all health-related fields. HSES moves beyond 'one health' and 'eco-health', expecting to identify emerging properties and determinants of health that may arise from a systemic view ranging across scales from molecules to the ecological and socio-cultural context, as well from the comparison with different disease endemicities and health systems (Zinsstag *et al.*, 2011).

Challenges to using the SES framework include the difficulty in evaluating all components empirically, the integration of knowledge and theories from different disciplines, the variety of possible explanations and competing priorities of actors within the system (including those involved in developing the framework to represent the system), the uncertainties of social and ecological processes (Baumgärtner *et al.*, 2008). Some of these challenges may be overcome by the adoption of common methodology, terminology and frameworks (McGinnis and Ostrom, 2014), whilst others may be inherent characteristics of complex system and require acknowledgement but can be incorporated into the system (Ford, 2010). Empirical methods that can be applied to SES frameworks are discussed in Section 6.3.5.

6.3.4 Agrarian system and food supply chain analyses

6.3.4.1 Agrarian system analysis

Issued from the French tradition of comparative agricultural studies (Cochet, 2012, 2015; Cochet *et al.*, 2007; Dufumier, 2007), the agrarian system analysis can be considered a heterodox approach characterized by an inherent rejection of some fundamental neoclassical paradigms, such as the rational choice scheme, and by an empirical-holistic method focused on the actual causes of the behaviours of the economic agents and institutions, rather than on the application of deductive theories (Colin and Crawford, 2000). The concept of agrarian system is the main operational tool for this type of analysis, applicable at the regional (or meso-economic; Section 6.3.4.2) level at different scales (village, municipality, district, etc.). Following a structuralist approach, an agrarian system is defined as the theoretical representation of a type of agriculture historically constituted and geographically localized, composed of a distinctive cultivated ecosystem (or agro-ecosystem) and a specific social productive system (Mazoyer and Roudart, 2006).

The agro-ecosystem (which includes the natural characteristics of the examined areas as modified by human activities) and the social productive system (which provides all the means and organization necessary to the agricultural production and to the agro-ecosystem fertility regeneration) are both analysed through sub-systems. The social productive system, for example, includes the agricultural holdings, the farming systems (i.e. the organization of production at the level of the agricultural holdings, including cropping systems, animal production systems, means of production, human work, farm management, etc.), the system of interrelations and exchanges taking place within the agricultural holdings and with the external agents, up to embrace all the relevant elements of the social and economic structure of which the agrarian system is part (i.e. institutions, social organization, markets, agro-food supply chains, etc.)(Mazoyer and Roudart, 2006). The identification of sub-systems and all their components implies the setting of the agrarian system boundaries that are functional to the specific aims of each study. The analysis consists in studying the organization and the functioning of all the system components and the complex of their feedbacks.

Productivity and sustainability of agrarian systems depend on the maintenance of fertility in the exploited agro-ecosystems and on the technical and organizational capacities of society. The historical and spatial contextualization and the concurrence of environmental resource exploitation, technology, and social organization to the definition of productivity levels in a given society match the Marxist concept of 'mode of production' and allowed a similar application of the agrarian system concept to historical and geographical studies on agriculture (e.g. Devienne, 2011; Ducourtieux, 2015; Dufumier, 2006a,b; Le Coq *et al.*, 2001; Mazoyer and Roudart, 2006).

As an analytical tool of comparative agriculture, the agrarian system concept followed the application of this discipline to agricultural development studies (Dufumier, 2007), by elaborating specific methods of economic analysis and using techniques of rural participative appraisal for the evaluation of projects and policies. A distinctive feature of this approach is the regional diagnostic of farming systems (Cochet, 2015; Cochet and Devienne, 2006; Dufumier, 1996), where the historical and functional analysis of the various components of a regional agrarian system is completed by a survey operated through interviews to farmers and local experts that bring to define a typology of the agricultural holdings. This is a basis for a characterization of the farming systems, which are represented through mathematical models showing their technical and economic performances and allowing evaluations about the perspectives of the different types of holdings under different scenarios.

6.3.4.2 Food supply chain analysis

Supply chain, also often referred as 'value chain', is a concept widely used in the economic analysis of the agri-food industry that, despite some important conceptual differences (Aragrande and Argenti, 2001; Lebailly, 1990; Temple *et al.*, 2011; Terpend, 1997), can be intended as the most common English translation of the French *filière*. In such terms, the supply chain can be defined as a set of business activities involved in the production, processing and distribution of a given product or a given type of products (e.g. dairy products, meat products, fish, etc.). The supply chain business units are connected directly or indirectly through technical and economic links (Labonne, 1987; Raikes *et al.*, 2000; Shaffer, 1973; Terpend, 1997).

The technical links refer to the operations performed by the different units along the supply chain necessary to bring the product (e.g. an agricultural commodity) to the final consumption stage: provisions of raw materials and technical means, transportation, processing, quality and health control, packaging, labelling, conservation, stocking, marketing, wholesale, retail, delivering, etc. Economic links denote the commercial and contractual relations between supply chain operators (e.g. supplier-customer, farmer cooperatives and other producer organizations for product processing or marketing, marketing boards, supply chain agreements, etc.) that define the integration among the different operational units and the different industries involved.

The supply chain horizontal relationships qualify the links among the business units belonging to the same economic sector (e.g. among agricultural producers, or among industrial processors). The vertical relationships indicate the links occurring between firms classified in different sectors (e.g. between farmers and industrial processors and between the latter and the retailers). The complementary or side activities surround the basic vertical structure of the supply chain by supplying technical inputs and services (e.g. financial services, energy, communication, informatics, transportation, packaging, advertising, etc.) to the main sectors. Vertical and horizontal links define the basic structure and organization of the supply chain, schematically represented in Figure 6.11.

Despite the lack of a rigid theoretical formalization (Labonne, 1987) and the methodological eclecticism which characterize the studies in this field (Temple *et al.*, 2011), supply chain analysis is a very flexible and intuitive method to study complex economic phenomena related to the production of good and services. Differently from many economic studies, mainly based on isolated firms or on specific industries, the supply chain concept focuses on a system of relationships that, within the context of specific products and spatial dimensions, cross-cuts the traditional sectoral classifications of economic activities (primary production, manufactures, services) without falling in the generalization of the whole (macro) economic system. The supply chain actually identifies a meso-economic level (or meso-system), where multiple aspects (technical, economic, social) and complex relationships can be understood.

The supply chain analysis can be applied to complex phenomena such as social contexts, habits, regulatory frameworks, and relationships among operators. This may occur on the demand side, with the inclusion of social and cultural factors in modelling demand behaviour in different types of society (Malassis, 1979), as well as on the supply side from the perspective of one single enterprise, up to a global perspective, with the concepts of value chain (Porter,

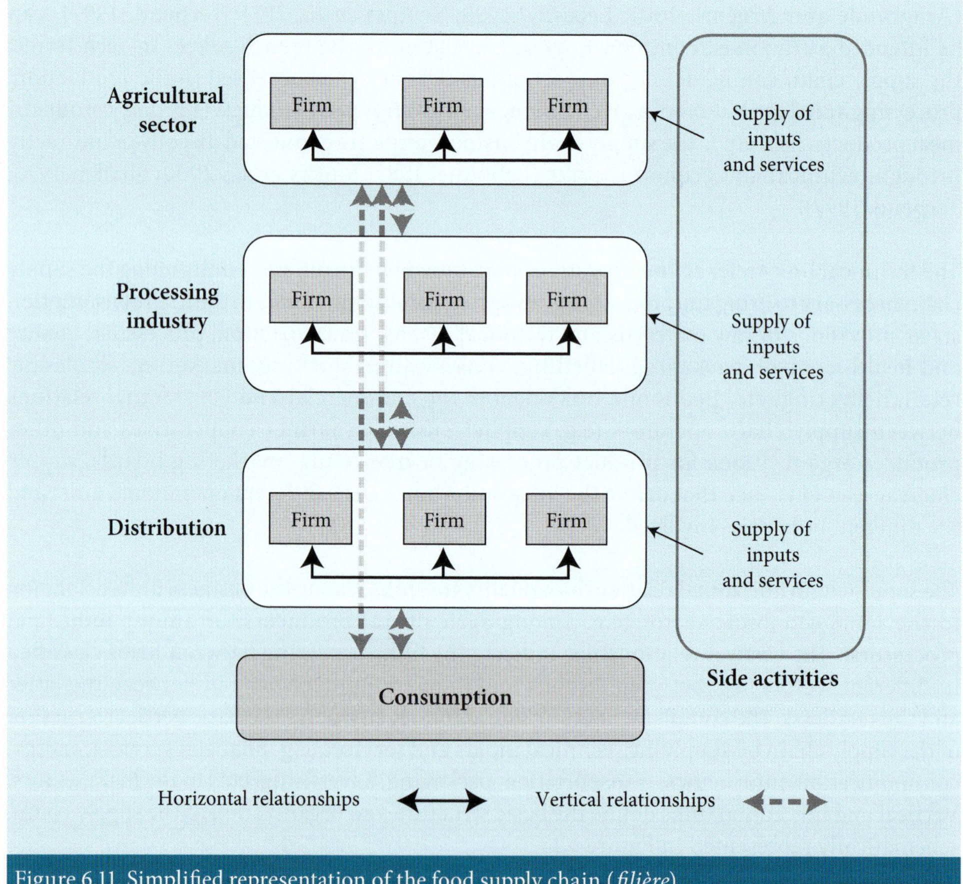

Figure 6.11. Simplified representation of the food supply chain (*filière*).

1985), global commodity chain (Raikes *et al.*, 2000), and global value chain (Gereffi *et al.*, 2005) revealing the means of worldwide governance of production in modern post-industrial economies by huge industrial groups. All these methods share a common inspiration to the basic concept of supply chain.

Supply chain management is a key issue for food safety and environmental protection (Canali *et al.*, 2017; Hammoudi *et al.*, 2015; Lang *et al.*, 2009; Nesheim *et al.*, 2015), which deeply involves One Health and related economic evaluations. On this perspective, the food supply chain analysis is a necessary tool to map the flows of materials and values in all the different activities connected to farming. This is crucial in the present-day context, when most of the income derived from food consumption in both developed and developing economies is not created by the farm sector, but in the upstream and downstream industries. In the United States, for example, out of $100 spent by consumers for food in 2012, only about $17 were destined to the farm sector, which contributed for less than 10% to the total added value generated along the food supply chain (Nesheim *et al.*, 2015). These figures explain the economic relevance of the activities siding animal and crop production and how health concerns in the food sector may have huge impacts on economy and society. Food scares have provided clear examples of the consequences of human, animal, and environmental health issues affecting the food supply chain, by showing that the collateral effects on economic activities may be costlier than the direct burden of the diseases (Adinolfi *et al.*, 2016; Aragrande and Canali, 2017; Buzby *et al.*, 1998; Hassouneh *et al.*, 2010, 2012; Hussain and Dawson, 2013; Livanis and Moss, 2005; Lloyd *et al.*, 2006).

6.3.4.3 Strength and weaknesses for One Health evaluations

A major limitation of supply chain analysis is seen in its prevailing qualitative and descriptive nature, as well as in the lack of adequate quantitative tools to fully account and hierarchize, for example, the impacts of specific animal health initiatives on the great variety of stakeholders involved (Rich and Perry, 2011; Rich *et al.*, 2011). Similar observations could be also advanced for the agrarian system analysis. However, such aspects can be considered inherent to these analytical approaches, which were not specifically conceived for health-related economic evaluations, but actually bring researchers to understand the wider context of animal and environmental health issues and identify, for each specific problem, the interactions that need to be deeply examined and the quantitative methods that are necessary to integrate the analysis (Jarvis and Valdes-Donoso, 2015; Rich *et al.*, 2011).

6.3.5 Dynamic transmission modelling

Dynamic models are mathematical models that account for time-dependent changes in the state of a system. Schlüter *et al.* (2014) defined dynamic model as a 'formal, theory- or empirically-based simplified mechanistic representation of the structure and processes of a real-world entity considered relevant to answer a specific question about the development of the system over time.' Based on this definition they are useful tools to study the change in SESs over time, particularly in situations for which time-series data are not available and experimentation is difficult.

Dynamic models have two distinct roles, prediction and understanding, which are related to the model properties of accuracy and transparency (Keeling and Danon, 2009). Predictive models usually require a high degree of accuracy, whereas transparency is a more important quality of models used to improve our understanding. Predictive models can be powerful tools in specific situations, guiding difficult policy decisions, where a trade-off between two or more alternative strategies or policies exist. Models can also be used to understand how dynamic systems behave in the real world, and how various complexities affect the dynamics. They provide an ideal world in which research can be conducted under 'experimental' conditions; individual factors can be examined in isolation and where every facet of the problem is recorded in perfect detail.

Using mathematical analysis to simulate interactions between the social and ecological components of the system, dynamic models facilitate the exploration of the consequences of salient relationships in the system. In turn, they can determine the system's sustainability and inform effective management strategies to improve the system being analysed. As such, models of SES are thought experiments for hypothesis generation and testing, particularly for exploring potential future development paths of a system under a given set of assumptions and potential pressures. Their usefulness for understanding the dynamics of SES is highest when they are part of a larger process of empirical and theoretical SES research (Baumgärtner *et al.*, 2008).

Dynamic modelling approaches commonly used for systems analysis include compartmental, network, agent-based models and system dynamics. In compartmental models, individuals in the population are divided into subgroups and the changes in the number of individuals in each subgroup are tracked over time based on different states. For example, mathematical analysis and modelling is central to infectious disease epidemiology and are used both in the generation and testing of hypotheses and the design of practical strategies for disease control (Grassly and Fraser, 2008; Keeling and Danon, 2009). The dynamics of infectious diseases among people, animals and plants result from the transmission of a pathogen either directly between hosts or indirectly through the environment, vectors, or intermediate hosts. The efficiency of transmission depends on the infectiousness of the infected host or hosts and the susceptibility of uninfected individuals who are exposed to infection. Infectiousness is a function of the biological, behavioural, and environmental context within which the pathogen is circulating.

Dynamic transmission studies constitute the backbone for dynamic economic modelling of One Health initiatives (Zinsstag *et al.*, 2005, 2009, 2017). The criteria that define an appropriate mathematical model with which to address a One Health question should be based on the principle of parsimony (chose the simplest model that explains the data), and the ability of the model to answer the question of interest (Grassly and Fraser, 2008). Therefore, the decision of which modelling approach to use and what level of detail to incorporate in the model should be based on the model's purpose. The usefulness of the model relies precisely on its ability to represent the key components of the system and their interactions, while ignoring the less important ones. There are many criticisms on the use and misuse of dynamic modelling and their limitations must be made explicit. Dynamic models of SESs are often based on assumptions about human behaviour or ecological dynamics that are uncertain but potentially have a substantial effect on model outcomes. Sources of uncertainty include: (1) lack of knowledge about people's decision-making processes; (2) how people value future benefits; and (3) the various processes that form a considerable component of natural resource dynamics, such as climatic variation (Schlüter *et al.*, 2014). Nevertheless, they provide useful tools in exploration of One Health scenarios, helping to improve our understanding and help to predict possible outcomes of One Health actions.

6.3.6 Bio-economic modelling

Bio-economic models define interrelations between economic and bio-physical variables with the aim to support decision-making in the management of biological resources. Typically, such models focus on how to maximize gains from economic exploitation of one or more species or ecosystems while maintaining the exploitation sustainable: e.g. the populations of the exploited organisms grow at given rates, depending on the carrying capacity of the ecosystems, and the exploitation should allow the maximum of profit compatible with those rates (Van der Ploeg *et al.*, 1987). Sectors like fishery, aquaculture, forestry, hunting, wild fauna, crop and animal farming, food security, agro-ecosystems, biodiversity, and environmental services have experienced the development of these tools (Brown, 2000; Brown and Hammack, 1972; Eggert, 1998; Janssen and Van Ittersum, 2007; Knowler, 2002; Kragt, 2012; Llorente and Luna, 2016; Mouysset *et al.*, 2011; Rewe and Kahi, 2012).

Bio-economic models are developed to respond to specific situations and problems, therefore the criteria for classifications may end up being extremely varied. Brown (2000) analyses the bio-physical or socio-economic components found in such models and identifies a first type of bio-economic models in biological process models enhanced with some economic components, e.g. accounting equations allowing cost-benefit assessments related to scenarios or management strategies simulated by the model. A second and opposite type is represented by economic optimization models that incorporate bio-physical features. In this case, the prevailing component of the model is socio-economic: examples are the models that account the multiple objectives of one decision-making unit (e.g. an agricultural holding) under different resource endowments and constraints. Integrated bio-economic models represent a convergence between the two previous types. They try to embody both to a large extent: the complex rationale of the economic optimisation models and the process simulation capacity of the biological process models (Brown, 2000).

Despite dating back several decades, integrating the two main components of bio-economic models remains extremely challenging, because of the inherent complexity. Table 6.4, reproduced from a literature review (Kragt, 2012), summarizes the main characteristics of bio-economic models from different sectors. The author of this study reported the scarcity of integrated studies and the lack of scientific foundation in many ecosystem evaluations. Economic optimization models are blamed for providing too simplistic representations of the underlying of natural processes at the origin of environmental transformations. On the other side, bio-economic integrated models represent complex systems, and the attempts to reduce complexity to make them more understandable and usable for stakeholders and decision makers, leading to the emergence of trade-offs that researchers are not always willing to resolve (Nielsen *et al.*, 2017). In this case, bio-scientists and ecologists often oppose the

Table 6.4. Identification of main characteristics of bio-economic models from different sectors based on a literature review (reproduced from Kragt, 2012).

	Forestry	Fisheries	Agriculture	Non-market values
Type of resource assessed	Single- or multi-species forest strands	Single- or multi-species fisheries	Representative farm systems	Variable
Modelling techniques	Linear/dynamic programming	Linear/dynamic programming, accounting	Accounting, regression, linear/dynamic programming	Environmental valuation techniques
Spatial scales and dynamics	Timber production per ha	Vary with habitat of study species	Paddock, whole-farm	Vary with resource under valuation
Temporal scales and dynamics	Steady state harvest models based on annual or seasonal changes in harvest activities			Typically, 15-30 year impacts
Bio-physical analysis	Mechanistic biological growth functions. Environmental conditions exogenous. Limited accounting for externalities or multiple ecosystem benefits			Environmental scenarios based on econometric considerations or expert opinion
Socio-economic analysis	Maximise net present value of profits from forestry, fish harvesting, crop and livestock production			Maximize net present value of allocating environmental resources across users and non-users

'irreducible complexity of ecosystem functioning', the idea of 'whole' as different from the sum of components, and holistic viewpoints to the reductionist approaches of economists (Kragt, 2012; Wam, 2010).

Herd models simulate livestock population dynamics under different parameters (e.g. mortality and fertility rates) to reproduce the effects of diseases or the outcome of disease control measures (Shaw, 2003). They can be classified as stochastic, when the parameters follow probability distributions, or deterministic, when the variables are taken as average values. Static models describe a heard steady state (e.g. James, 1995), while dynamic models track output changes over time resulting from animal health interventions (e.g. Doran, 2000). They are also distinguished for the prevailing targets, e.g. to assess the output value (economic models) or manage feed resources by considering a variety of biological parameters (bioeconomic models, e.g. Von Kaufmann *Et Al.*, 1990).

Progress in integrated bio-economic modelling requires improvements in the capacity of managing complexity by scientists and the interdisciplinary collaboration between bio-physical and socio-economic research (Kragt *et al.*, 2016), which is also the main challenges of One Health. The convergence between the One Health vision and the developments needed to the evolution of bio-economic modelling potentially makes it one of the most promising fields for economic research within the One Health context. On this perspective, a number of studies already gives relevant examples of models integrating environmental, epidemiological and socioeconomic variables in animal disease or welfare management (e.g. Collins and Part, 2013; Fenichel *et al.*, 2010, 2012; Getaneh *et al.*, 2017; Grace *et al.*, 2017; Horan *et al.*, 2008; Kingwell, 2002; Rich, 2007; Shwiff *et al.*, 2013; Sikhweni, 2014; Tschopp *et al.*, 2012).

6.3.7 Multicriteria decision analysis

Traditional tools for economic evaluation tend to focus on a single outcome measure or criterion. Different outcomes can then be assigned a monetary value to make them comparable and facilitate a synthetic judgement. As discussed in Section 6.2, this approach has limitations and may not be feasible or appropriate in all cases. MCDA offers an alternative approach, which can be particularly well-suited for the evaluation of One Health interventions. With reference to a set of alternative interventions examined, an MCDA can express, through a single synthetic judgement, a complex set of assessments related to all the criteria chosen for the evaluation by taking into account the priorities established among them. Criteria can include environmental sustainability, wildlife preservation, inequality of health impacts, or economic costs among others. Furthermore, MCDA allows for the consideration of both quantitative and qualitative information. This framework is consistent in principle with a decision maker perspective but can integrate several perspectives.

MCDA can be used as an additional step, where the results of the economic evaluation are considered alongside other criteria, such as ethical or political concerns. Alternatively, MCDA can constitute an integral part of the economic evaluation. Tiwari *et al.* (1999) apply an MCDA framework for environmental-economic decision making in irrigated agricultural lowlands. This study incorporates measures of monetized NPV from several perspectives (government, farmers, societal), alongside a range of non-monetized environmental sustainability criteria. In the area of health economics this MCDA is still not widely used, although some authors have advocated for a move towards this methodology, which can better reflect the range of concerns that decision makers face. Baltussen and Niessen (2006) propose a multi-criteria framework for economic evaluation integrating efficiency measures as well as other criteria such as potential to reduce health inequalities of vulnerable populations, or to respond to life-threatening situations. To improve knowledge integration in the governance of One health initiatives, Hitziger *et al.* (2018) indicate MCDA as a main tool to create a participative convergence among interests, preferences and values of the multiple actors involved and identify shared priorities for collective action.

Acknowledgements

This chapter is based upon work from COST Action 'Network for Evaluation of One Health' (TD1404), supported by COST (European Cooperation in Science and Technology).

References

Ackerman, F. and Heinzerling, L., 2005. Priceless: on knowing the price of everything and the value of nothing. The New Press, New York, NY, USA.

Adinolfi, F., Di Pasquale, J. and Capitanio, F., 2016. Economic issues on food safety. Ital J Food Saf 5: 5580.

Anand, S. and Hanson, K., 1997. Disability-adjusted life years: a critical review. J Health Econ 16: 685-702.

Anderson, E., 1988. Values, risks, and market norms. Philos Public Aff 17: 54-65.

Aragrande, M. and Argenti, O., 2001. Studying food supply and distribution systems to cities in developing countries and countries in transition: methodological and operational guide (revised version) (No. DT/36-01E), food into cities. FAO, Rome, Italy.

Aragrande, M. and Canali, M., 2017. Animal health and price transmission along livestock supply chains. OIE Rev Sci Tec 36: 87-96.

Arrow, K.J., 1951. Social choice and individual values. John Wiley & Sons, New York, NY, USA.

Arrow, K.J., 1970. Political and economic evaluation of social effects and externalities. In: Margolis, J. (ed.) The analysis of public output. Columbia University Press for National Bureau of Economic Research, New York / London, USA / UK, pp. 1-30.

Arrow, K.J., 1994. Methodological individualism and social knowledge. Am Econ Rev 84: 1-9.

Attfield, R., 1998. Existence value and intrinsic value. Ecol Econ 24: 163-168.

Babo Martins, S. and Rushton, J., 2014. Cost-effectiveness analysis: adding value to assessment of animal health, welfare and production. Rev Sci Tech Off Int Epiz 33: 681-689.

Baio, G. and Dawid, A.P., 2008. Probabilistic sensitivity analysis in health economics (No. 292), research report. University Colllege London, London, UK. Available at: http://discovery.ucl.ac.uk/1472742.

Baltussen, R. and Niessen, L., 2006. Priority setting of health interventions: the need for multi-criteria decision analysis. Cost Eff Resour Alloc 4: 14.

Barker, C. and Green, A., 1996. Opening the debate on DALYs. Health Policy Plan 11: 179-183.

Baumgärtner, S., Becker, C., Frank, K., Müller, B. and Quaas, M., 2008. Relating the philosophy and practice of ecological economics: the role of concepts, models, and case studies in inter- and transdisciplinary sustainability research. Ecol Econ 67: 384-393.

Berbel, J., Martin-Ortega, J. and Mesa, P., 2011. A cost-effectiveness analysis of water-saving measures for the Water Framework Directive: the case of the Guadalquivir River Basin in Southern Spain. Water Resour Manag 25: 623-640.

Birch, S. and Gafni, A., 1992. Cost effectiveness/utility analyses: do current decision rules lead us to where we want to be? J Health Econ 11: 279-296.

Birch, S. and Gafni, A., 2004. Evidence based health economics: answers in search of questions? In: Kristiansen, I.S. and Mooney, G. (eds.) Evidence-based medicine in its place. Routledge, Abingdon/ New York, UK/USA, pp. 163.

Birol, E., Asare-Marfo, D., Ayele, G., Mensa-Bonsu, A., Ndirangu, L., Okpukpara, B., Roy, D. and Yakhshilikov, Y., 2010. Investigating the role of poultry in livelihoods and the impact of avian flu on livelihoods outcomes in Africa (No. 01011). International Food Policy Research International, Washington, DC, USA.

Bleichrodt, H., 1995. QALYs and HYEs: under what conditions are they equivalent? J Health Econ 14: 17-37.

Blundell, R. and Costa Dias, M., 2005. Evaluation methods for non-experimental data. Fisc Stud 21: 427-468.

Boardman, A.E. and Greenberg, D.H., 1998. Discounting and the social discount rate. In: Thompson, F. and Green, M.T. (eds.) Handbook of public finance. Marcel Dekker, New York, NY, USA, pp. 269-318.

Bond, A., Pope, J., Morrison-Saunders, A. and Retief, F., 2016. A game theory perspective on environmental assessment: what games are played and what does this tell us about decision making rationality and legitimacy? Environ Impact Assess Rev 57: 187-194.

Bowling, A., 1995. Measuring disease: a review of disease-specific quality of life measurement scales. Open University Press, Buckingham, UK. Available at: https://tinyurl.com/y8vfsolk.

Bresnahan, B.W. and Dickie, M., 1995. Averting behavior and policy evaluation. J Environ Econ Manage 29: 378-392.

Brid Gleeson, H., 2007. House values, incomes, and industrial pollution. J Environ Econ Manage 54: 100-112.

Briggs, A., 1995. Handling the uncertainty in the results of economic evaluation (No. 32), OHE Briefing, London, UK.

Briggs, A., 2001. Handling uncertainty in economic evaluation. In: Drummond, M. and McGuire, A. (eds.) Economic evaluation in health care: merging theory with practice. Oxford University Press, Oxford, UK, pp. 172-214.

Brouwer, W.B.F., Culyer, A.J., Van Exel, N.J.A. and Rutten, F.F.H., 2008. Welfarism vs. extra-welfarism. J Health Econ 27: 325-338.

Brown, D.R., 2000. A review of bio-economic models. Paper prepared for the Cornell African Food Security and Natural Resource Management (CAFSNRM) Program. Ithaca, NY, USA.

Brown, G.M.J. and Hammack, J., 1972. A preliminary investigation of the economics of migratory waterflow. In: Krutilla, J.V. (ed.) Natural environments: studies in theoretical and applied analysis. Johns Hopkins University Press, Baltimore, MD, USA, pp. 171-204.

Buzby, J.C., Fox, J.A., Ready, R.C. and Crutchfield, S.R., 1998. Measuring consumer benefits of food safety risk reductions. J Agric Appl Econ South Agric Econ Assoc 30: 69-82.

Canali, M., Östergren, K., Amani, P., Aramyan, L., Sijtsema, S., Korhonen, O., Silvennoinen, K., Moates, G., Waldron, K. and O'Connor, C., 2014. Drivers of current food waste generation, threats of future increase and opportunities for reduction. FUSIONS Project, Wageningen UR, Wageningen, the Netherlands.

Carson, R.T., 2000. Contingent valuation: a user's guide. Environ Sci Technol 34: 1413-1418.

Clarke, P.M., 1998. Cost-benefit analysis and mammographic screening: a travel cost approach. J Health Econ 17: 767-787.

Claxton, K. and Culyer, A.J., 2006. Wickedness or folly? The ethics of NICE's decisions. J Med Ethics 32: 373-377.

Claxton, K.P., 2005. Cost consequences: implicit, opaque and anti scientific. BMJ 329: 1233.

Coast, J., 2004. Is economic evaluation in touch with society's health values? BMJ 329: 1233-1236.

Cochet, H. and Devienne, S., 2006. Fonctionnement et performances économiques des systèmes de production agricole: une démarche à l'échelle régionale. Cah Agric 15: 578-583.

Cochet, H., 2012. The systeme agraire concept in francophone peasant studies. Geoforum 43: 128-136.

Cochet, H., 2015. Comparative agriculture. Springer Netherlands, Dordrecht, the Netherlands.

Cochet, H., Devienne, S. and Dufumier, M., 2007. L'agriculture comparée, une discipline de synthèse? Écon Rural 297-298: 99-112.

Colander, D.C., 2000. The complexity vision and the teaching of economics. Edward Elgar Publishing, Glos, UK.

Colin, J. and Crawford, E.W., 2000. Economic perspectives in agricultural system analysis. Rev Agric Econ 22: 192-216.

Collins, L.M. and Part, C.E., 2013. Modelling farm animal welfare. Animals 3: 416-441.

Cullino, R., 1996. An estimate of environmental and defensive expenditures. In: Musu, I. and Siniscalco, D. (eds.) National accounts and the environment. Springer Netherlands, Dordrecht, the Netherlands, pp. 253-272.

Currie, J., Davis, L., Greenstone, M. and Walker, R., 2015. Environmental health risks and housing values: evidence from 1,600 toxic plant openings and closings. Am Econ Rev 105: 678-709.

De Scitovszky, T., 1941. A note on welfare propositions in economics. Rev Econ Stud 9: 77-88.

Devienne, S., 2011. Les États-Unis: une agriculture puissante, mais des espaces ruraux de moins en moins agricoles. In: Guibert, M. and Jean, Y. (eds.) Dynamiques des espaces ruraux dans le monde. Armand Colin, Paris, France, pp. 270-296.

Devleesschauwer, B., Havelaar, A.H., Maertens de Noordhout, C., Haagsma, J.A., Praet, N., Dorny, P., Duchateau, L., Torgerson, P.R., Van Oyen, H. and Speybroeck, N., 2014a. Calculating disability-adjusted life years to quantify burden of disease. Int J Public Health 59: 565-569.

Devleesschauwer, B., Havelaar, A.H., Maertens de Noordhout, C., Haagsma, J.A., Praet, N., Dorny, P., Duchateau, L., Torgerson, P.R., Van Oyen, H. and Speybroeck, N., 2014b. DALY calculation in practice: a stepwise approach. Int J Public Health 59: 571-574.

Diamond, G.A. and Kaul, S., 2009. Cost, effectiveness, and cost-effectiveness. Circ Cardiovasc Qual Outcome 2: 49-54.

Dolan, P., 2000. The measurement of health-related quality of life for use in resource allocation decisions in health care. Handb Heal Econ 1: 1723-1760.

Donaldson, C., Mugford, M., Vale, L., Farquhar, C., Brown, P., Jefferson, T., Demicheli, V., Ament, A., Evers, S., Goossens, M., De Vet, H., Van Tulder, M., Sonbo Kristiansen, I., Gosden, T., Birch, S., Drummond, M., Currie, G. and Mans, B., 2002. Evidence-based health economics: from effectiveness to efficiency in systematic review. John Wiley & Sons, New York, NY, USA, 180 pp.

Dopfer, K., Foster, J. and Potts, J., 2004. Micro-meso-macro. J Evol Econ 14: 263-279.

Doran, M., 2000. Socio-economics of trypanosomosis. Implications for control strategies within the Common Fly-Belt of Malawi, Mozambique, Zambia and Zimbabwe. Bovine trypanosomosis in Southern Africa. Vol. 3. Regional tsetse and trypanosomiasis control project for Southern Africa, Harare, Zimbabwe.

Drummond, M.F., Sculpher, M.J., Claxton, K., Stoddart, G.L. and Torrance, G.W., 2015. Methods for the economic evaluation of health care programmes, 4th edition. Oxford University Press, Oxford, UK.

Drummond, M.F., Sculpher, M.J., Torrance, G.W., O'Brein, B.J. and Stoddart, G.L., 2005. Methods for the economic evaluation of health care programmes, 3rd edition. Oxford University Press, New York, NY, USA.

Ducourtieux, O., 2015. Agriculture in the forest. Ecology and rationale of shifting cultivation. In: Peh, K.S.H., Corlett, R.T. and Bergeron, Y. (eds.) Routledge handbook of forest ecology. Routledge, Abingdon / New York, UK / USA, pp. 573-587.

Dufumier, M., 1996. Les projets de développement agricole: manuel d'expertise. Karthala, Paris, France.

Dufumier, M., 2006a. Biodiversité et agricultures paysannes des Tiers-Mondes. Ann Georgr 651: 550.

Dufumier, M., 2006b. Diversité des exploitations agricoles et pluriactivité des agriculteurs dans le Tiers Monde. Cah Agric 15: 584-588.

Dufumier, M., 2007. Agriculture comparée et développement agricole. Rev Tiers Monde 191: 611.

Dürr, S., Meltzer, M.I., Mindekem, R. and Zinsstag, J., 2008. Owner valuation of rabies vaccination of dogs, Chad Emerg Infect Dis 14: 1650-1652.

Edlin, R., McCabe, C., Hulme, C., Hall, P. and Wright, J., 2015. Outputs from probabilistic sensitivity analysis. Cost effectiveness modelling for health technology assessment. Springer International Publishing, Cham, Switzerland, pp. 163-175.

Eggert, H., 1998. Bioeconomic analysis and management. Environ Resour Econ 11: 3-4.

Elbasha, E.H., Fitzsimmons, T.D. and Meltzer, M.I., 2000. Costs and benefits of a subtype-specific surveillance system for identifying *Escherichia* coli O157:H7 outbreaks. Emerg Infect Dis 6: 293-293.

Fayers, P., Bottomley, A., EORTC Quality of Life Group and Quality of Life Unit, 2002. Quality of life research within the EORTC-the EORTC QLQ-C30. European Organisation for Research and Treatment of Cancer. Eur J Cancer 38: S125-133.

Fenichel, E.P., Horan, R.D. and Hickling, G.J., 2010. Management of infectious wildlife diseases: bridging conventional and bioeconomic approaches. Ecol Appl 20: 903-914.

Ferrero, G. and Neri, S., 2017. Monetary policy in a low interest rate environment. Banca d'Italia, Occasional Papers 392.

Ford, A. and Frederick, A., 2010. Modeling the environment. Island Press, Washington, DC, USA.

Foster, J., 2005. From simplistic to complex systems in economics. Cambridge J Econ 29: 873-892.

Freeman III, A.M., Herriges, J.A. and Kling, C.L., 2014. The measurement of environmental and resource values. Theory and methods, 3rd edition. RFF Press - Taylor and Francis and RFF Press - Routledge, Abingdon / New York, UK / USA.

Garber, A. and Phelps, C., 1997. Economic foundations of cost-effectiveness analysis. J Health Econ 16: 1-31.

Garnett, T., Roos, E. and Little, D., 2015. Lean, green, mean, obscene? What is efficiency and is it sustainable? Animal production and consumption reconsidered. University of Oxford, Oxford, UK.

Gereffi, G., Humphrey, J. and Sturgeon, T., 2005. The governance of global value chains. Rev Int Polit Econ 12: 78-104.

Getaneh, A.M., Mekonnen, S.A. and Hogeveen, H., 2017. Stochastic bio-economic modeling of mastitis in Ethiopian dairy farms. Prev Vet Med 138: 94-103.

Gittinger, J.P., 1985. Economic analysis of agricultural projects, 2nd edition. World Bank, Washington, DC, USA.

Gold, M.R., Stevenson, D. and Fryback, D.G., 2002. HALYs and QALYs and DALYs, oh my: similarities and differences in summary measures of population health. Annu Rev Publ Health 23: 115-134.

Görlach, B., Interwies, E. and Newcombe, J., 2006. How are we performing? The role of ex-post cost-effectiveness-analyses in European environmental policies. Berlin Conference on the Human Dimensions of Global Environmental Change, Berlin, Germany, pp. 25.

Grace, D., Lindahl, J., Wanyoike, F., Bett, B., Randolph, T. and Rich, K.M., 2017. Poor livestock keepers: ecosystem-poverty-health interactions. Philos Trans R Soc Lond B Biol Sci 372(1725): 20160166.

Grassly, N.C. and Fraser, C., 2008. Mathematical models of infectious disease transmission. Nat Rev Microbiol 6: 477-487.

Guyatt, G.H., Bombardier, C. and Tugwell, P.X., 1986. Measuring disease-specific quality of life in clinical trials. CMAJ 134: 889-895.

Halasa, Y.A., Shepard, D.S., Wittemberg, E., Fonseca, D.M., Farajollahi, A., Healy, S.P., Gaugler, R., Strickman, D.A. and Clark, G.G., 2012. Willingness-to-pay for an area-wide integrated pest management program to control the Asian tiger mosquito in New Jersey. J Am Mosq Control Assoc 28: 225-236.

Hammoudi, A., Grazia, C., Surry, Y. and Traversac, J.-B., 2015. Food safety, market organization, trade and development. Springer International Publishing, Cham, Switzerland.

Hanley, N., Mourato, S. and Wright, R.E., 2002. Choice modelling approaches: a superior alternative for environmental valuation? J Econ Surv 15: 435-462.

Hansson, H. and Lagerkvist, C.J., 2015. Identifying use and non-use values of animal welfare: evidence from Swedish dairy agriculture. Food Policy 50: 35-42.

Hansson, H. and Lagerkvist, C.J., 2016. Dairy farmers' use and non-use values in animal welfare: determining the empirical content and structure with anchored best-worst scaling. J Dairy Sci 99: 579-592.

Hansson, S.O., 2007. Philosophical problems in cost-benefit analysis. Econ Philos 23: 163-183.

Harrington, W., Mcconnell, V. and Ando, A., 1999. The enhanced I/M program in Arizona: costs, effectiveness, and a comparison with pre-regulatory estimates (No. 99-37). Resources for the Future, Washington, DC, USA.

Häsler, B., Cornelson, L., Bennani, H. and Rushton, J., 2014. A review of the metrics for One Health benefits. Sci Tech Rev Off Int des Epizoot 33: 453-464.

Häsler, B., Gilbert, W., Jones, B.A., Pfeiffer, D.U., Rushton, J. and Otte, M.J., 2012. The economic value of One Health in relation to the mitigation of zoonotic disease risks. Curr Top Microbiol Immunol 365: 127-151.

Hassouneh, I., Radwan, A., Serra, T. and Gil, J.M., 2012. Food scare crises and developing countries: the impact of avian influenza on vertical price transmission in the Egyptian poultry sector. Food Policy 37: 264-274.

Hassouneh, I., Serra, T. and Gil, J.M., 2010. Price transmission in the Spanish bovine sector: the BSE effect. Agric Econ 41: 33-42.

Heckman, J.J., Ichimura, H. and Todd, P.E., 1997. Matching as an econometric evaluation estimator: evidence from evaluating a job training programme. Rev Econ Stud 64: 605-654.

Heinzerling, L., Ackerman, F. and Massey, R., 2005. Applying cost-benefit to past decisions: was environmental protection ever a good idea? Adm Law Rev 57: 155-192.

Hicks, J.R., 1939. The foundations of welfare economics. Econ J 49: 696-712.

Hicks, J.R., 1943. The four consumer's surpluses. Rev Econ Stud 11: 31-41.

Hitziger, M., Esposito, R., Canali, M., Aragrande, M., Häsler, B. and Rüegg, S.R., 2018. Knowledge integration in One Health policy formulation, implementation and evaluation. Bull World Health Organ 96: 211-218.

HM Treasury, 2003. The green book: appraisal and evaluation in Central Government Treasury Guidance. HM Treasury, London, UK.

Hodgson, G.M., 2000. What is the essence of institutional economics? J Econ Issues 34: 317-329.

Holling, C.S. and Gunderson, L.H., 2002. Resilience and adaptive cycles. In: Gunderson, L.H. and Holling, C.S. (eds.) Panarchy: understanding transformations in human and natural systems. Island Press, Washington, DC, USA, pp. 25-62.

Holston, K., Laubach, T. and Williams, J.C., 2017. Measuring the natural rate of interest: international trends and determinants. J Int Econ 108: S59-S75.

Horan, R.D., Wolf, C.A. and Fenichel, E.P., 2012. Dynamic perspectives on the control of animal disease: merging epidemiology and economics. In: Zilberman, D., Otte, J., Roland-Holst, D. and Pfeiffer, D. (eds.) Health and animal agriculture in developing countries. Springer, New York, NY, USA, pp. 101-118.

Horan, R.D., Wolf, C.A., Fenichel, E.P. and Mathews, K.H., 2008. Joint management of wildlife and livestock disease. Environ Resource Econ 41: 47-70.

Hussain, M.A. and Dawson, C.O., 2013. Economic impact of food safety outbreaks on food businesses. Foods 2: 585-589.

Hutton, G., 2008. Economic evaluation of environmental health interventions to support decision making. Environ Health Insight 2: 137-155.

Ichimura, H. and Taber, C., 2000. Direct estimation of policy impacts (No. 254). NBER Technical Working Papers, Cambridge, MA, USA.

Jacklin, P.B., Roberts, J.A., Wallace, P., Haines, A., Harrison, R., Barber, J.A., Thompson, S.G., Lewis, L., Currell, R., Parker, S. and Wainwright, P., 2003. Virtual outreach: economic evaluation of joint teleconsultations for patients referred by their general practitioner for a specialist opinion. BMJ 327(7406): 84.

Jakubiak-Lasocka, J. and Jakubczyk, M., 2014. Cost-effectiveness versus cost-utility analyses: what are the motives behind using each and how do their results differ? A Polish example. Value Health Reg Issues 4: 66-74.

James, A.D., 1995. Methods for the assessment of disease on livestock productivity. Proceedings of the Society for Veterinary Epidemiology and Preventive Medicine. Society for Veterinary Epidemiology and Preventive Medicine, Edinburgh, UK, pp. 156-161.

Jan, S., 1998. A holistic approach to the economic evaluation of health programs using institutionalist methodology. Soc Sci Med 47: 1565-1572.

Janssen, S. and Van Ittersum, M.K., 2007. Assessing farm innovations and responses to policies: a review of bio-economic farm models. Agric Syst 94: 622-636.

Jarvis, L.S. and Valdes-Donoso, P., 2015. A selective review of the economic analysis of animal health management. J Agric Econ 69: 201-225.

Jeuland, M., Lucas, M., Clemens, J. and Whittington, D., 2010. Estimating the private benefits of vaccination against cholera in Beira, Mozambique: a travel cost approach. J Dev Econ 91: 310-322.

Jevons, W.S., 1871. The theory of political economy, 3rd edition. Macmillan and Co., London, UK.

Johannesson, M., 1995. The relationship between cost-effectiveness analysis and cost-benefit analysis. Soc Sci Med 41: 483-489.

Kaldor, N., 1939. Welfare propositions of economics and interpersonal comparisons of utility. Econ J 49: 549-552.

Kaufman, R. and Watkins, R., 1996. Costs-consequences analysis. Hum Resour Dev Q 7: 87-100.

Kaufman, R., Watkins, R., Sims, L., Crispo, N.S., Hall, J.C. and Sprague, D.E., 1997. Cost-consequences analysis: a case study. Perform Improv Q 10: 7-21.

Keeling, M.J. and Danon, L., 2009. Mathematical modelling of infectious diseases. Br Med Bull 92: 33-42.

Kelleher, J.C. and MacCormack, J.J., 2004. Internal rate of return: a cautionary tale. McKinsey Finance 2004: 71-75.

Kelman, S., 1981. Cost-benefit analysis: an ethical critique. AEI J Gov Soc Regul 5: 33-40.

Kingwell, R., 2002. Sheep animal welfare in a low rainfall Mediterranean environment: a profitable investment? Agric Syst 74: 221-240.

Kirkley, A. and Griffin, S., 2003. Development of disease-specific quality of life measurement tools. Arthrosc J Arthrosc Relat Surg 19: 1121-1128.

Klose, T., 1999. The contingent valuation method in health care. Health Policy 47: 97-123.

Knowler, D., 2002. A review of selected bioeconomic models with environmental influences in fisheries. J Bioeconomic 4: 163-181.

Kragt, M.E., 2012. Bioeconomic modelling: integrating economic and environmental systems? In: Seppelt, R., Voinov, A.A., Lange, S. and Bankamp, D. (eds.) Managing resources of a limited planet: pathways and visions under uncertainty. International Congress on Environmental Modelling and Software, sixth Biennial Meeting. International Environmental Modelling and Software Society, Leipzig, Germany, pp. 1628-1636.

Kragt, M.E., Pannell, D.J., Mcvittie, A., Stott, A.W., Ahmadi, B.V. and Wilson, P., 2016. Improving interdisciplinary collaboration in bio-economic modelling for agricultural systems. Agric Syst 143: 217-224.

Krugman, P., 1996. What economists can learn from evolutionary theorists. Available at: http://web.mit.edu/krugman/www/evolute.html.

Kula, E., 2006. Social discount rate in cost-benefit analysis – the British experience and lessons to be learned. Milan European Economy Workshops, Milan, Italy.

Labonne, M., 1987. Sur le concept de la filière en économie agro-alimentaire. In: Kermel-Torrès, D., Roca, P.-J., Bruneau, M. and Courade, G. (eds.) Terres, comptoirs et silos: des systèmes de production aux politiques alimentaires. ORSTOM, Paris, France, pp. 137-149.

Lang, T., Barling, D. and Caraher, M., 2009. Food policy: Integrating health, environment and society. Oxford University Press, Oxford, UK.

Laurila-Pant, M., Lehikoinen, A., Uusitalo, L. and Venesjärvi, R., 2015. How to value biodiversity in environmental management? Ecol Indic 55: 1-11.

Le Coq, J.F., Dufumier, M. and Trébuil, G., 2001. History of rice production in the Mekong delta. 3rd Euroseas Conference. September 6-8, 2001. London, UK, pp. 27.

Lebailly, P., 1990. Concept de filière, économie agro-alimentaire et développement. Tropicult 81: 9-14.

Leischow, S.J. and Milstein, B., 2006. Systems thinking and modeling for public health practice. Am J Public Health 96: 403-405.

Leung, L., 2016. Health economic evaluation: a primer for healthcare professionals. Prim Heal Care Open Access 6: 1-8.

Livanis, G. and Moss, C.B., 2005. Price transmission and food scares in the U.S. beef sector. American Agricultural Economics Association Meetings, Providence, RI, pp. 19.

Llorente, I. and Luna, L., 2016. Bioeconomic modelling in aquaculture: an overview of the literature. Aquac Int 24: 931-948.

Lloyd, T.A., McCorriston, S., Morgan, C.W. and Rayner, A.J., 2006. Food scares, market power and price transmission: the UK BSE crisis. Eur Rev Agric Econ 33: 119-147.

Lock, K., Smith, R.D., Dangour, A.D., Keogh-Brown, M., Pigatto, G., Hawkes, C., Fisberg, R.M. and Chalabi, Z., 2010. Health, agricultural, and economic effects of adoption of healthy diet recommendations. Lancet 376: 1699-1709.

Lyon, R.M., 1990. Federal discount rate policy, the shadow price of capital, and challenges for reforms. J Environ Econ Manage 18: S29-S50.

Malassis, L., 1979. Économie agro-alimentaire. Tome 1: Économie de la consommation et de la production agro-alimentaire. Cujas, Paris, France.

Marshall, A., 1890. Principles of economics. MacMillian and Co., London, UK.

Mathers, C.D., Salomon, J.A., Murray, C.J.L. and Lopez, A.D., 2003. Alternative summary measures of average population health. In: Murray, C.J.L. and Evans, D.B. (eds.) Health systems performance assessment: debates, methods and empiricism. World Health Organization, Geneva, Switzerland, pp. 319-334.

Mazoyer, M. and Roudart, L., 2006. A history of world agriculture: from the neolithic age to the current crisis. Monthly Review Press, London, UK.

McClelland, G.H., Schulze, W.D., Lazo, J.K., Waldman, D.M., Doyle, J.K., Elliott, S.R., Irwin, J.R. and Carlin, A., 1992. Methods for measuring non-use values, a contingent valuation study of groundwater cleanup. US Environmental Protection Agency. Available at: https://tinyurl.com/y8jth4jj.

McGinnis, M.D. and Ostrom, E., 2014. Social-ecological system framework: initial changes and continuing challenges. Ecol Soc 19: 30.

McInerney, J., 2004. Animal welfare, economics and policy. Farm and Animal Health Economics Division of Defra, Exeter, UK.

Menard, C. and Shirley, M.M. (eds.), 2005. Handbook of new institutional economics. Springer-Verlag, Berlin/Heidelberg, Germany.

Mindekem, R., Lechenne, M.S., Naissengar, K.S., Oussiguéré, A., Kebkiba, B., Moto, D.D., Alfaroukh, I.O., Ouedraogo, L.T., Salifou, S. and Zinsstag, J., 2017. Cost description and comparative cost efficiency of post-exposure prophylaxis and canine mass vaccination against rabies in N'Djamena, Chad Front Vet Sci 4: 38.

Morrow, R.H. and Bryant, J.H., 1995. Health policy approaches to measuring and valuing human life: conceptual and ethical issues. Am J Public Health 85: 1356-1360.

Moscardini, A.O., Lawler, K. and Loutfi, M., 1999. Using system dynamic's methodology to teach macroeconomics. In: Cavana, R.Y., Vennix, J.A.M., Rouwette, E.A.J.A., Stevenson-Wright, M. and Candlish, J. (eds.) Systems thinking for the next millennium. 17th International Conference of the System Dynamics Society and 5th Australian and New Zealand Systems Conference. July 20-23, 1999. System Dynamics Society, ANZSYS, Wellington, New Zealand.

Mouysset, L., Doyen, L., Jiguet, F., Allaire, G. and Leger, F., 2011. Bio economic modeling for a sustainable management of biodiversity in agricultural lands. Ecol Econ 70(4): 617-626.

Murray, C.J.L. and Acharya, A.K., 1997. Understanding DALYs. J Health Econ 16: 703-730.

Narrod, C., Zinsstag, J. and Tiongco, M., 2012. A One Health framework for estimating the economic costs of zoonotic diseases on society. Ecohealth 9: 150-162.

Nash, J., 1951. Non-cooperative games. Ann Math 54: 286.

Nesheim, M.C., Oria, M. and Yih, P.T. (eds.), 2015. A framework for assessing effects of the food system. National Academies Press, Washington, DC, USA.

Nielsen, J.R., Thunberg, E., Holland, D.S., Schmidt, J.O., Fulton, E.A., Bastardie, F., Punt, A.E., Allen, I., Bartelings, H., Bertignac, M., Bethke, E., Bossier, S., Buckworth, R., Carpenter, G., Christensen, A., Christensen, V., Da-Rocha, J.M., Deng, R., Dichmont, C., Doering, R., Esteban, A., Fernandes, J.A., Frost, H., Garcia, D., Gasche, L., Gascuel, D., Gourguet, S., Groeneveld, R.A., Guillén, J., Guyader, O., Hamon, K.G., Hoff, A., Horbowy, J., Hutton, T., Lehuta, S., Little, L.R., Lleonart, J., Macher, C., Mackinson, S., Mahevas, S., Marchal, P., Mato-Amboage, R., Mapstone, B., Maynou, F., Merzéréaud, M., Palacz, A., Pascoe, S., Paulrud, A., Plaganyi, E., Prellezo, R., van Putten, E.I., Quaas, M., Ravn-Jonsen, L., Sanchez, S., Simons, S., Thébaud, O., Tomczak, M.T., Ulrich, C., Van Dijk, D., Vermard, Y., Voss, R. and Waldo, S., 2017. Integrated ecological-economic fisheries models-evaluation, review and challenges for implementation. Fish Fisheries 19(1): 1-29.

Nirmala, G.M., 2014. Averting expenditure – measure of willingness to pay. J Environ Earth Sci 4: 28-32.

Nuijten, M.J.C. and Dubois, D.J., 2011. Cost-utility analysis: current methodological issues and future perspectives. Front Pharmacol 2: 29.

O'Neil, J., 1993. Ecology, policy, and politics: human well-being and the natural world. Routledge, London, UK.

Ostrom, E., 2007. A diagnostic approach for going beyond panaceas. Proc Natl Acad Sci USA 104: 15181-15187.

Ostrom, E., 2008. Doing institutional analysis: digging deeper than markets and hierarchies. In: Handbook of new institutional economics. Springer, New York, NY, USA, pp. 819-848.

Ostrom, E., 2009. A general framework for analyzing sustainability of social-ecological systems. Science 325: 419-422.

Otte, M.J., Nugent, R. and McLeod, A., 2004. Transboundary animal diseases: assessment of socio-economic impacts and institutional responses (No. 9). Livestock Policy Discussion Paper. FAO, Rome, Italy.

Pareto, V., 1894. Il massimo di utilità dato dalla libera concorrenza. G Degli Econ 9: 48-66.

Pearce, D., Atkinson, G. and Mourato, S., 2006. Cost-benefit analysis and the environment. Recent developments. OECD, Paris, France.

Pigou, A.C., 1920. The economics of welfare. MacMillan, London, UK.

Pinstrup-Andersen, P. and Watson, D.D., 2011. Food policy for developing countries: the role of government in global, national, and local food systems. Cornell University Press, Ithaca, NY, USA.

Porter, M.E., 1985. Competitive advantage: creating and sustaining superior performance. Free Press and Collier MacMillan, New York, NY, USA.

Portney, P.R., 1981. Housing prices, health effects, and valuing reductions in risk of death. J Environ Econ Manage 8: 72-78.

Radzicki, M.J., 1990. Methodologia oeconomiae et systematis dynamis. Syst Dyn Rev 6: 123-147.

Radzicki, M.J., 2009. System dynamics and its contribution to economics and economic modelling. Complex systems in finance and econometrics. Springer, New York, NY, pp. 727-737.

Raftery, J., 1999. Methodological limitations of cost-effectiveness analysis in health care: implications for decision making and service provision. J Eval Clin Pract 5: 361-366.

Raikes, P., Friis Jensen, M. and Ponte, S., 2000. Global commodity chain analysis and the French filière approach: comparison and critique. Econ Soc 29: 390-417.

Remme, M., Vassall, A., Lutz, B., Luna, J. and Watts, C., 2014. Financing structural interventions: going beyond HIV-only value for money assessments. AIDS 28: 425-434.

Rewe, T.O. and Kahi, A., 2012. Design and application of bio-economic modelling in livestock genetic improvement in Kenya: a review. Egert J Sci Technol 12: 113-119.

Rich, K.M. and Perry, B.D., 2011. The economic and poverty impacts of animal diseases in developing countries: new roles, new demands for economics and epidemiology. Prev Vet Med 101: 133-147.

Rich, K.M., 2007. New methods for integrated models of animal disease control. Selected paper prepared for the 2007 American Agricultural Economics Association Meetings. July 29- August 1, 2007. Portland, OR, USA.

Rich, K.M., Ross, R.B., Baker, A.D. and Negassa, A., 2011. Quantifying value chain analysis in the context of livestock systems in developing countries. Food Policy 36: 214-222.

Robbins, L., 1932. An essay on the nature and significance of economic science. Macmillan and Co., London, UK.

Robinson, R., 1993. Cost-effectiveness analysis. Bio-Medical J 307: 793-795.

Rojo-Gimeno, C., Postma, M., Dewulf, J., Hogeveen, H., Lauwers, L. and Wauters, E., 2016. Farm-economic analysis of reducing antimicrobial use whilst adopting improved management strategies on farrow-to-finish pig farms. Prev Vet Med 129: 74-87.

Roth, F., Zinsstag, J., Orkhon, D., Chimed-Ochir, G., Hutton, G., Cosivi, O., Carrin, G. and Otte, J., 2003. Human health benefits from livestock vaccination for brucellosis: case study. Bull World Health Organ 81: 867-876.

Rutherford, M., 2001. Institutional economics: then and now. J Econ Perspect 15: 173-194.

Sagoff, M., 1988. Some problems with environmental economics. Environ Ethics 10: 55-74.

Samuelson, P.A., 1942. Constancy of the marginal utility of income. In: Lange, O., McIntyre, F. and Yntema, T.O. (ed.) Studies in mathematical economics and econometrics. University of Chicago Press, Chicago, IL, USA, pp. 75-91.

Schelling, E., Bechir, M., Ahmed, M.A., Wyss, K., Randolph, T.F. and Zinsstag, J., 2007. Human and animal vaccination delivery to remote nomadic families, Chad. Emerg Infect Dis 13: 373-379.

Schelling, E., Zinsstag, J., 2015. Transdisciplinary research and One Health. In: Zinsstag, J., Schelling, E., Waltner-Toews, D., Whittaker, M. and Tanner, M. (eds.) One Health: the theory and practice of integrated health approaches. CABI, Wallingford, UK, pp. 366-373.

Schlüter, M., Hinkel, J., Bots, P.W.G. and Arlinghaus, R., 2014. Application of the SES framework for model-based analysis of the dynamics of social-ecological systems. Ecol Soc 19: 36.

Schreiner, J.A. and Hess, S., 2017. The role of non-use values in dairy farmers' willingness to accept a farm animal welfare programme. J Agric Econ 68: 553-578.

Screpanti, E. and Zamagni, S., 2005. An outline of the history of economic thought. Oxford University Press, Oxford, UK.

Sen, A., 1970. The impossibility of a Paretian liberal. J Polit Econ 78: 152-157.

Sen, A., 1999. The possibility of social choice. Am Econ Rev 89: 349-378.

Shaffer, J.D., 1973. On the concept of subsector studies. Am J Agric Econ 55: 333.

Shaw, A.P.M., 2003. Economic guidelines for strategic planning of tsetse and trypanosomiasis control in West Africa. FAO, Rome, Italy.

Shaw, A.P.M., Rushton, J., Roth, F. and Torgenson, P.R., 2017. DALYs, dollars and dogs: how best to analyse the economics of controlling zoonoses. Rev Sci Tech l'OIE 36: 147-161.

Shwiff, S., Aenishaenslin, C., Ludwig, A., Berthiaume, P., Bigras-Poulin, M., Kirkpatrick, K., Lambert, L. and Bélanger, D., 2013. Bioeconomic modelling of raccoon rabies spread management impacts in Quebec, Canada. Transbound Emerg Dis 60: 330-337.

Sikhweni, N.P., 2014. Bio-economic analysis of foot-and-mouth disease transmission between wildlife and livestock populations in Limpopo Province, South Africa. University of Pretoria, Pretoria, South Africa.

Stern, N., 2008. The economics of climate change. Am Econ Rev 98: 1-37.

Stokols, D., 1996. Translating social ecological theory into guidelines for community health promotion. Am J Heal Promot 10: 282-298.

Sugden, R. and Williams, A., 1978. The principles of practical cost-benefit analysis. Oxford University Press, Oxford, UK, 288 pp.

Sunstein, C.R., 2005. Cost-benefit analysis and the environment. Ethics 115: 351-385.

Taylor, M. and Filby, A., 2009. What is sensitivity analysis? Hayward Medical Communications. Available at: http://www.whatisseries.co.uk/what-is-sensitivity-analysis/.

Temple, L., Lançon, F., Palpacuer, F. and Paché, G., 2011. Actualisation du concept de filière dans l'agriculture et l'agroalimentaire. Écon Soc 33: 1785-1797.

Tengs, T.O., 2004. Cost-effectiveness versus cost-utility analysis of interventions for cancer: does adjusting for health-related quality of life really matter? Value Heal 7: 70-78.

Terpend, N., 1997. Guide pratique de l'approche filière. Le cas de l'approvisionnement et de la distribution des produits alimentaires dans les villes (No. DT/18-97F). Aliments dans les villes. FAO, Rome, Italy.

Thompson, S. and Valentinov, V., 2017. The neglect of society in the theory of the firm: a systems-theory perspective. Cambridge J Econ 41: 1061-1085.

Tiwari, D.N., Loof, R. and Paudyal, G.N., 1999. Environmental-economic decision-making in lowland irrigated agriculture using multi-criteria analysis techniques. Agric Syst 60: 99-112.

Torgerson, P.R. and Rüegg, S., Devleesschauwer, B., Abela-Ridder, B., Havelaar, A.H., Shaw, A.P.M., Rushton, J., Speybroeck, N., 2018. zDALY: an adjusted indicator to estimate the burden of zoonotic diseases. One Health 5: 40-45.

Tschopp, R., Hattendorf, J., Roth, F., Choudhoury, A., Shaw, A., Aseffa, A. and Zinsstag, J., 2012. Cost estimate of bovine tuberculosis to Ethiopia. In: Mackenzie, J.S., Jeggo, M., Daszak, P. and Richt, J.A. (eds.) One Health: the human-animal-environment interfaces in emerging infectious diseases. The concept and examples of a One Health approach. Springer, New York, NY, USA, pp. 249-268.

Van der Ploeg, S.W.F., Braat, L.C. and Van Lierop, W.F.J., 1987. Integration of resource economics and ecology. Ecol Modell 38: 171-190.

Vázquez Rodríguez, M.X. and León, C.J., 2004. Altruism and the economic values of environmental and social policies. Environ Resour Econ 28: 233-249.

Venkatachalam, L., 2004. The contingent valuation method: a review. Environ Impact Assess Rev 24: 89-124.

Von Kaufmann, R., MacIntire, J. and Itty, P., 1990. ILCA Bio-economic herd model for microcomputer. Technical reference guide. ILCA, Addis Ababa, Ethiopia.

Von Neumann, J. and Morgenstern, O., 1953. Theory of games and economic behavior, 3rd edition. Princeton University Press, Princeton. NJ, USA.

Walker, D. and Fox-Rushby, J., 2001. Allowing for uncertainty in economic evaluations: qualitative sensitivity analysis. Health Policy Plan 16: 435-443.

Walras, L., 1874. Éléments d'économie politique pure, ou Théorie de la richesse sociale. Corbaz, Paris, France.

Wam, H.K., 2010. Economists, time to team up with the ecologists! Ecol Econ 69: 675-679.

Weatherly, H., Drummond, M., Claxton, K., Cookson, R., Ferguson, B., Godfrey, C., Rice, N., Sculpher, M. and Sowden, A., 2009. Methods for assessing the cost-effectiveness of public health interventions: key challenges and recommendations. Health Policy 93: 85-92.

Weinstein, M.C., Torrance, G. and McGuire, A., 2009. QALYs: the basics. Value Heal 12: S5-S9.

Weintraub, W.S. and Cohen, D.J., 2009. The limits of cost-effectiveness analysis. Circ Cardiovasc Qual Outcomes 2: 55-58.

Weyler, E. and Gandjour, A., 2011. Empirical validation of patient versus population preferences in calculating QALYs. Health Serv Res 46: 1562-1574.

Wheat, I.D., 2007. The feedback method of teaching macroeconomics: is it effective? Syst Dyn Rev 23: 391-413.

Wilkinson, D., 1999. Cost-benefit analysis versus cost-consequences analysis. Perform Improv Q 12: 71-81.

World Bank, 2007. Cost-benefit analysis: evaluation criteria (or: 'Stay away from the IRR') (No. 40619). Knowledge brief for bank staff. World Bank, Washington, DC, USA.

Wynn, G., 2002. The cost-effectiveness of biodiversity management: a comparison of farm types in extensively farmed areas of Scotland. J Environ Plan Manag 45: 827-840.

Zhuang, J., Liang, Z., Lin, T. and De Guzman, F., 2007. Theory and practice in the choice of social discount rate for cost-benefit analysis: a survey. ERD Working Paper, Manila, Philippines.

Zinsstag, J., Choudhury, A., Roth, F. and Shaw, A., 2015a. One Health economics. In: Zinsstag, J., Schelling, E., Waltner-Toews, D., Whittaker, M. and Tanner, M. (eds.) One Health: the theory and practice of integrated health approaches. CABI, Wallingford, UK, pp. 134-145.

Zinsstag, J., Dürr, S., Penny, M.A., Mindekem, R., Roth, F., Menendez Gonzalez, S., Naissengar, S. and Hattendorf, J., 2009. Transmission dynamics and economics of rabies control in dogs and humans in an African city. Proc Natl Acad Sci USA 106: 14996-15001.

Zinsstag, J., Lechenne, M., Laager, M., Mindekem, R., Naïssengar, S., Oussiguéré, A., Bidjeh, K., Rives, G., Tessier, J., Madjaninan, S., Ouagal, M., Moto, D.D., Alfaroukh, I.O., Muthiani, Y., Traoré, A., Hattendorf, J., Lepelletier, A., Kergoat, L., Bourhy, H., Dacheux, L., Stadler, T. and Chitnis, N., 2017. Vaccination of dogs in an African city interrupts rabies transmission and reduces human exposure. Sci Transl Med 9: eaaf6984.

Zinsstag, J., Mahamat, M.B. and Schelling, E., 2015b. Measuring added value from integrated methods. In: Zinsstag, J., Schelling, E., Waltner-Toews, D., Whittaker, M. and Tanner, M. (eds.) One Health: the theory and practice of integrated health approaches. CABI, Wallingford, UK, pp. 53-59.

Zinsstag, J., Roth, F., Orkhon, D., Chimed-Ochir, G., Nansalmaa, M., Kolar, J. and Vounatsou, P., 2005. A model of animal-human brucellosis transmission in Mongolia. Prev Vet Med 69: 77-95.

Zinsstag, J., Schelling, E., Waltner-Toews, D. and Tanner, M., 2011. From 'one medicine' to 'one health' and systemic approaches to health and well-being. Prev Vet Med 101: 148-156.

Zinsstag, J., Waltner-Toews, D. and Tanner, M., 2015c. Theoretical issues of One Health. In: Zinsstag, J., Schelling, E., Waltner-Toews, D., Whittaker, M. and Tanner, M. (eds.) One Health: the theory and practice of integrated health approaches. CABI, Wallingford, UK, pp. 16-25.

Chapter 7

One Health governance: knowledge integration in One Health policy formulation, implementation and evaluation

Photo: Martin Hitziger

This chapter was published before as Martin Hitziger, Roberto Esposito, Massimo Canali, Maurizio Aragrande, Barbara Häsler and Simon R Rüegg, 2018. Knowledge integration in One Health policy formulation, implementation and evaluation. Bulletin of the World Health Organization 96: 211-218. https://doi.org/10.2471/BLT.17.202705. Republished with permission.

Martin Hitziger[1*], Roberto Esposito[2], Massimo Canali[3], Maurizio Aragrande[3], Barbara Häsler[4] and Simon R. Rüegg[1]

[1]Epidemiology Section, Vetsuisse Faculty, University of Zurich, Winterthurerstrasse 270, 8057 Zürich, Switzerland; [2]External Relation Office, National Institute of Health, Via Giano della Bella 34, 00199 Rome, Italy; [3]Department of Agricultural and Food Sciences, University of Bologna, Viale Giuseppe Fanin 44, 40127 Bologna, Italy; [4]Department of Pathobiology and Population Sciences, Veterinary Epidemiology Economics and Public Health Group, Royal Veterinary College, Hawkshead Lane, North Mymms, Hatfield, Hertfordshire, AL9 7TA, United Kingdom; martin.hitziger2@uzh.ch

Abstract

The One Health concept covers the interrelationship between human, animal and environmental health and requires multistakeholder collaboration across many cultural, disciplinary, institutional and sectoral boundaries. Yet, the implementation of the One Health approach appears hampered by shortcomings in the global framework for health governance. Knowledge integration approaches, at all stages of policy development, could help to address these shortcomings. The identification of key objectives, the resolving of trade-offs and the creation of a common vision and a common direction can be supported by multicriteria analyses. Evidence-based decision-making and transformation of observations into narratives detailing how situations emerge and might unfold in the future can be achieved by systems thinking. Finally, transdisciplinary approaches can be used both to improve the effectiveness of existing systems and to develop novel networks for collective action. To strengthen One Health governance, we propose that knowledge integration becomes a key feature of all stages in the development of One-Health-related policies. We suggest several ways in which such integration could be promoted.

Keywords: One Health, knowledge integration, governance, systems thinking, multicriteria analyses, transdisciplinarity

7.1 Introduction

The concept of One Health initially arose from integrated research on zoonoses, (Woods and Bresalier, 2014; Zinsstag *et al.*, 2015) but now covers all of the interconnections between human, animal and environmental health. The concept is a collaborative, interdisciplinary and intersectoral multi-institutional approach, linking many different forms of knowledge and expertise (CDCP, 2018; Cork *et al.*, 2014; Rüegg *et al.*, 2017). One Health is represented by a complex biological and social system that involves multiple actors and processes and their interactions over time, at local, national and global levels (Rüegg *et al.*, 2017). To date, relatively little attention has been given to the epistemological, institutional, political and social factors associated with the implementation of a One Health approach (Lebov *et al.*, 2017; Woods and Bresalier, 2014) This is illustrated by the almost complete lack of literature on One Health governance.

There is an existing framework for global health governance: a combination of the formal and informal institutions, rules and processes that influence global decisions on health policy (Okello *et al.*, 2015; Shiroyama *et al.*, 2012). Ideally, such a framework should transcend national boundaries, embrace multisectoral and interdisciplinary approaches and engage with the whole wide range of relevant actors (Dodgson *et al.*, 2002). In reality, however, the current framework is affected by fragmentation of health interests, programmes and sectors, a general lack of societal participation and by professional focus on very limited areas of expertise, so-called professional silos (Galaz *et al.*, 2015; Lee *et al.*, 2009). The dysfunctionality of the current framework, in terms of the core elements of the One Health concept, emphasizes the need for a dedicated framework for One Health governance (Lee and Brumme, 2013).

It has been suggested that some of the current framework's shortcomings could be overcome by the development of coordinated supranational bodies, the promotion of specialized training and career opportunities and the creation of dedicated funding mechanisms (Dodgson *et al.*, 2002; Frankson *et al.*, 2016; Lee and Brumme, 2013; Queenan *et al.*, 2017) We suggest that the framework may also be strengthened by improving the integration of its management (Cork *et al.*, 2014; De Savigny and Adam, 2009; Okello *et al.*, 2015) and by integrating knowledge at all stages of any related policy development (Boyle *et al.*, 2001; Chaffin *et al.*, 2014; Dietz *et al.*, 2003). In 2012, knowledge integration was listed as one of the United States National Cancer Institute's key recommendations for improving 21st century epidemiology (Khoury *et al.*, 2013).

Since 2014, about 230 experts and representatives of governments and nongovernmental organizations, from the fields of environmental, public and veterinary health and associated sciences, have come together in the Network for the Evaluation of One Health (NEOH, 2018). This network's main aim is to develop standards for assessing integration in One Health. Since 2016, this work has been enhanced by a core group of experts on complex systems, governance and knowledge integration. This paper summarizes the results of this group's investigation of knowledge integration in governance, as a mechanism for multi-institutional learning to improve the governance and coordination of One Health implementation in the absence of hierarchical chains of command.

7.2 Coordination and governance

In policy cycles, multiple rounds of agenda setting, policy formulation, decision-making, implementation and evaluation lead to the creation, implementation and revision of policies (Jann and Wegrich, 2007). We believe that, in terms of the interdisciplinary, intersectoral and multi-institutional One Health approach, knowledge integration at every stage of policy development, in every policy cycle, could strengthen the coordination and governance of One Health implementation. Although some integration of knowledge from different disciplines, institutions and sectors can, and does, take place intuitively, in many circumstances, we believe that it needs to become a regular, routine and institutionalized process at project, programme and policy levels (Assmuth and Lyytimäki, 2015; Chaffin *et al.*, 2014; Shiroyama *et al.*, 2012) In the development of health policies, knowledge assessment is often confined to the last, that is evaluation, stage of each policy cycle (El Allaki *et al.*, 2012). We believe that, to optimize the coordination and governance of the One Health approach, knowledge integration should be central at every stage of policy development.

In its broadest sense, knowledge integration has been defined as the building of shared and meaningful syntheses between distinct mental models, based on a recognition and explanation of the relevant differences between the models (Jahn *et al.*, 2012; Körner *et al.*, 2016) Rather than seeking consensus, knowledge integration can be used to build a common framework that allows an understanding of the links between the knowledge of multiple individuals. Such integration has been likened to the weaving of multiple perspectives into a central vision or a search for coherence and correspondence (Klein, 2008; Wickson *et al.*, 2006). The fostering of effective knowledge integration in a policy cycle is a multidimensional challenge because it requires the integration of cognitive concepts, organizational and social interests and perspectives and communicative and cultural factors. The relevant literature distinguishes target knowledge from systems and transformation knowledge. Target, or normative, knowledge relates to objectives and interests, while systems, or descriptive, knowledge relates to perspectives on factual processes. Transformation, or prescriptive, knowledge relates to the transformation of the current version of a system towards a more desired version (Pohl and Hirsch Hadorn, 2007). The integration of these three forms of knowledge throughout a policy cycle can be facilitated by three different approaches: multicriteria analyses for target knowledge, systems thinking for systems knowledge and transdisciplinary approaches for transformation knowledge.

7.3 Multicriteria analyses

The key to integrating target knowledge is to understand the often-conflicting interests, preferences and values of the multiple actors, as a first step to mediation, negotiation and, ultimately, collective action (Scholz and Tietje, 2002) Multicriteria analyses can assist such integration because they elicit and structure value systems in a way that accommodates a multiplicity of information sources and types (Von Winterfeldt and Edwards, 1986) Such analyses can incorporate any objective that has relevance to the point of view under consideration, rely on non-monetary units and apply valuation methods that are independent of pricing mechanisms. This makes these analyses particularly suited for priority setting in implementation of the One Health approach, which typically involves equity, intergenerational justice and non-marketed goods (Baltussen and Niessen, 2006; Bots and Hulshof, 2000) When combined with systems analysis for strategic, long-term assessments, multicriteria analyses offer a flexible yet systematic method of valuation that can bridge the gap between governance and action (Montibeller and Franco, 2011; Munaretto *et al.*, 2014). Like systems thinking, multi-criteria analyses are a rigorous set of methods. They are based on the multicriteria utility theory. Readers can learn about their main characteristics and find further literature in Box 7.1.

7.4 Systems thinking

Systems knowledge refers to an understanding of the complex interactions, between the many actors and processes in the fields of human, animal and environmental health that emerge and feedback over long time scales. To integrate such knowledge, the management discipline known as systems thinking can be used. Systems thinking can assist human thought by permitting the analytical inference of dynamic consequences, from complex nets of long causal chains that often have feedback loops and unintended effects. System thinking also allows information from multiple sources, e.g. quantitative data, expert knowledge and stakeholders' experiential insight, to be combined systematically (Dörner, 1996; Rasmussen *et al.*, 1995) These different sources of information are complementary because of missing data, methodological differences and interest-based selective perception, even among members of the same scientific team (Bergmann *et al.*, 2010; Scholz and Steiner, 2015). By using all of the available relevant information to understand the possible outcomes of policy interventions and by linking diverse bodies of relatively abstract information with the narratives that guide everyday experience, systems thinking can reduce uncertainty in complex governance problems (De Savigny and Adam, 2009; Lane, 2016; Rosenhead and Mingers, 2001).

Box 7.1. Multicriteria analyses for One Health impact evaluations.

This box outlines of a multi-criteria tool to provide a comprehensive assessment of impacts in One Health initiatives. It is thus complementary to Chapter 3, which provides a method to assess One Health initiatives with regard to set-up and project design. In that chapter, six main metrics are used. They are intended to apply to any One Health initiative, and their integration in a spider diagram implicitly assigns equal weight to each of them. Chapter 4-6 describe a wealth of metrics to measure impacts in the medical, social, environmental, and economic sectors. In contrast to Chapter 3, it is, however, impossible for any initiative to address all of them, and it is unrealistic to refine a small set of impact metrics that any One Health initiative would be expected to address. Instead, each initiative will need to prioritise its efforts with a particular emphasis towards a small number of disciplinary/sectoral, interdisciplinary/intersectoral and OH metrics to monitor. The overarching challenge is to provide a rigorous approach to assessing incommensurable information that is sufficiently flexible to accommodate each project's individual objectives, while being capable of integrating information from a wide variety of sources.

Multi-criteria methods are based on a multi-attribute extension of expected utility theory (Von Neumann and Morgenstern, 1953; Von Winterfeldt and Edwards, 1986). They are concerned with building and aggregating individual values or utilities into composite indices, to enable evaluations that do not exclusively rely on monetary scales. Multi-criteria methods were developed as a prospective tool for comparing different pathways of action against incommensurable or conflicting objectives (Baron, 2008; Keeney, 1982; Von Winterfeldt and Edwards, 1986). Applying them to monitor ongoing initiatives or retrospectively evaluate and compare project impacts requires to increase complexity: (1) to monitor developments over time, and (2) to distinguish between impacts caused by the initiative versus changes of the monitored metrics under consideration due to contextual circumstances. In case the aim is to monitor complex impacts beyond a single initiative, such as a programme, an additional level of complexity is (3) to define a single, comprehensive value system that is able to accommodate the different initiatives. Therefore, we propose a methodology which is based on the following steps:

1. Definition of a value system: Elicitation of a hierarchy of objectives and metrics (see Chapter 3), which reflect the aims of the One Health initiatives under consideration. The value system needs to comprehensively cover the full breadth of expected impacts across all initiatives[1]. The used metrics can differ between the initiatives to a certain degree, as long as they assess similar aspects of the same overall objectives. As emphasized by Keeney (1982) doing this step before commencing the initiative is a useful tool to plan and structure the activities efficiently, and to avoid a psychological anchoring bias that results from framing the analysis in terms of the status quo. It should, however be reviewed before conducting the evaluation, to accommodate for unexpected impacts or novel developments. For methodological and practical reasons, it is important to structure the evaluation problem with a 'requisite' evaluation model (Phillips, 1984). This term describes a (a) sufficiently comprehensive, but (b) minimally complex set of objectives and metrics that satisfyingly reflects the systems model and the theory of change of the initiatives from the perspective of the decision maker that is interested in the evaluation:

>>>

Box 7.1. Continued.

   a. To assure comprehensive coverage of all aspects relevant to the decision maker's preferences, the selection of objectives should be wide and each should be operationalized by metrics that satisfyingly reflect the decision maker's preferences and his theory of change.
   b. To assure a minimally complex value system, it is crucial that objectives and metrics be mutually independent. Specifically, it should exclude interactions – reaching one objective, or scoring highly/lowly on any specific metric should not automatically result in reaching or failing any other objective, or scoring highly or lowly on any other metric. Neglecting this condition runs the risk to over-represent certain aspects and thus introduce a bias in the overall multi-attribute utility (step 5).
2. Eliciting weights: This step elicits subjective weights of the objectives that form the evaluated value system. The weights entirely depend on the purpose of the evaluation, and the perspective from which it is done (internal monitoring, retrospective comparison, different stakeholder perspectives).
3. Estimation of the impact that the initiative has on each metric: This requires a comparison of ex ante and ex post states of the system and is equivalent to any conventional evaluation procedure. Many of the metrics presented in Chapters 4-6 already come with defined methodologies for their assessment. Where this is not the case, the full breadth of quantitative and qualitative data collection methods of any scientific disciplinary or interdisciplinary analysis can be integrated. Particular emphasis should be placed on gathering perspectives from involved stakeholders, for which transdisciplinary workshops, focus group discussions, or soft operations research methods are particularly useful. Regardless of the applied method of data collection and metric assessment, the end result of this step should be a two-fold short summary for each metric: (a) a concise grasp of its development over time, and (b) a thorough and critical reflection on the impact that is attributable to the initiative, contrasted to impacting forces attributable to external or contextual developments.
4. Scoring the impacts: Scoring transforms impact data into single-attribute utilities or preference values. It reduces complexity that originates from different metrics for each objective. Thus, it allows to merge several metrics that assess the same objective, and also allows to deal with various initiatives that might assess the same objective with slightly different metrics. In comparative evaluations, the best and the worst performing initiative's scores are usually set as positive and negative benchmarks. In evaluations of individual projects, benchmarks need to be defined separately. This step is tightly linked to data derived from step 2. Transforming this data into preference orders does, however, involve a degree of subjectivity. This subjectivity becomes stronger the more the metrics differ.
5. Calculating multi-attribute utilities: Transforming single-attribute utilities into a multi-attribute utility results in a preference ranking (in comparative evaluations), or a rating of the achieved impact as compared to a defined benchmark (in evaluations of individual initiatives). This is done by means of a decision rule. While different decision rules are mathematically possible, the most common one is additive (Von Neumann and Morgenstern, 1953; Von Winterfeldt and Edwards, 1986), assuming independence of the individual objectives (step 1). Taking into account human limits of mathematical reasoning, it allows mathematically consistent ‚rational' aggregation of single-attribute utilities and avoids the challenges that come with elicitation and measurement of non-linear conceptual entities[2]. It calculates an overall utility u from the sum of the utilities u_i of the individual objectives scores as a function on their metric scores x_i, each multiplied with the relative weight k_i of the objective.

$$u(x_1, \ldots, x_n) = \sum_{i=1}^{n} k_i u_i(x_i)$$

>>>

Box 7.1. Continued.

A wide range of methods are available to aid the elicitation of scores and weights (Keeney, 1982; Von Winterfeldt and Edwards, 1986). They use different techniques to facilitate reflective thinking, and critical scrutiny of preference building. They also offer rigorous approaches to induce preference ratings that take into account psychological biases and modes of thought. Many of them come in computer-aided packages that allow straightforward calculation of scores and weights from basic data and preference elicitations. Multi-criteria decision analyses are thus rigorous and replicable procedures with two effects. On the one hand, they contribute to reflection and critical scrutiny of preferences. On the other hand, they allow to incorporate any objective of relevance to the decision maker, and to flexibly accommodate any piece of available data in comprehensive evaluations. Based on a thorough procedure, they transparently render the objectives and trade-offs explicit that are inherent in any evaluation or decision.

[1] In this box, we use the term objective to qualitatively describe a change of the situation that the One Health initiative is envisioned to bring about. An impact is the actual change of the situation brought about through the initiative. Quantitative metrics and qualitative indicators measuring the degree to which that change has happened. For simplicity, we use the term ‚metric' to collectively refer to quantitative and qualitative indicators.

[2] Non-linear decision models are sometimes used in very specific contexts that allow a crisp, quantitative, and mathematically valid calculation of non-linear relations between interdependent metrics. One example is multiplicative aggregation of length of life and quality of life metrics in medical science (Ara and Wailoo, 2012; Weinstein *et al.*, 2003). However, these can also be grasped in composite qualitative indicators and scored accordingly (steps 1 and 4). Considering the wealth of objectives and metrics used in One Health projects, as well as the overall data quality available, we suggest as a rule of thumb to stick to additive decision rules.

7.5 Transdisciplinarity

Transdisciplinary approaches, which are sometimes called boundary management, are designed to build a bridge, at the science-policy interface and between potentially diverse knowledge systems, by facilitating communication, mediation and translation across cultural, disciplinary, institutional and/or sectoral divides. Although multiple analytic methods may be employed, (Bergmann *et al.*, 2010; Scholz and Tietje, 2002) the distinctive characteristics of such approaches are mainly sociocultural and aim to foster collective action towards societal transformations (Cash *et al.*, 2003; Lang *et al.*, 2012; Scholz and Steiner, 2015) They include the selection of actors that legitimately represent the interest groups of relevance to the research problem. Co-leadership helps to ensure the equitable representation of interests and perspectives and to mitigate power differentials. The joint negotiation and definition of research objectives and hypotheses is a crucial step in building mutual understanding and enabling successful collaborations. Linking narratives and experiential perceptions with conceptual or explanatory systems knowledge is a central challenge. This challenge can be overcome by careful consideration and the development of a deep understanding in experiential encounters, by repeatedly exposing the different bodies of knowledge to each other and by working towards joint outputs. The sustained commitment of the varied stakeholders needs to be supported by strong leadership, trust building and conflict management (Bergmann *et al.*, 2010; Pohl and Hirsch Hadorn, 2007; Scholz and Tietje, 2002). Transdisciplinary approaches may make three crucial contributions to societal transformations. First, they create social contexts for successful knowledge integration, even where such contexts do not occur naturally. Second, as a result of their collaborative and interactive nature, they tend to produce knowledge that is generally perceived as credible, legitimate and salient. Finally, by fostering collaboration among societal and scientific partners, they can build trust and networks that are independent of any hierarchical chains of command.

7.6 Case studies

We believe that the effective implementation of the One Health strategy, as an interdisciplinary and intersectoral approach that links different forms of knowledge and expertise across multiple institutions, depends on knowledge integration. Six case studies support this view: three general One Health initiatives and three integrated health initiatives that included multicriteria analyses, systems thinking or a transdisciplinary approach (Table 7.1).

7.8 Integration of target knowledge

The integration of target knowledge has been fostered by including stakeholder perspectives in agenda setting and decision-making, through either explicit co-leadership and negotiation (Hitziger *et al.*, 2017; Mbabu *et al.*, 2014; Paternoster *et al.*, 2017) or changes of perspective in collaborative work assignments (Hitziger *et al.*, 2017; Lane *et al.*, 2016; Sripa *et al.*, 2015) In Quebec, Canada, a rigorous multicriteria analysis, of Lyme disease surveillance and control strategies was used to support the public health authorities' decision-making and programme direction (Aenishaenslin *et al.*, 2013). In the latter investigation, a participative approach that involved health professionals and other stakeholders from governmental and nongovernmental organizations was used to compare several surveillance strategies in terms of their likely animal, environmental and public health and socioeconomic impacts. The stakeholder group provided input during the definition of management strategies, the assessment of objectives and their relative importance and the scoring of the strategies in terms of their likely attainment of the objectives. Since stakeholders represented their institutional perspectives, the process presumably assured the balanced representation of each of the relevant institutional viewpoints. The analyses allowed preference rankings of several possible intervention strategies for the management of Lyme disease, facilitated a better understanding of the conflicts between the key objectives and the relevance of such conflicts to each stakeholder group, and apparently improved each stakeholder group's appreciation of the preferences and priorities of the other stakeholder groups. In short, the analyses contributed to resolving trade-offs and setting a common vision and direction. While multicriteria analyses have mostly been focused on the early stages of policy development, e.g. agenda setting and policy formulation, they have important evaluative elements and can build consensus, to strengthen collective action, during policy implementation (Table 7.1).

Table 7.1. Comparison of integration of three types of knowledge in six initiatives.

Initiative, country, study period	General details	Integration		
		Systems knowledge	Target knowledge	Transformation knowledge
One Health initiatives				
West Nile virus surveillance, Italy, from 2013 (Paternoster *et al.*, 2017)	Inter-institutional working groups of local and regional authorities in human, animal and environmental health, covering Emilia-Romagna, Lombardy and Piedmont. Implementation of integrated surveillance of birds, horses, humans and mosquitos, including sampling protocols, technical procedures, data-sharing agreements and public information campaigns.	Comprehensive conceptual framework, multispecies sampling protocols, data sharing and linking of information in interdisciplinary groups allowed for integration of systems knowledge. Dissemination to the general public promoted via seminars and educational activities.	Shared leadership fostered integration of target knowledge. Objectives and targets, for overall initiative and individual expert teams, were well defined. Lack of funding for specific targets demonstrated the incomplete alignment of objectives between central and local levels. Institutional set-up lacked flexibility for adaptation.	Strong institutional backing and complex and competent actor network facilitated legitimacy, implementation and resilience. Joint field activities created a team spirit and fostered communication. Annual plenary meetings improved effectiveness. However, public involvement and accessibility of transformation knowledge were considered limited.
Opisthorchiasis control in Lawa province, Thailand, from 2005 (Sripa *et al.*, 2015)	Longstanding research track at local university complemented with community-based integrated surveillance, parasite sampling in fish, human screening, medical treatment and education campaigns targeted at public and schools. Linked to international helminth control programme.	Research on opisthorchiasis endemicity and human prevalence. Collaboration with community members for data collection and dissemination fostered integration of local systems knowledge. The need for a more integrated surveillance approach, to understand transmission dynamics, was recognized.	An iterative approach, to facilitate mutual learning in local communities, resulted in an increasingly broad scope and comprehensive objectives. High level of local commitment and collaboration indicated a strong alignment of target knowledge between initiative and all local actors and stakeholders.	Transformation knowledge integrated via collaboration with, and capacity building in, local hospitals. Education strategies for communities and schools aimed to foster transformation knowledge among general public. Production of manuals should allow replication of approach.

>>>

Table 7.1. Continued.

Initiative, country, study period	General details	Integration		
		Systems knowledge	Target knowledge	Transformation knowledge
Strategic plan for implementing One Health, Kenya, from 2011 (Mbabu *et al.*, 2014)	Establishment of interministerial committees and task forces in charge of programme development, e.g. a national influenza task force, a zoonosis technical working group, One Health zoonotic disease units at central and peripheral levels and a One Health task force covering central and eastern Africa. Establishment of One Health offices within disease units and a national One Health secretariat.	Joint situation analyses of zoonotic diseases and the adoption of a One Health approach in routine and/or emergency activities fostered a shared understanding of systems knowledge.	Development of a One Health strategy/action plan strengthened common vision and direction at operational/institutional level. Inadequate funding for coordinated activities and lack of political will indicated insufficient alignment of objectives between initiative and high-level decision-makers.	A lack of institutional arrangements for coordination and collaboration between the line agencies and operational departments indicated that networks for collective action needed to be strengthened.
Other initiatives[1]				
Review of complex intersectoral services for child protection, the United Kingdom, 2010-2011 (Lane *et al.*, 2016)	Analysis of entire child-protection system to review and improve service provision at national level. Collaborative integration of evidence with stakeholders across entire chain of interests and responsibilities: affected individuals, charities, family proceedings courts, local institutions, national department of education and professionals.	Authorities and stakeholders jointly defined 60 relevant variables, and provided evidence on their relations, interactions and feedback loops. There was integration of systems knowledge through personal interactions, facilitated by joint building and analysis of system dynamics models. Group understanding developed in joint model analysis and validation.	Actor targets were analysed, as determinants of system behaviour. Target knowledge was integrated via analysis of system dynamics, as determined by organizational targets.	Integration of transformation knowledge supported by joint definition and analysis of scenarios for transforming the activities and structures of the child-protection sector. Trust, networks and collaborative capacities for implementation were strengthened across hierarchies and sectors.

>>>

Table 7.1. Continued.

Initiative, country, study period	General details	Integration		
		Systems knowledge	Target knowledge	Transformation knowledge
One Health surveillance and control, Canada, 2010-2012 (Aenishaenslin *et al.*, 2013)	Analysis of integrated Lyme disease surveillance and control strategies to support decision-making and programme direction of public health authorities in Quebec. Collaboration with five national and regional authorities in agriculture, environment and public health. Actor perspectives on 11 strategic option's effects on 16 target criteria were analysed under emerging and epidemic outbreak scenarios.	Focus groups, expert interviews and literature review facilitated integration of systems knowledge by joint problem definition and performance assessment of strategic options.	Target knowledge was integrated by defining targets in dialogue, discussion and reflection, by the elicitation and systematic analysis of stakeholder institution's perspectives on target weights for animal, environmental and public, health, economic, operational, social and strategic impacts and surveillance, and by joint reflection on, and validation of the resulting multicriteria assessments.	Supported integrated transformation knowledge through joint elaboration of strategic options and target criteria, indicators and scales that were relevant to pertinent authorities. Participation in research, analysis and data analysis built collaborative capacities, networks for implementation and trust.
Intercultural collaboration for integrated health, Guatemala, 2012-2015 (Hitziger *et al.*, 2017)	Analyses of impacts, of a facilitated transdisciplinary approach, on trust, networks and mutual learning among biomedical doctors and traditional Maya healers. All in a country where structural violence hampers the development of integrative health systems. Collaborative referral designed to integrate different health systems in patients' health-seeking pathways.	Integration of systems knowledge facilitated among practitioners via joint design and validation of empirical research on barriers to integrative health services. Group understanding was developed through workshop techniques and the changes in perspective that occurred during joint fieldwork.	Integration of target knowledge was supported through increased understanding of the perspectives of other participating actors and via negotiation of the characteristics and objectives of the transdisciplinary approach.	Integration of transformation knowledge was supported by the strengthening of collaborative capacities, by an improved understanding of the viewpoints of other participating actors, viewpoints that, in many other projects, remain hidden because of the segregation of institutions and sectors, and by the joint development, implementation and assessment of pilot models for institutional and operational transformation.

[1] Initiatives that applied specific knowledge integration approaches in fields of relevance to One Health.

7.9 Integration of systems knowledge

The integration of systems knowledge has been used in the joint definition of broad conceptual bases for the collection and assessment of evidence (Aenishaenslin *et al.*, 2013; Hitziger *et al.*, 2017; Mbabu *et al.*, 2014; Paternoster *et al.*, 2017) and in facilitating group understanding of the evidence collected via collaborative data analysis and validation (Table 7.1) (Aenishaenslin *et al.*, 2013; Mbabu *et al.*, 2014; Paternoster *et al.*, 2017). In the United Kingdom of Great Britain and Northern Ireland, a comprehensive intersectoral review of the activities, culture, effectiveness, policies and social relations within the child-protection sector demonstrated how One Health governance could be supported by structured and rigorous systems thinking (Lane *et al.*, 2016). This review engaged a reference group of relevant stakeholders, e.g. representatives from charities, the civil service and other government departments, an adoptive mother and young people who had been through the child-protection system themselves, and drew on evidence from databases, written sources and individual stakeholders' perspectives. The collaborative development of causal loop diagrams, with 60 variables, facilitated both a better understanding of the systemic outcomes of interdependent decision-making processes and a comparative assessment of potential policy interventions. The recommendations drawn from this review's results were largely accepted by the commissioning government authority and triggered substantial policy changes (Lane *et al.*, 2016). Systems thinking can therefore transforms complex mixtures of individual observations into coherent narratives that state how situations emerge and how they may unfold in the future. While systems thinking has mostly focused on the evaluation stage of policy development, it usually includes target knowledge, as a determinant of behaviour, and its participative nature can also build trust and foster mutual learning between decision-makers and scientists.

7.10 Integration of transformation knowledge

Most One Health and related initiatives rely on a multi-institutional network of actors. This network often contributes to the integration of transformation knowledge in two ways: via the institutional support provided by relevant decision-makers (Aenishaenslin *et al.*, 2013; Lane *et al.*, 2016; Paternoster *et al.*, 2017) and via the collaboration of individuals who have a broad range of implementation-related skills and expertise in many specialist fields (Aenishaenslin *et al.*, 2013; Lane *et al.*, 2016; Paternoster *et al.*, 2017; Sripa *et al.*, 2015). The potential usefulness of transdisciplinary approaches for coordinating and managing such interdisciplinary, intersectoral and intercultural collaboration, even in challenging societal contexts, was illustrated by a collaboration in Guatemala (Hitziger *et al.*, 2017). The main aim of this collaboration was to bridge the gaps between the knowledge systems of biomedical doctors and those of traditional Maya healers and, in so doing, promote collaboration and mutual learning between the two groups. After facilitating joint patient diagnosis and subsequent treatment reconstruction, the collaboration was deemed useful and relevant by both groups of subjects and appears to have reduced the long-standing prejudices held by each group towards the other. Scientific institutions that, in terms of these prejudices, were perceived as neutral acted as intermediaries and helped ensure the credibility of the results. The process provided multiple opportunities for the building of mutual trust, via dialogue and experiential exchange and also triggered reflection, by pointing out the shortcomings of the current health systems, and appears to have educated all of the participants. In short, it developed and/or strengthened the networks for collective action. While the Guatemalan study focused on the implementation stage of policy development, the transdisciplinary approaches also had effects on agenda setting, by influencing the actors' target knowledge and on evaluation, by enabling process assessments that were more inclusive of the divergent knowledge systems (Table 7.1). Furthermore, the new networks and increased levels of trust helped achieve consensus and collective action at every stage of policy development.

7.11 Discussion

We believe that knowledge integration is both an integral element of successful One Health governance systems and a prerequisite for the effective implementation of the One Health approach. The combined use of multicriteria analyses, systems thinking and transdisciplinary approaches (Figure 7.1) could contribute to more systematic and successful collaborations within and across existing institutions and form a procedural backbone for converting the aspirations of the One Health concept into institutional processes. In general, the aim of multicriteria analyses, systems thinking and transdisciplinary approaches is to create, maintain and inform collective action by broad coalitions of societal partners. If successfully implemented over extended time spans, they could contribute to the building of trust, networks and institutions that are not primarily dependent on any existing hierarchical structures of government.

Although multicriteria analyses, systems thinking and transdisciplinary approaches mainly focus on different, crucial aspects of One Health governance, they are complementary and overlapping rather than mutually exclusive. They provide methods to resolve trade-offs and set a common vision and a common direction across disciplines, institutions and sectors. They serve as toolbox for systemic monitoring and feedback to transform observations into narratives detailing how situations emerge and might unfold in the future. Finally, they contribute to the development and/or strengthening of networks for collective action towards a common vision. Potentially, therefore, as a decisive element in policy development, knowledge integration could help resolve the main shortcomings of the current global framework for health governance, by managing complexity and shaping interactions between actors and institutions towards joint learning (Boyle *et al.*, 2001; Chaffin *et al.*, 2014). Knowledge integration could also be used to complement educational and institutional measures for improving the implementation of the One Health approach (Queenan *et al.*, 2017). We therefore propose that policy cycles relevant to One Health should aim at knowledge integration and make the best possible use of multicriteria analyses, systems thinking and transdisciplinary approaches. Whenever they are used as elements of the implementation of the One Health approach, the processes involved in knowledge integration should be reported explicitly in the associated scientific articles. Ideally, such reporting should be based on standardized criteria and systematic evaluation frameworks, like the one proposed by the Network for Evaluation of One Health (NEOH, 2018; Rüeg *et al.*, 2017). To develop and improve best practices in One Health, the practitioners and scientists in active One Health networks should be educated on knowledge integration and encouraged to discuss their ideas with those of more established governance actors, ideally in programmes supported by permanent professional associations or organizations. Finally, attention should be directed towards developing and implementing efficient technical mechanisms to facilitate stakeholder involvement and brokering at all levels of health governance, from local to global level.

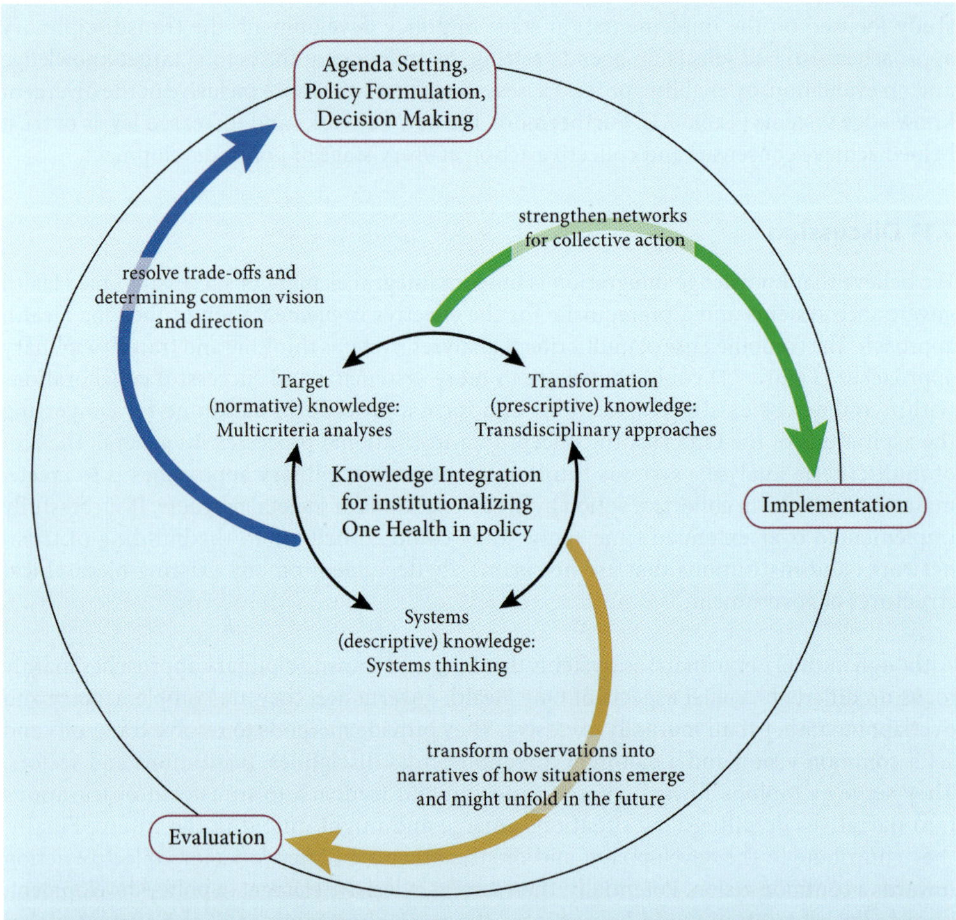

Figure 7.1. Potential uses of knowledge integration within One Health policy cycles. The outer and inner circles indicate the stages within each round of policy development (Jann and Wegrich, 2007) and the approaches for integrating target, transformation and systems knowledge (Pohl and Hirsch Hadorn, 2007), respectively. The text on the arrows summarizes the main benefits of knowledge integration for each stage of policy development.

Acknowledgements

This chapter is based upon work from COST Action 'Network for Evaluation of One Health' (TD1404), supported by COST (European Cooperation in Science and Technology).

References

Aenishaenslin, C., Hongoh, V., Cissé, H.D., Hoen, A.G., Samoura, K., Michel, P., Waaub, J.-P. and Bélangeret, D., 2013. Multi-criteria decision analysis as an innovative approach to managing zoonoses: results from a study on Lyme disease in Canada. BMC Publ Health 13(1): 897.

Ara, R. and Wailoo, A., 2012. Using health state utility values in models exploring the cost-effectiveness of health technologies. Value Health 15(6): 971-974.

Assmuth, T. and Lyytimäki, J., 2015. Co-constructing inclusive knowledge within converging fields: environmental governance and health care. Environ Sci Pol 51: 338-350.

Baltussen, R. and Niessen, L., 2006. Priority setting of health interventions: the need for multi-criteria decision analysis. Cost Eff Resour Alloc 4(1): 14.

Baron, J., 2008. Thinking and deciding, 4th edition. Cambridge University Press, New York, NY, USA.

Bergmann, M., Jahn, T., Knobloch, T., Krohn, W., Pohl, C. and Schramm, E., 2010. Methods for transdisciplinary research: a primer for practice. Campus Verlag, Frankfurt, Germany, 295 pp.

Bots, P. and Hulshof, J.A.M., 2000. Designing multi-criteria decision analysis processes for priority setting in health policy. J Multi-Criteria Decis Anal 9(1-3): 56-75.

Boyle, M., Kay, J. and Pond, B., 2001. Monitoring in support of policy: an adaptive ecosystem approach. In: Munn, T. (ed.) Encyclopedia of global environmental change. Vol. 4. Wiley, New York, NY, USA, pp. 116-137.

Cash, D.W., Clark, W.C., Alcock, F., Dickson, N.M., Eckley, N., Guston, D.H., Jäger, J. and Mitchell, R.B., 2003. Knowledge systems for sustainable development. Proc Natl Acad Sci USA 100(14): 8086-8091.

Centers for Disease Control and Prevention (CDCP), 2018. One Health. Centers for Disease Control and Prevention, Atlanta, GA, USA. Available at: https://tinyurl.com/y7jvwxdd.

Chaffin, B.C., Gosnell, H. and Cosens, B.A., 2014. A decade of adaptive governance scholarship: synthesis and future directions. Ecol Soc 19(3): 56.

Cork, S.C., Geale, D.W. and Hall, D.C., 2015. One Health in policy development: an integrated approach to translating science into policy. In: Zinstag, J., Schelling, E., Waltner-Toews, D., Whittaker, M. and Tanner, M. (eds.) One Health: the theory and practice of integrated health approaches. CAB International, Wallingford, UK, pp. 304-317.

De Savigny, D. and Adam, T. (eds.), 2009. Systems thinking for health systems strengthening. World Health Organization, Geneva, Switzerland. Available at: https://tinyurl.com/y8q6jarj.

Dietz, T., Ostrom, E. and Stern, P.C., 2003. The struggle to govern the commons. Science 302(5652): 1907-1912.

Dodgson, R., Lee, K. and Drager, N., 2002. Global health governance, a conceptual review. World Health Organisation, Geneva, Switzerland. Available at: https://tinyurl.com/y982oqgu.

Dörner, D., 1996. The logic of failure: recognizing and avoiding error in complex situations. Metropolitan Books, New York, NY, USA.

El Allaki, F., Bigras-Poulin, M., Michel, P. and Ravel, A.A., 2012. Population health surveillance theory. Epidemiol Health 34: e2012007.

Frankson, R., Hueston, W., Christian, K., Olson, D., Lee, M., Valeri, L., Hyatt, R., Annelli, J. and Rubin, C., 2016. One Health core competency domains. Front Publ Health 4: 192.

Galaz, V., Leach, M., Scoones, I. and Stein, C., 2015. The political economy of One Health research and policy. STEPS Working Paper 81. STEPS Centre, Brighton, UK. Available at: https://tinyurl.com/y7vrk3ut.

Hitziger, M., Berger, Gonzalez, M., Gharzouzi, E., Ochaíta Santizo, D., Solis Miranda, R., Aguilar Ferro, A.I., Vides-Porras, A., Heinrich, M., Edwards, P. and Krütli, P., 2017. Patient-centered boundary mechanisms to foster intercultural partnerships in health care: a case study in Guatemala. J Ethnobiol Ethnomed 13(1): 44.

Jahn, T., Bergmann, M. and Keil, F., 2012. Transdisciplinarity: between mainstreaming and marginalization. Ecol Econ 79: 1-10.

Jann, W. and Wegrich, K., 2007. Theories of the policy cycle. In: Fischer, F., Miller, G.J., Sidney, M.S. (eds.) Handbook of public policy analysis: theory, politics and methods. CRC Press, Boca Raton, FL, USA, pp. 43-62.

Keeney, R., 1982. Value-focused thinking: a path to creative decision making. Harvard University Press, Cambridge, MA, USA.

Khoury, M.J., Lam, T.K., Ioannidis, J.P., Hartge, P., Spitz, M.R., Buring, J.E., Chanock, S.J., Croyle, R.T., Goddard, K.A., Ginsburg, G.S., Herceg, Z., Hiatt, R.A., Hoover, R.N., Hunter, D.J., Kramer, B.S., Lauer, M.S., Meyerhardt, J.A., Olopade, O.I., Palmer, J.R., Sellers, T.A., Seminara, D., Ransohoff, D.F., Rebbeck, T.R., Tourassi, G., Winn, D.M., Zauber, A. and Schully, S.D., 2013. Transforming epidemiology for 21st century medicine and public health. Cancer Epidemiol Biomark Prev 22(4): 508-516.

Klein, J.T., 2008. Evaluation of interdisciplinary and transdisciplinary research: a literature review. Am J Prev Med 35(2): S116-123.

Körner, M., Lippenberger, C., Becker, S., Reichler, L., Müller, C., Zimmermann, L., Rundel, M., Baumeister, H., 2016. Knowledge integration, teamwork and performance in health care. J Health Organ Manag 30(2): 227-243.

Lane, D., 2016. 'Till the muddle in my mind have cleared awa': can we help shape policy using systems modelling? Syst Res Behav Sci 33(5): 633-650.

Lane, D.C., Munro, E., Husemann, E., 2016. Blending systems thinking approaches for organisational analysis: reviewing child protection in England. Eur J Oper Res 251(2): 613-623.

Lang, D.J., Wiek, A., Bergmann, M., Stauffacher, M., Martens, P., Moll, P., Swilling, M., Thomas, C.J., 2012. Transdisciplinary research in sustainability science: practice, principles, and challenges. Sustain Sci. 7(1): 25-43.

Lebov, J., Grieger, K., Womack, D., Zaccaro, D., Whitehead, N., Kowalcyk, B., MacDonald, P.D.M., 2017. A framework for One Health research. One Health 24(3): 44-50.

Lee, K. and Brumme, Z.L., 2013. Operationalizing the One Health approach: the global governance challenges. Health Pol Plan 28(7): 778-785.

Lee, K., Koivusalo, M., Ollila, E., Labonté, R., Schuftan, C. and Woodward, D., 2009. Global governance for health. In: Labonté, R., Schrecker, T., Packer, C. and Runnels, V. (eds.) Globalization and health: pathways, evidence and policy. Routledge, London, UK, pp. 289-316.

Mbabu, M., Njeru, I., File, S., Osoro, E., Kiambi, S., Bitek, A., Ithondeka, P., Kairu-Wanyoike, S., Sharif, S., Gogstad, E., Gakuya, F., Sandhaus, K., Munyua, P., Montgomery, J., Breiman, R., Rubin, C. and Njenga, K., 2014. Establishing a One Health office in Kenya. Pan Afr Med J 19: 106.

Montibeller, G. and Franco, L.A., 2011. Raising the bar: strategic multi-criteria decision analysis. J Oper Res Soc 62(5): 855-867.

Munaretto, S., Siciliano, G. and Turvani, M.E., 2014. Integrating adaptive governance and participatory multicriteria methods: a framework for climate adaptation governance. Ecol Soc 19(2): 74.

Network for Evaluation of One Health (NEOH), 2018. COST Action TD1404. COST Association, Brussels, Belgium. Available at: http://neoh.onehealthglobal.net.

Okello, A., Vandersmissen, A. and Welburn, S.C., 2015. One Health into action: integrating global health governance with national priorities in a globalized world. In: Zinstag, J., Schelling, E., Waltner-Toews, D., Whittaker, M. and Tanner, M. (eds.) One Health, the theory and practice of integrated health approaches. CAB International, Wallingford, UK, pp. 283-303.

Paternoster, G., Tomassone, L., Tamba, M., Chiari, M., Lavazza, A., Piazzi, M., Favretto, A.R., Balduzzi, G., Pautasso, A. and Vogler, B.R., 2017. The degree of One Health implementation in the West Nile virus integrated surveillance in northern Italy, 2016. Front Publ Health 5(5): 236.

Phillips, L., 1984. A theory of requisite decision models. Acta Psychol 56: 29-48.

Pohl, C. and Hirsch Hadorn, G., 2007. Principles for designing transdisciplinary research. Oekom Verlag, Munich, Germany.

Queenan, K., Garnier, J., Rosenbaum Nielsen, L., Buttigieg, S., De Meneghi, D., Holmberg, M., Zinsstag, J., Rüegg, S., Häsler, B. and Kock, R., 2017. Roadmap to a One Health agenda 2030. Perspect Agric Vet Sci Nutr Nat Resour 12(14): 1-17.

Rasmussen, J., Nixon, P., Warner, F., 1990. Human error and the problem of causality in analysis of accidents. Philos Trans R Soc Lond B Biol Sci 327(1241): 449-460.

Rosenhead, J. and Mingers, J., 2001. Rational analysis for a problematic world revisited. Wiley, Chichester, UK.

Rüegg, S.R., McMahon, B.J., Häsler, B., Esposito, R., Nielsen, L.R., Ifejika Speranza, C., Ehlinger, T., Peyre, M., Aragrande, M., Zinsstag, J., Davies, P., Mihalca, A.D., Buttigieg, S.C., Rushton, J., Carmo, L.P., De Meneghi, D., Canali, M., Filippitzi, M.E., Goutard, F.L., Ilieski, V., Milićević, D., O'Shea, H., Radeski, M., Kock, R., Staines, A. and Lindberg, A., 2017. A blueprint to evaluate One Health. Front Publ Health 16(5): 20.

Scholz, R.W. and Steiner, G., 2015. The real type and ideal type of transdisciplinary processes: part I – theoretical foundations. Sustain Sci 10(4): 527-544.

Scholz, R.W. and Tietje, O., 2002. Embedded case study methods: itegrating quantitative and qualitative knowledge. Sage Publications, Thousand Oaks, CA, USA.

Shiroyama, H., Yarime, M., Matsuo, M., Schroeder, H., Scholz, R. and Ulrich, A.E., 2012. Governance for sustainability: knowledge integration and multi-actor dimensions in risk management. Sustain Sci 7, Suppl. 1: 45-55.

Sripa, B., Tangkawattana, S., Laha, T., Kaewkes, S., Mallory, F.F., Smith, J.F. and Wilcox, B.A., 2015. Toward integrated opisthorchiasis control in northeast Thailand: the Lawa project. Acta Trop 141B: 361-367.

Von Neumann, J. and Morgenstern, O., 1953. Theory of games and economic behaviour. Princeton University Press, Princeton, NJ, USA.

Von Winterfeldt, D. and Edwards, W., 1986. Decision analysis and behavioural research. Cambridge University Press, Cambridge, UK.

Weinstein, M.C., O'Brien, B., Hornberger, J., Jackson, J., Johannesson, M. and McCabe, C., 2003. Principles of good practice for decision analytic modeling in health-care evaluation. Report of the ISPOR Task Force on Good Research Practices – Modeling Studies. Value Health 6(1): 9-17.

Wickson, F., Carew, A.L. and Russell, A.W., 2006. Transdisciplinary research: characteristics, quandaries and quality. Futures 38(9): 1046-1059.

Woods, A. and Bresalier, M., 2014. One Health, many histories. Vet Rec 174(26): 650-654.

Zinsstag, J., Schelling, E., Waltner-Toews, D., Whittaker, M. and Tanner, M. (eds.), 2015. One Health, the theory and practice of integrated health approaches. CAB International, Wallingford, UK.

Epilogue

For our handbook, 'Integrated approaches to health: a handbook for the evaluation of One Health', many One Health researchers and practitioners joined forces, minds and experiences. With the support of versed evaluators, we developed new methods and tools for the evaluation of One Health.

Once the evaluation framework, tools and protocols were developed, we conducted a series of training schools, short-term scientific missions, and workshops to give people the skills and confidence to apply the methods and tools in case studies and thereby test the novel approach. None of the NEOH members were professional evaluators, and it was a shared learning journey for all – thus, we applied a One Health approach to develop a One Health evaluation framework.

In the process, the evaluation framework and tools were applied to existing One Health initiatives from different NEOH member countries, which complied with the following criteria: (1) the initiative matched the topic area; (2) the initiative was an One Health initiative; (3) data were available or could be generated; (4) resources for evaluation were available or could be found; (5) the initiative was relevant for Europe; and (6) the initiative was inter- or transdisciplinary. A total of 12 case studies were selected and each evaluation was led by a case study leader and a small, often international team. Topics included infectious zoonotic diseases, vector borne diseases, education, animal welfare and non-communicable diseases such as obesity. The evaluations were conducted over a timeframe of about one year. Most case studies focused particularly on Elements 1 and 3 of the evaluation framework described in Chapter 3. Through iterative cycles, with repeated feedback from the case study teams to the handbook authors, the evaluation tools and protocols were refined and further developed. Finally, the case studies were published in a dedicated Frontiers Research Topic 'Concepts and experiences in framing, integration and evaluation of One Health and EcoHealth' https://www.frontiersin.org/research-topics/5479) to serve as examples and document the evolutionary process of the NEOH framework.

The most important finding across the case studies was that the protocols were useful to reflect on the implementation of the One Health process (what we coined One Health-ness), i.e. how well the initiatives were doing in the six aspects defined. They also provided first evidence that the context of an initiative is crucial for its capacity to address these aspects. Moreover, the case studies demonstrated that size and scope as well as the time period considered in the evaluation, strongly influence the theory of change and consequently how an evaluation is framed. Because of the lack of benchmarking data to date, we do not know yet what the optimal level of One Health-ness is to achieve the best outcomes. However, the NEOH approach helped the evaluators to broaden their scope, let go of pre-defined ideas, reflect on the system, and stimulate interesting discussions with a range of stakeholders. Thus, with this work, we have made a small, but important step towards the assessment of One Health initiatives, thereby helping to move One Health from the often criticised sphere of being 'theoretical', 'intellectual', or 'conceptual' to something much more tangible, practical, and applied. Implicitly, we claim that One Health is an approach based on systems thinking, using One Health thinking, working, and planning in an enabling environment of sharing, learning and systemic organisation, with the aim to achieve outcomes critical for sustainable development – ecology, society, and economy.

In a final step of NEOH, the case studies and articles from peer-reviewed literature were scored using the presented protocols in a desk-based meta-study to generate evidence on the usefulness of our approach and to identify factors that are critical for the added value of One Health initiatives. But this is not the end of our journey. We welcome you to join forces, and apply the handbook in your work to move towards benchmarking of One Health initiatives.

Sara Savić, Simon Rüegg and Barbara Häsler

Glossary

Actor	Actors are a subgroup of stakeholders (see there for distinction) such as 'any individual, group or organisation who acts, or takes part' in the context of the OH initiative.
Agency	Agency refers to the possibility of people to shape actions and societal structures in which they are embedded in such a way that the members of a society have equal chances to bring their views to social, economic and material expressions.
Anthropocentric	Assuming that humans are the centre or ultimate end of the universe.
Association	Group of individuals who enter into an agreement to accomplish a purpose.
Boundaries (system boundaries)	Operational delimitations of the system in different dimensions (e.g. geographical, time, governance) that can affect the system aims or indicators.
Component	Systems are composed of a set of interacting or interdependent components that form a complex whole. Components may be tangible (e.g. humans, animals, forests, lakes) or intangible (e.g. cultural behaviours, values, norms, language expressions) and are linked by interactions.
Constraint	Condition that narrows the system.
Context	The system or socio-ecological system within which the initiative is aiming to evoke change towards a health outcome.
Cross-sectoral	Where the actions of one industry sector impacts on one or more other sectors.
Disability-adjusted life year (DALY)	One DALY can be thought of as one lost year of 'healthy' life. The sum of these DALYs across the population, or the burden of disease, can be thought of as a measurement of the gap between current health status and an ideal health situation where the entire population lives to an advanced age, free of disease and disability.
Dimension	Systems are organised in hierarchical order. Hierarchies depend on a fundamental quality that defines this order. Examples

The reference for detailed definitions can be found in the text of the handbook.

	for dimensions are life with its different organisational levels; within the semantic space (dimension) expands the hierarchy of meanings of words; within the dimension of faith various beliefs are organised within larger clusters, but also governance, time, geographical space, and many more are dimensions. This is a specific use of the term in Chapter 3. The use of dimension in other chapters refers to a component, an aspect, a feature, or a facet.
Discipline	Field of study, a branch of knowledge.
Driver	Element of a system that has a major or critical effect on the associated elements or the entire system.
Evaluation design	A plan for conducting an evaluation.
Feedback	The process in which part of the output of a [One Health] system is returned to its input.
Governance	The exercise of political, economic and administrative authority in the management of a country's affairs at all levels.
Governance system (GS)	Governance systems are a further core subsystem of a social-ecological system and represent the system that is managing specific resource systems. In contrast to Ostrom[1], we do not differentiate between users and resource units, because users may represent a resource from e.g. a disease perspective.
Government	The person or persons authorized to administer the laws, the ruling power and the administration.
Impact	Positive and negative, primary and secondary long-term effects produced by a development intervention, directly or indirectly, intended or unintended.
Inequality	Health inequality is the generic term used to designate differences, variations, and disparities in the health achievements of individuals and groups.

[1] Ostrom, E., 2009. A general framework for analyzing sustainability of social-ecological systems. Science 325: 419-422.

Inequity	Health inequity refers to those inequalities in health that are deemed to be unfair or stemming from some form of injustice. The crux of distinguishing 'between equality and equity is that the identification of health inequities entails normative judgment premised upon (1) one's theories of justice; (2) one's theories of society; and (3) one's reasoning underlying the genesis of health inequalities. Because identifying health inequities involves normative judgment, science alone cannot determine which inequalities are also inequitable, nor what proportion of an observed inequality is unjust or unfair'.
Institutional memory	Stored knowledge within organisations. Requires ongoing transmission of memories between members of the organisation.
Interdisciplinary (ID)	The interdisciplinary approach involves the integration of perspectives, concepts, theories, and methods to address a common challenge.
Inter-sectoral	Aims/challenges common to several sectors. Effects from one sector to another are not necessarily significant.
Knowledge	Knowledge can be differentiated into system-, target- and transformation knowledge, where system knowledge comprises the analytical aspect (e.g. what are the relations between society and nature?), target knowledge embodies the normative aspect (e.g. what kind of social values are needed?) and transformation knowledge represents the political aspects of knowledge (e.g. what practical strategies should be adopted?).
Leadership	The essence of effective leadership: • Creating alignment around shared objectives and strategies to attain them. • Increasing enthusiasm, optimism, confidence and excitement about the work. • Helping people to appreciate each other, and to learn how to resolve differences constructively. • Helping people to co-ordinate activities, continuously improve and collectively learn about better ways to work together. • Kindness and courage.
Learning	The acquisition of knowledge or skill. In the context of One Health the term is extended to cover all mental and physical changes of organisms to improve the interactions with their environment.
Level	Used as synonym to scale.
Logic model	Logic models graphically illustrate the components (inputs, activities, outputs, outcomes, impacts) of a programme in a structured, logical and sequential way.
Method	A way of proceeding or doing something, especially a systematic or regular one.
Metric	Parameter or measure of quantitative assessment used for measurement, comparison or to track performance or production.
Ministry	A governmental organization, headed by a minister that is usually meant to manage a specific sector of public administration.

Multidisciplinary (MD) The multi-disciplinary approach is typically understood as the sequential or additive combination of ideas or methods.

Multisectoral Collaboration of several sectors, such as human and animal health, agriculture, food processing industry, etc. See also 'sector' for a definition.

Network for Evaluation of One Health (NEOH) A network funded by the European Cooperation in Science and Technology (TD1404) with the aim to enable future quantitative evaluations of OH activities and to further the evidence base by developing and applying a science-based evaluation protocol in a community of experts.

One Health (OH) OH emphasizes the commonalities of human, animal, plant, and environmental health. In this perspective, it can be regarded as an 'umbrella' term that captures integrative approaches to health across these highly interlinked components.

One Health initiative (OH initiative) Any initiative, such as research projects, developmental programs, policy, etc. that relies on the concept of OH as described above. In a generic way, a OH initiative aims at generating change in a social-ecological system (context) towards improved health of humans, animals and/or ecosystems. We do not refer to the pro bono Kahn-Kaplan-Monath-Woodall-Conti 'One Health Initiative' at http://www.onehealthinitiative.com.

Outcome The likely or achieved short-term and medium-term effects of an OH initiative's outputs.

Outcome mapping An approach used for planning and assessing programmes that focus on change and social transformation. It provides a set of tools to design and gather information on the outcomes, defined as behavioural changes, of the change process.

Output The products, capital goods and services which result from a OH initiative; may also include changes resulting from the intervention which are relevant to the achievement of outcomes.

Paradigm A theory providing a unifying explanation for a set of phenomena in some field, which serves to suggest methods to test the theory and develop a fuller understanding of the topic, and which is considered useful until it is replaced by a newer theory providing more accurate explanations or explanations for a wider range of phenomena.

Participation The act or state of taking part in an activity, or sharing in common with others.

Programme An organised set of financial, organisational and human resources mobilised to achieve an objective or set of objectives in a given timeframe. A programme is delimited in terms of a schedule and a budget and its objectives are defined beforehand. It is always under the responsibility of an authority or several authorities who share the decision-making. Programmes are generally broken down into measures and projects.

Project	The complex of actions, which have a potential for resulting in a [physical] change [in the environment].
Quality-adjusted life year (QALY)	A measure of the state of health of a person or group in which the benefits, in terms of length of life, are adjusted to reflect the quality of life. One QALY is equal to 1 year of life in perfect health. QALYs are calculated by estimating the years of life remaining for a patient following a particular treatment or intervention and weighting each year with a quality-of-life score (on a 0 to 1 scale). It is often measured in terms of the person's ability to carry out the activities of daily life, and freedom from pain and mental disturbance.
Resilience	Adaptability of a system upon disturbances to allow it to recover and remain sustainable.
Resource system (RS)	Resource systems are core subsystems of a social-ecological system such as forested areas, wildlife, water systems, national parks, etc. We extend the idea of Ostrom[2] and consider social systems as resource systems too, e.g. health care system, local community, food chains, etc. They 'provide' or host resource units such as trees, shrubs, susceptible persons, traders, food items, etc. which contribute to the system.
Resource units (RU)	Resource units are product or component of the resource system and represent a link of the resource system to other components. In contrast to Ostrom[2], we do not differentiate between users and resource units, because users may represent a resource from e.g. a disease perspective.
Restrictions	Human action to account for the 'conditions' and 'constrain'.
Scale	Identical to level. Systems are organised in hierarchical order. This hierarchy implies that different levels of the hierarchy can be in the focus of attention. As an example in the hierarchy of life, one can look at individuals, populations, communities or ecosystems, i.e. different scales of the same quality (life).
Sector	A sector is an area of activity aimed at benefits to society, characterised by common processes and institutions. Examples include agriculture, health, transportation, education and environment. Sub-sectors would be units within the sector; for example, in agriculture these could be livestock, crops, agro-forestry, fishing and aquaculture.
Space	Here used as synonym to dimension. Extension, considered independently of anything which it may contain; that which makes extended objects conceivable and possible. Specifically: Geographical area where the system is happening: place, region, state, nation, also international space. Or a minor dimension, if it is involving individuals or populations.

[2] Ostrom, E., 2009. A general framework for analyzing sustainability of social-ecological systems. Science 325: 419-422.

Sustainability	The continuation of benefits from an intervention. The probability of continued long-term benefits. The resilience to risk of the net benefit.
Stakeholder	Any individual, group or organisation who may affect, be affected by, or perceive themselves to be affected by, a decision or activity. See 'actor' for distinction.
System, social-ecological system (SES)	A system is a set of interacting, interrelated or independent components that form a complex and unified whole. Human made systems are usually conceived to achieve a defined aim. However, this may not be the case for social-ecological systems (SES), which were defined as a hierarchy of subsystems and internal variables at multiple levels analogous to organisms composed of organs, organs of tissues, tissues of cells, etc. The core subsystems of a SES are resource systems, resource units, governance systems and users.
Task	An activity that is accomplished within a defined period of time or terminated with a deadline.
Team	A team is a group of individuals who work together to produce products or deliver services for which they are mutually accountable. Team members share goals and are mutually held accountable for meeting them, they are interdependent in their accomplishment, and they affect the results through their interactions with one another. Because the team is held collectively accountable, the work of integrating with one another is included among the responsibilities of each member.
Theory of change (TOC)	The TOC explains all the different pathways that might lead to the desired effect of an initiative. It not only shows the outputs, outcomes and impact of an initiative, but also requires outlining (and explaining) the causal linkages. Each effect is shown in a logical relationship to all the others.
Transdisciplinary (TD)	The transdisciplinary approach entails not only the integration of approaches, but also the creation of fundamentally new conceptual frameworks, hypotheses, and research strategies that synthesize diverse approaches and ultimately extend beyond them to transcend pre-existing disciplinary boundaries. The term transdisciplinarity refers to scholarship that transgresses the boundaries between academia and communities outside academia. By doing so, it enables inputs and scoping across scientific and non-scientific stakeholder communities and facilitates a systemic way of addressing a challenge.
Wicked problem	Problem that cannot be solved with a linear approach. It has causes that seem incomprehensible and solutions that seem uncertain, and often requires to transcend conventions and to question current practice.

About the editors

Dr Simon Rüegg, leader of Working Group 1 in the Network for Evaluation of One Health (together with Jakob Zinsstag), is an animal health researcher from the University of Zürich, Switzerland. After his veterinary medicine degree, he completed a doctorate (DVM) and a PhD in epidemiology, biostatistics and molecular diagnostics of tick-borne equine piroplasmoses. Throughout field projects in southern Mongolia and as a veterinary practitioner he gained a solid understanding of practical aspects of veterinary medicine and epidemiology. From 2013 he worked as researcher at the Veterinary Public Health Institute in Bern investigating the cost-effectiveness of surveillance for vector-borne diseases. In September 2013, he joined the Veterinary Epidemiology Group in Zürich as senior research assistant, lecturer and statistics consultant. His research interest is the application of the theory of complex adaptive systems to health questions, in particular its impact on medical decision making; the relation between scientific, emotive, financial, ethical and social aspects; the sustainability of medical practices; and how to promote new integrated approaches to health.

Dr Simon Rüegg, Section of Epidemiology, Vetsuisse Faculty, University of Zürich, Winterthurerstrasse 270, 8057 Zürich, Switzerland.

Dr Barbara Häsler, chair of the Network for Evaluation of One Health, is a veterinary researcher and senior lecturer in Agrihealth at the Royal Veterinary College, UK, with expertise in animal health economics, evaluation, food systems and One Health. Her main area of interest is the integration of economic, social and epidemiological aspects in animal disease mitigation to inform practical and feasible solutions for disease management in livestock food systems. She is particularly committed to research that focuses on the interconnectedness of the health of people, animals and the environment and the use of systems-based approaches that help to create co-benefits for various sectors and populations in a sustainable way. She

holds a PhD in animal health economics, a postgraduate certificate in veterinary education and a postgraduate certificate in economics from the University of London, UK, and a doctorate in veterinary epidemiology and economics and a diploma of veterinary medicine from the University of Bern, Switzerland.

Dr Barbara Häsler, Department of Pathobiology and Population Sciences, Veterinary Epidemiology Economics and Public Health Group, Royal Veterinary College, Hawkshead Lane, North Mymms, Hatfield, Hertfordshire, AL9 7TA, United Kingdom.

Prof. Jakob Zinsstag, leader of Working Group 1 in the Network for Evaluation of One Health (together with Simon Rüegg), is a professor of epidemiology at the Swiss Tropical and Public Health Institute (Swiss TPH) and the University of Basel, Switzerland. He graduated with a doctorate in veterinary medicine from the Veterinary Faculty of the University of Bern in 1986 and holds a PhD in tropical animal production. After graduation, he worked in rural practice and as post-doctoral fellow on trypanosomiasis research at the Swiss Tropical Institute. From 1990 to 1993, he led a livestock helminthosis project for the University of Bern in The Gambia. From 1994 to 1998, he headed the Centre Suisse de Recherches Scientifiques in Abidjan, Côte d'Ivoire. Since 1998, he has been in charge of a research group at the Swiss TPH investigating the interface of human, animal and environmental health with the focus on the health of nomadic people and the control of zoonoses under the paradigm of One Health. He is a long-standing and passionate advocate for One Health.

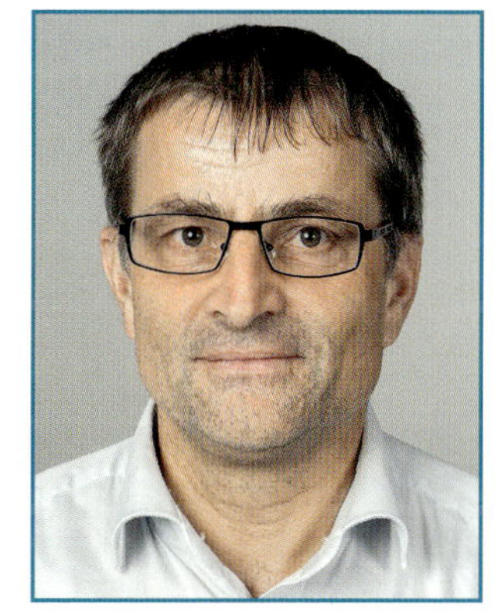

Prof. Jakob Zinsstag, Department of Epidemiology and Public Health, Swiss Tropical and Public Health Institute, University of Basel, P.O. Box, 4002 Basel, Switzerland.